Differential Topology
and
Quantum Field Theory

Differential Topology
and
Quantum Field Theory

Charles Nash

St Patrick's College
Maynooth
Ireland

ACADEMIC PRESS
An imprint of Elsevier
Amsterdam   Boston   London   New York   Oxford   Paris
San Diego   San Francisco   Singapore   Sydney   Tokyo

ACADEMIC PRESS
An imprint of Elsevier
84 Theobald's Road
London WC1X 8RR, UK

United States Edition published by
ACADEMIC PRESS
An imprint of Elsevier Science
San Diego, CA 92101

A catalogue for this book is available from the British Library
ISBN 0–12–514076–2 (paperback)

Typeset using T$_{\text{E}}$X
Transferred to Digital Printing 2008

Preface

The twentieth century has been witness to a great burgeoning of mathematics and physics. In the early part of the century the new physical theories of relativity and quantum mechanics made extensive use of the machinery of differential geometry and Hilbert spaces. Some time later quantum field theory began to pose difficult mathematical problems. As a partial response to this the subject of axiomatic quantum field theory was born. The main thrust of this approach was to tackle the formidable problems of quantum field theory head on using the most powerful mathematical tools available; the bulk of these tools being drawn from analysis. More recently there has been considerable evidence that the way forward in these problems is considerably illuminated if, in addition to analysis, one uses differential topology. There have also been advances in the differential topology of dimensions three and four which have drawn extensively on physical sources, the principal source being Yang–Mills theories. Thus there has been a genuinely two sided interaction between the worlds of mathematics and mathematical physics. This book is intended as an informal introduction to some of these mathematical and physical ideas. It should be of use to graduate students and other research workers taking an interest in this material for the first time.

September 1990 *Charles Nash*

For Jeanine Niamh Marie–Thérèse Nash

27th September 1977—15th February 1980

Contents

CHAPTER III

CHAPTER IV

CHAPTER V

CHAPTER VI

Infinite Dimensional Groups   **170**

CHAPTER VII

Morse Theory   **192**

CHAPTER VIII

Instantons and Monopoles   **216**

CHAPTER XII

CHAPTER I

A Topological Preliminary

§ 1. From homeomorphism to diffeomorphism

In this chapter we survey, in an informal way, some of the principal techniques used in topology. In so doing we both establish some of the basic notation that we will employ in subsequent chapters and give an account of certain standard results and constructions that will be in frequent use. Topology sets itself, as its initial goal, the task of classifying all topological spaces up to homeomorphism—two topological spaces X and Y being homeomorphic if there exist between them continuous maps α and α^{-1}:

$$\alpha : X \longrightarrow Y \qquad \alpha \quad continuous$$
$$\alpha^{-1} : Y \longrightarrow X \qquad \alpha^{-1} \quad continuous \qquad (1.1)$$
$$\alpha \circ \alpha^{-1} = \alpha^{-1} \circ \alpha = I$$

However this is far too ambitious and, except in certain special cases, the goal is not reached.

For example, if one works in a low dimension such as 2 and one restricts the spaces to be compact closed orientable surfaces, M say, then one knows that M is homeomorphic to a sphere with g handles; and two such M are homeomorphic if and only if they possess the same number of handles. Such M are of course the celebrated Riemann surfaces and the number g is called the genus.

In higher dimensions simple classifications of this kind do not exist—this is due in part to the fact that in dimension 2 topological phenomena are somewhat restricted e.g. if a deformation construction involving an M of dimension 2 encounters a sub-manifold N as an obstacle then necessarily $\dim N = 1$ or 2 and there may simply not be enough room to avoid the obstacle. On the other hand, if $\dim M \gg 2$ then a sub-manifold of dimension

1 or 2 may well be movable to some 2-dimensional subspace of M which is in a region not involved in the deformation.

Another illustration is provided by the theory of knots: a knot is regarded as a map from the circle S^1 to $\mathbf{R}^3$; many knots are non-trivial but all of them can be untied if one simply increases the dimensions by replacing $\mathbf{R}^3$ by $\mathbf{R}^4$. More generally, suppose that dim $M = n$, then at what value of n should M be deemed a topological space of higher dimension? The answer seems to be when n is about 5 and we shall say more about this matter in § 6 of this chapter. All this suggests that we should study a much smaller class of spaces than the class of all topological spaces.

To this end let us now lower our sights somewhat and announce our intention to study the smaller class of spaces which are known as *topological* manifolds. As the name suggests a topological manifold M of dimension n is a (Hausdorff) space which is covered by patches U_α each of which is homeomorphic to some open set in $\mathbf{R}^n$, and for which the usual transition functions $g_{\alpha\beta}$ between a pair of patches U_α and U_β are homeomorphisms. Now even this class of spaces is rather abstractly defined and sparsely provided with concrete mathematical structures. We can do a little to remedy this. If we bear in mind that algebraic topology often makes use of combinatorial and triangulation methods to calculate invariants such as homology groups, then one can require that the transition functions $g_{\alpha\beta}$ and $g_{\alpha\beta}^{-1}$ are piecewise-linear or PL. We then have a *piecewise-linear* manifold of dimension n. Now the physicist or differential topologist would like to make use of calculus. Thus we take one further step and require the transition functions $g_{\alpha\beta}$ to be diffeomorphisms, i.e. both $g_{\alpha\beta}$ and $g_{\alpha\beta}^{-1}$ are infinitely differentiable or C^∞. Thus M has now become a *differentiable* manifold of dimension n.

It is important to realise that the distinctions made in describing these three types of manifold are not purely academic: It might be thought that, given a manifold M, there are always three mutually homeomorphic realisations of M as topological, piecewise-linear, or differentiable, which one could denote by M_{TOP}, M_{PL} and M_{DIFF} respectively. This is not so—there exist topological manifolds which admit no PL structure, i.e. M_{TOP} exists but not M_{PL}, also there exist topological and PL manifolds which admit no differentiable structure, i.e. both M_{TOP} and M_{PL} exist but not M_{DIFF}. Further when M_{DIFF} does exist it may not be unique in the sense that the differentiable coordinates with which it endows M are not unique, i.e. one can have two M_{DIFF} which are homeomorphic but not diffeomorphic. In fact the existence of these results is a sort of raison d'être for the singling out of *differential topology* as an individual branch of the subject Topology.

The particular properties of manifolds that are described here might be thought unlikely to have any relevance for quantum field theory. However, this turns out to be far from true and we shall have occasion to return to this subject both in this chapter and in subsequent chapters.

Despite the above discussion we shall usually deal with differentiable manifolds. Also in accordance with standard practice we shall simply refer to a differentiable manifold as a manifold. If the manifold is not differentiable we shall explicitly say so. In the main we wish to make full use of calculus on manifolds to calculate geometric and topological quantities such as curvatures and de Rham cohomology groups respectively.

§ 2. Some algebraic topology: homotopy

Recall that the general idea in algebraic topology is to attach to topological spaces invariants which can be used to distinguish the spaces one from another. These invariants can be numbers, such as the dimension of the space or its Euler–Poincaré characteristic. They can also be mathematical objects furnished with an algebraic structure like homotopy groups or homology groups, etc. However, whatever these invariants are, they are *topological* invariants—in other words, the invariants for a particular space are unchanged if this space is traded for another to which it is homeomorphic.

We now summarise in brief some of the principal invariants that we shall have a need for. One of the most powerful notions available is that of *homotopy*. While the notion of homeomorphism, viewed as an equivalence relation on the class of all topological spaces, divides these spaces into equivalence classes, homotopy generates equivalence classes that contain continuous maps: Given two spaces X and Y, let $Map(X,Y)$ denote the set of continuous maps from X to Y and let us consider two maps α and β so that

$$\alpha \in Map(X,Y), \qquad \beta \in Map(X,Y) \qquad (1.2)$$

Then α is said to be homotopic to β, which we write as

$$\alpha \simeq \beta \qquad (1.3)$$

if there is a continuous map

$$F : [0,1] \times X \longrightarrow Y \qquad (1.4)$$

such that for $x \in X$ we have

$$F(0,x) = \alpha(x) \quad \text{and} \quad F(1,x) = \beta(x) \qquad (1.5)$$

the map F is then referred to as a homotopy from α to β. Thus homotopy divides $Map(X, Y)$ up into equivalence classes. The set of all these equivalence classes is denoted by

$$[X, Y] \tag{1.6}$$

and the equivalence class in $[X, Y]$ to which a given map $\alpha \in Map(X, Y)$ belongs is denoted in the standard fashion by $[\alpha]$. If we view $[X, Y]$ as function of Y with X fixed then it is immediately verifiable that the resulting collection of equivalence classes is a topological invariant of Y—this is because the use of a homeomorphism, h say, to change Y for a homeomorphic copy Y' induces a canonical replacement of a homotopy F from α to β, by the homotopy $h \circ F$ from $h \circ \alpha$ to $h \circ \beta$. Thus one can canonically identify $[X, Y]$ and $[X, Y']$.

Continuing on our discussion of invariance properties of topological spaces we come to the definition of *homotopy type*. A space X is said to be of the same homotopy type as another space Y if there exists a continuous map α

$$\alpha : X \longrightarrow Y$$

such that α, though not necessarily possessing an inverse, must have a *homotopic inverse*, β say. This simply means that if I_X and I_Y denote the identity maps on X and Y respectively then for α and β together we have the properties

$$\begin{aligned} \alpha : X &\longrightarrow Y \\ \beta : Y &\longrightarrow X \\ \alpha \circ \beta \simeq I_Y, \qquad &\beta \circ \alpha \simeq I_X \end{aligned} \tag{1.7}$$

We shall write this as

$$X \simeq Y \tag{1.8}$$

and since this is clearly an equivalence relation we often refer to $X \simeq Y$ as the statement that X is homotopy equivalent to Y.

Topological invariants such as $[X, Y]$, where one of X and Y is fixed and the other varies, are also known as *invariants of homotopy type*. Suppose, to be definite, that X is fixed, then this terminology means that not only do we have

$$[X, Y] = [X, Y'] \qquad \text{for } Y \text{ homeomorphic to } Y' \tag{1.9}$$

but we also have

$$[X, Y] = [X, Y'] \qquad \text{for } Y \text{ of the same homotopy type as } Y' \tag{1.10}$$

The most commonly encountered topological invariants in algebraic topology are invariants of homotopy type. One should realise, however, that to ask for an invariant of homotopy type of X is to ask for less than an invariant of the homeomorphism class of X. For example, the space $\mathbf{R}^n$ can be contracted to a point and so is of the same homotopy type as a point; and this has the well known consequence that its homotopy and homology groups coincide with those of a point and are thus trivial. On the other hand, $\mathbf{R}^n$ $n > 0$, is not homeomorphic to a point, and indeed $\mathbf{R}^n$ is homeomorphic to $\mathbf{R}^m$ if and only if $n = m$. Thus the dimension of a space X is an example of a topological invariant which is *not* an invariant of homotopy type.

Returning to the consideration of $[X, Y]$, a particularly important choice for the fixed space X is the circle S^1. Then one is considering homotopy classes of loops on X and one knows that this leads naturally to the fundamental group $\pi_1(X)$. However, before we get to the fundamental group we observe that the loops on X have no obvious composition law with which to make them a group. Indeed to obtain such a law we restrict the loops to have a point in common—the point where they all begin and end. This simple but powerful notion is that of endowing every space with a base point (often denoted by $*$), and requiring that maps between pairs of spaces preserve base points. Because of its importance we expand on this notion below.

To formalise the notion of base points and their preservation in the way just described we introduce the space of maps

$$Map_0(X, Y) \qquad\qquad (1.11)$$

which is the space of continuous maps between X and Y in which the base point of X is mapped into that of Y; we often refer to such maps as *based maps*. Thinking homotopically about $Map_0(X, Y)$ leads us straightaway to introduce the space

$$[X, Y]_0 \qquad\qquad (1.12)$$

which is the space of homotopy classes of $Map_0(X, Y)$. The previous space $[X, Y]$ is also sometimes called the space of *free* homotopy classes, and a free homotopy is one which is not required to preserve base points. An understanding of the difference between these two kinds of homotopy classes can be obtained on introduction of the fundamental group—we do this in the next section and we shall return to this matter there.

A pair $\{X, *\}$ consisting of a space X together with its base point $*$ is frequently referred to as a *pointed space*. Now if we dispense with base points temporarily, then, given three spaces X, Y and Z, there is a natural bijection

$$i : Map(X, Map(Y, Z)) \longrightarrow Map(X \times Y, Z) \qquad (1.13)$$

To construct this bijection recall that, set theoretically, the set of all maps (continuous or not) from X to Y is given by

$$Y^X \qquad (1.14)$$

Thus if we consider X, Y and Z we have the obvious bijections

$$Z^{(Y \times X)} \simeq Z^{(X \times Y)} \simeq (Z^X)^Y \qquad (1.15)$$

This means that to establish the bijection i of 1.13 above one simply has to check that the set theoretic version 1.15 still goes through when the various maps are made continuous.

Since our interest in groups, such as the fundamental group, leads us to study spaces with base points we would like to have a similar bijection for pointed spaces. Such a bijection exists provided we introduce a method for correctly handling the base points involved. The only space occurring in 1.13 whose base point is not naturally determined is $X \times Y$. We have to devise a sensible way of assigning a base point to $X \times Y$. This standard construction has two parts: first construct what is called the reduced join of X and Y and is written as

$$X \vee Y \qquad (1.16)$$

This is simply the disjoint union of X and Y with the two base points identified. Then take the Cartesian product $X \times Y$ and quotient out $X \vee Y$, i.e. form what is called the reduced product $X \wedge Y$ which is defined by

$$X \wedge Y = \frac{X \times Y}{X \vee Y} \qquad (1.17)$$

$X \wedge Y$ is also referred to as the smash product because, in taking the quotient, $X \vee Y$ is smashed to a point, this point becoming the base point of $X \wedge Y$. In any case our object is achieved in that $X \wedge Y$ has a base point which is naturally determined by the base points x_0 and y_0 of X and Y respectively. Having completed our two part construction we can give the analogue, for pointed spaces, of 1.13: This is that there is a bijection i_* of the form

$$i_* : Map_0(X, Map_0(Y, Z)) \longrightarrow Map_0(X \wedge Y, Z) \qquad (1.18)$$

One very often encounters $X \wedge Y$ in the form $S^1 \wedge X$ where S^1 is the circle and X is arbitrary; $S^1 \wedge X$ is then called the suspension or reduced suspension of X and is usually denoted by SX, i.e. one has

$$S^1 \wedge X = SX$$

A further piece of notation is that $Map_0(S^1, X)$, which is the space of based maps from S^1 to X, is also denoted by ΩX and referred to as the space of (based) loops on X. So we have that

$$Map_0(S^1, X) = \Omega X \qquad (1.19)$$

If the X happens to be a sphere S^n, say, then it is easy enough to check, cf. Whitehead [1], that the suspension of X is (homeomorphic to) a sphere of one higher dimension. Thus one has

$$S^1 \wedge S^n = SS^n \simeq S^{n+1} \qquad (1.20)$$

Finally let us think of the symbols S and Ω as being *operators* which act on spaces, or, to give them their topological name, *functors*. We shall then see that they are formal adjoints of one another. In short we have the canonical bijection (this follows immediately from 1.18 with $Y = S^1$)

$$Map_0(SX, Y) \stackrel{\rightarrow}{\leftarrow} Map_0(X, \Omega Y) \qquad (1.21)$$

from which one deduces immediately that for the corresponding homotopy classes we also have

$$[SX, Y]_0 \stackrel{\rightarrow}{\leftarrow} [X, \Omega Y]_0 \qquad (1.22)$$

thus displaying the adjoint relationship of S and Ω just mentioned above.

It is well known that a group is an extremely valuable structure in mathematics and this is so also in algebraic topology. A useful way to approach the various groups such as homotopy groups and cohomology groups that arise in algebraic topology, is to think of them as groups structures possessed by the homotopy classes $[X, Y]_0$: this can happen in two ways, the first occurs when Y possesses an appropriate structure, P say, the second when X possesses an appropriate structure, P' say; moreover these two situations are dual to one another. We now give a short explanation of what we mean by the structures P and P'.

Beginning with P notice that if the space Y is actually a group G then it is immediately possible to give a group structure not just to $[X, Y]_0$ but to its precursor $Map_0(X, Y)$: to do this set $Y = G$, and let $\alpha, \beta \in Map_0(X, G)$, then we denote their group product by $\alpha \bullet \beta$ whose definition is

$$\alpha \bullet \beta : X \longrightarrow G$$
$$x \longmapsto \alpha(x) * \beta(x) \tag{1.23}$$

where $*$ stands for the product in G. This product $\bullet$ then induces a product in $[X, G]_0$ (which for convenience we also denote by $\bullet$) which renders it a group. However, viewing the product $*$ in G as a map

$$* : G \times G \longrightarrow G$$

one sees that, to obtain a group structure on $[X, G]_0$, it is not necessary that the map $*$ be a group product, it is sufficient that it be a map $* : G \times G \to G$ which descends to a group product on the homotopy classes $[X, G]_0$. Given $[X, Y]$, the existence of such a map $* : Y \times Y \to Y$ that renders $[X, Y]_0$ a group is referred to by saying that Y is a homotopy associative H-space, or simply a H-space—the rôle that the words 'homotopy associative' play is that they remind one that, though the group product on $[X, Y]_0$ must of course be associative, it need not be so on $Map_0(X, Y)$; on $Map_0(X, Y)$ it need only be associative up to homotopy. Thus Y possesses the property that we called P above if it is a H-space, and note that then $[X, Y]_0$ is a group for *any* X.

Turning now to P' is to take the dual approach, thus the *multiplication* map $*$ is replaced by a *comultiplication* map $*'$ by which we mean a map of the form

$$*' : X \longrightarrow X \vee X \tag{1.24}$$

(Note that $X \vee X$ occurs in the above where one would expect $X \times X$, this is to maintain our requirements for base points.) In any case, granted such a comultiplication $*'$ then it necessarily induces a product law $\bullet'$ on $[X, Y]_0$ where $\bullet'$ is given by

$$\bullet' : [X, Y]_0 \times [X, Y]_0 \longrightarrow [X, Y]_0 \tag{1.25}$$

Finally if this product on $[X, Y]_0$ renders it a group then our object is achieved and X is called a homotopy associative *coH-space*, or simply a coH-space. We note also that, in a similar fashion to the situation for H-spaces, when X is a coH-space then $[X, Y]_0$ is a group for *any* Y.

We have seen above that a simple example of a H-space is obtained when $[X, Y]_0$ has $Y = G$ with G a group. An example of a coH-space is obtained when $[X, Y]_0$ has $X = SZ$ for any space Z, with SZ denoting the suspension of Z. For the details cf. Husemoller [1]. This completes our summary of the property P'.

To end this section we point out that it is perfectly possible for the properties P and P' to be present simultaneously, i.e. given a particular $[X, Y]_0$, it may be the case that Y is a H-space *and* X is a coH-space, for example one could choose $Y = G$ and $X = SZ$. When this is so it would appear that there are two possible group structures on $[X, Y]_0$—in fact it is an elementary calculation, cf. Spanier [1], that both these group structures coincide, and that the resulting single group structure is *always commutative*. As we shall observe in the next section this means that the fundamental group of a topological group G (e.g. G is some Lie group) is always commutative.

§ 3. Homotopy groups

The first homotopy group of a pointed space X is the fundamental group $\pi_1(X)$ and is defined by

$$\pi_1(X) = [S^1, X]_0 \tag{1.26}$$

Thus the fundamental group of X is the homotopy classes of based loops on X. Because $\pi_1(X)$ is a set of homotopy classes of maps, rather than the maps themselves, then one does not distinguish loops which are homotopic to one another. The group product is constructed as follows: let α and β be based loops so that $\alpha, \beta \in Map_0(S^1, X) = \Omega X$, denote their corresponding homotopy classes in $[S^1, X]_0$ by $[\alpha]$ and $[\beta]$ respectively, then the product is written as $[\alpha][\beta]$. Suppose $[\alpha][\beta] = [\gamma]$ then the loop γ is defined by

$$\gamma(t) = \begin{cases} \alpha(2t) & 0 \leq t \leq \frac{1}{2} \\ \beta(2t - 1) & \frac{1}{2} \leq t \leq 1 \end{cases} \tag{1.27}$$

where the variable t is the coordinate on the circle which we have taken to be of unit circumference. Also the well known process of checking that the product defined above is both invertible and associative up to homotopy is precisely the verification that ΩX is (another example of) a H-space.

The higher homotopy groups are obtained by replacing the circle S^1 by the spheres S^n, $n = 2, 3, \ldots$ and are denoted by $\pi_n(X)$. Accordingly, if we put all the values of n together we obtain

$$\pi_n(X) = [S^n, X]_0 \quad n = 1, 2, \ldots \tag{1.28}$$

Making use of 1.22 we can deduce at once that

$$[S^n, X]_0 = [S^{n-1}, \Omega X]_0 \tag{1.29}$$

But since we have just seen that ΩX is a H-space then this implies that $[S^{n-1}, \Omega X]_0$ is a group and so the higher homotopy groups $\pi_n(X)$ $n \geq 2$ are indeed groups. The fundamental group $\pi_1(X)$ can be non-commutative, as is the case when X is a Riemann surface of genus $g \geq 2$. However, the higher homotopy groups $\pi_n(X)$ are always commutative. This follows also from the H and coH properties of $[S^n, X]$: we observe that, since $n \geq 2$, then we can write $\pi_n(X)$ in the form

$$\pi_n(X) = [S^n, X]_0 = [S^{n-1}, \Omega X]_0 = [SS^{n-2}, \Omega X]_0 \qquad (1.30)$$

so that $\pi_n(X)$ for $n \geq 2$ is expressible in the form $[X, Y]_0$ with $Y = \Omega X$ a H-space and $X = SS^{n-2}$ a coH-space; thus by our remark at the close of the last section the group structure on $[X, Y]_0 = \pi_n(X)$ is both unique and commutative. We can now use a similar argument to establish our earlier assertion that the fundamental group of a topological group is always commutative. Let $X = G$ be a topological group, then we write [1]

$$\pi_1(G) = [S^1, G]_0 = [SS^0, G]_0$$

and because G is a group, and hence a H-space, and $S^1 = SS^0$ is a coH-space, then $\pi_1(G)$ is necessarily commutative.

Next we return briefly to the distinction made in §2 between free and based homotopy classes. In summary, given the two sorts of classes $[X, Y]$ and $[X, Y]_0$, there is an action of $\pi_1(Y)$ on the based classes $[X, Y]_0$, and on forming the appropriate quotient one gets the free classes $[X, Y]$. We write this as

$$[X, Y]_0/\pi_1(Y) \simeq [X, Y] \qquad (1.31)$$

To understand all this suppose one has a map $\alpha \in Map_0(X, Y)$ that preserves base points and hence determines the class $[\alpha] \in [X, x_0; Y, y_0]_0$, where we have temporarily amended our notation to include the relevant base points. Then without loss of generality we can leave x_0 fixed and ask the question: How does $[X, x_0; Y, y_0]_0$ depend on the base point y_0? Well suppose we choose another base point y_1 then we must now compare $[X, x_0; Y, y_0]_0$ and $[X, x_0; Y, y_1]_0$. To do this join y_0 to y_1 by a path γ, say, then one can pass between the two sets of homotopy classes by going back and forth along γ. Thus we obtain a γ-dependent isomorphism i_γ between $[X, x_0; Y, y_0]_0$ and $[X, x_0; Y, y_1]_0$. However, given two paths γ and γ' between y_0 and y_1 then we shall have the same isomorphism provided γ and

---

[1] By analogy with the sphere S^n which has the equation $x_1^2 + \cdots x_{n+1}^2 = 1$, the sphere S^0 simply has the equation $x_1^2 = 1$ and so consists of a pair of points one of which is its base point.

γ' are *homotopic*. But the difference between γ and γ' is a *loop*, i.e. an element of $\pi_1(Y)$—thus we have an action of $\pi_1(Y)$ on $[X, x_0; Y, y_0]$. Passing to the quotient then gives the desired result. If it happens that $\pi_1(Y)$ is zero then the action is necessarily trivial, and so the free and the based homotopy classes coincide. Note, though, that we must assume that Y is *path-connected*, otherwise such γ's need not exist.

Homotopy groups, though simple and intuitive to describe, can be very difficult to calculate. Even for the spheres S^n the results are incomplete: Starting with S^1 we have

$$\pi_m(S^1) = \begin{cases} \mathbf{Z} & m = 1 \\ 0 & m > 1 \end{cases} \tag{1.32}$$

While for a general value of n, $n \geq 1$, it is also relatively easy, cf. Bott and Tu [1], to deduce that

$$\pi_m(S^n) = \begin{cases} 0 & m < n \\ \mathbf{Z} & m = n \end{cases} \tag{1.33}$$

However for $m > n$ the picture is more complicated. For example $\pi_m(S^2)$ is not known for all m. It is known though, (cf. again Bott and Tu [1]) that

$$\pi_m(S^n) = \begin{cases} \text{a finite group} & m > n, \quad \text{and } n \text{ odd} \\ \text{a finite group} & m > n, \quad \text{and } n \text{ even but } m \neq 2n - 1 \\ \mathbf{Z} \times F & m > n, \quad n \text{ even and } m = 2n - 1 \end{cases}$$
$$\tag{1.34}$$

where F is some finite group. Despite the intricacy evident in the above results there are some patterns in the values of $\pi_m(S^n)$. These are known as *stability* properties of homotopy groups and emerge when one considers $\pi_{m+n}(S^n)$ for large enough n. More precisely, if one calculates $\pi_{m+n}(S^n)$ for fixed m as n increases, then for $n \geq m + 2$ the result stabilises and one has

$$\pi_{m+n}(S^n) = \pi_{m+n+1}(S^{n+1}) = \cdots = G_m \quad \text{say}, \quad n \geq m + 2 \tag{1.35}$$

where G_m is obtained by setting $n = m + 2$ so that

$$G_m = \pi_{2m+2}(S^{m+2}) \tag{1.36}$$

The homotopy groups of the classical compact Lie groups G are somewhat better known than those of spheres. There is also a pattern in the $\pi_n(G)$ for appropriate n and G. This is the celebrated Bott periodicity and we shall encounter it in chapter 3 when we discuss K-theory.

A much more calculable set of invariants than the homotopy groups are the cohomology groups which we now consider.

§ 4. Cohomology and homology groups

Cohomology groups can be defined in a variety of ways. However, cohomology groups of X are invariants of homotopy type; one particular definition which brings this out this useful point is the definition via what are called *Eilenberg–MacLane spaces*. These latter are spaces with only one non-vanishing homotopy group. To be exact, an Eilenberg–MacLane space is denoted by $K(G, n)$ with G a group and n a positive integer. It has the property that

$$\pi_m\left(K(G,n)\right) = \begin{cases} G & \text{if } m = n \\ 0 & \text{otherwise} \end{cases} \tag{1.37}$$

We already have a simple example: from 1.32 we see that

$$S^1 = K(\mathbf{Z}, 1)$$

In any case given G and n there always exist such spaces, and, though not unique, they are unique up to homotopy equivalence. We use them to define cohomology groups as follows: the nth cohomology group of X with coefficients in the group G is denoted by $H^n(X; G)$ where

$$H^n(X; G) = [X, K(G, n)]_0 \tag{1.38}$$

In passing we note that because $\pi_n(X)$ is always commutative for $n \geq 2$ then the same holds for the group G in the Eilenberg–MacLane space $K(G, n)$ if $n \geq 2$. Now given a $K(G, n)$ it is evident that $\Omega K(G, n+1)$ satisfies

$$\pi_m\left(\Omega K(G, n+1)\right) = \begin{cases} G & \text{if } m = n \\ 0 & \text{otherwise} \end{cases} \tag{1.39}$$

In other words, $\Omega K(G, n+1)$ is also a $K(G, n)$. By our remark about uniqueness up to homotopy equivalence above we see that $\Omega K(G, n+1)$ and $K(G, n)$ are both of the same homotopy type—for that matter the spaces $\Omega^m K(G, n+m)$ $m = 1, 2, \ldots$ are also all $K(G, n)$'s. However, the point here is that since $K(G, n)$ can be expressed as ΩY for a suitable space Y, then $K(G, n)$ is a H-space, and so $[X, K(G, n)]_0$ is automatically endowed with a group structure as befits a cohomology group.

There is also the definition of cohomology via a complex: for this one needs a collection of spaces $E^n(X)$, say, together with linear maps d_n which connect pairs of spaces

$$\cdots \xrightarrow{d_{n-2}} E^{n-1}(X) \xrightarrow{d_{n-1}} E^n(X) \xrightarrow{d_n} E^{n+1}(X) \xrightarrow{d_{n+1}} \cdots \tag{1.40}$$

The complex $\{E^n(X), d_n\}$ must have the property that the composition of two consecutive maps is zero

$$d_n \circ d_{n-1} = 0$$

This crucial property implies that

$$Im\, d_{n-1} \subseteq ker\, d_n \qquad (1.41)$$

If $Im\, d_{n-1} = ker\, d_n$ for each n then the sequence 1.40 is called *exact*—this is false, in general, and the cohomology groups of the complex are a measure of how far the sequence is from being exact. The nth cohomology group of X associated to the complex above is then given by

$$H_E^n(X) = \frac{ker\, d_n}{Im\, d_{n-1}} \qquad (1.42)$$

For example, let X be a differentiable manifold; if d is taken to be the exterior derivative, and $E_n = \Omega^n(X)$, where $\Omega^n(X)$ are the n-forms on X, then the resulting complex is the de Rham complex from which derives the de Rham cohomology of X which is, of course, isomorphic to the real cohomology group $H^n(X; \mathbf{R})$:

$$\cdots \xrightarrow{d_{n-2}} \Omega^{n-1}(X) \xrightarrow{d_{n-1}} \Omega^n(X) \xrightarrow{d_n} \Omega^{n+1}(X) \xrightarrow{d_{n+1}} \cdots$$
$$H_E^n(X) = \frac{ker\, d_n}{Im\, d_{n-1}} = H_{de\, Rham}^n(X) \simeq H^n(X; \mathbf{R}) \qquad (1.43)$$

Passing to the dual will generate the corresponding *homology groups*. We denote the complex dual to $\{E^n, d_n\}$ by $\{E_n, \partial_n\}$; note that the maps now go in the opposite direction: i.e. from E_n to E_{n-1}. In summary we have

$$\cdots \xleftarrow{\partial_{n-2}} E_{n-1}(X) \xleftarrow{\partial_{n-1}} E_n(X) \xleftarrow{\partial_n} E_{n+1}(X) \xleftarrow{\partial_{n+1}} \cdots$$
$$\partial_{n-1} \circ \partial_n = 0, \qquad H_n^E(X) = \frac{ker\, \partial_{n-1}}{Im\, \partial_n} \qquad (1.44)$$

We can choose $\{E_n, \partial_n\}$ to be the singular chain complex for X—this means that E_n denotes the singular n-chains on X and ∂_n is the corresponding boundary map. The homology of the complex is the well known singular homology group, and is isomorphic to the integral cohomology group $H_n(X; \mathbf{Z})$.

Finally we recall a few well known and useful facts. Homology and cohomology groups are finite in number and can be non-trivial only for $0 \leq n \leq \dim X$. For cohomology the sum of all the groups $H^n_E(X)$ forms a *ring* which is denoted by $H^*_E(X)$. The product in the ring $H^*_E(X)$ is known as the cup product and is written as $\cup$. For example, if we take the de Rham complex, where the cup product $\cup$ coincides with the wedge product $\wedge$, then one has

$$H^*_{de\ Rham}(X) \simeq H^*(X; \mathbf{R}) = \bigoplus_{n \geq 0} H^n(X; \mathbf{R})$$

$$\cup : H^m(X; \mathbf{R}) \times H^n(X; \mathbf{R}) \longrightarrow H^{m+n}(X; \mathbf{R}) \qquad (1.45)$$

$$([\omega], [\nu]) \longmapsto [\omega] \cup [\nu] = [\omega \wedge \nu]$$

where, in the above, $[\omega]$ and $[\nu]$ represent cohomology classes in $H^m(X; \mathbf{R})$ and $H^n(X; \mathbf{R})$; these in turn are determined by ω and ν which are closed differential forms belonging to $\Omega^m(X)$ and $\Omega^n(X)$ respectively.

Homology and cohomology groups are usually much easier to calculate than homotopy groups—we have seen that the homotopy groups, $\pi_m(S^n)$, for spheres are still not fully known. In contrast, for the less refined homology and cohomology invariants, a simple calculation yields the result

$$H^m(S^n; \mathbf{Z}) \simeq H_m(S^n; \mathbf{Z}) = \begin{cases} \mathbf{Z} & \text{if } m = 0 \text{ or } n \\ 0 & \text{otherwise} \end{cases} \qquad (1.46)$$

There is, though, an important relation, between the homotopy and the homology of X, known as the Hurewicz isomorphism. This is the statement that the first non-vanishing homotopy and homology groups of a path connected X occur at the same dimension and are isomorphic. One should add to this that homology (and cohomology) groups are always commutative, but homotopy groups $\pi_n(X)$ are only necessarily commutative for $n \geq 2$; should it happen that the first non-vanishing homotopy group of X is a *non-commutative* $\pi_1(X)$ (e.g. X is a Riemann surface of genus 2 say), then the Hurewicz isomorphism statement above is modified to say that $H_1(X; \mathbf{Z})$ is isomorphic to the *commutative part* of $\pi_1(X)$—the commutative part of $\pi_1(X)$ is constructed by taking the quotient of $\pi_1(X)$ by its commutator subgroup $[\pi_1(X), \pi_1(X)]$. In short one has

$$H_1(X; \mathbf{Z}) \simeq \pi_1(X)/[\pi_1(X), \pi_1(X)]$$

§ 5. Fibre bundles and fibrations

The notion of a fibre bundle derives directly from naturally occuring structures in mathematics and in this section we just wish to recall, for the convenience of the reader, some of the standard properties, cf. Nash and Sen [1] and Husemoller [1] for more details.

One of the most basic examples is the tangent bundle TM to a differentiable manifold M. The tangent bundle TM is the collection of all the tangent spaces to M and is given by

$$TM = \bigcup_{p \in M} T_p M$$

where $T_p M$ is the tangent space to M at the point p. Every bundle has fibres, a total space, and a base space. For the tangent bundle the fibres are the vector spaces $T_p M$, the total space is TM, and the base space is M; when the fibres are vector spaces the bundle is often referred to as a vector bundle; and the fibres always have the same dimension. In this case the dimension is $n = \dim M$ and TM is called a vector bundle of rank n. One can go on to consider vector bundles of various ranks, for example new bundles can be constructed by carrying out the standard operations of linear algebra on the fibres of TM: one can form direct sums, tensor products, pass to the dual, and so on. Underlying such bundles is a principal bundle P with total space P, base M and fibre G where $G = GL(n, \mathbf{R})$, this group being the group which transforms the fibres of the various vector bundles into one another. Every fibre bundle is locally trivial, this means that, given a bundle E with fibre F, projection π and base M, there is a covering of the base by open sets $\{O_\alpha\}$ such that over each O_α the bundle is homeomorphic to a direct product, i.e.

$$\pi^{-1}(O_\alpha) \simeq O_\alpha \times F \tag{1.47}$$

An important property of bundles is their behaviour under the operation of pullback—recall that the notion of pullback is often used when dealing with the cohomology groups $H^n(M; G)$ of a space M—the principal ingredient in any pullback calculation is a map $f : N \to M$, then if $\omega \in H^n(M; G)$ one can 'pull back' ω to $f^*\omega$, so that $f^*\omega \in H^n(N; G)$. Thus any knowledge of the cohomology ring $H^*(M; G)$ on M induces, via the pullback, some knowledge of the cohomology ring $H^*(N; G)$ on N, this knowledge being expressible as the object $f^*H^*(M; G)$. In the case of bundles one pulls back bundles instead of cohomology groups. Thus if E is a bundle over M then f^*E is a bundle over N. Also, as is the case for cohomology, two homotopic maps produce identical pullbacks: i.e. if $f_1, f_2 : N \to M$ *and $f_1 \simeq f_2$ then the two pullback bundles are isomorphic*

$$f_1^*E \simeq f_2^*E \tag{1.48}$$

The notion of pullback also plays a key rôle in the classification of bundles. This is because there exist so called universal bundles E_G over a base space BG, known as a classifying space; and any bundle E with structure group G, over a space M, arises as a pullback f^*E_G for some map $f : M \to BG$ (this classification of bundles is treated in Nash and Sen [1], and at greater length in Husemoller [1] and Bott and Tu [1]). Since homotopic maps produce indistinguishable pullbacks, the set of (isomorphism classes) of bundles over M with group G is in one to one correspondence with the homotopy classes

$$[M, BG] \tag{1.49}$$

and as $[f] \in [M, BG]$ runs through all possible choices then f^*E_G runs through all possible bundles E over M with group G. In practice one needs some means of detecting the non-triviality of a bundle E—the construction of E via a classifying space provides us with just that: The point is that the cohomology of the classifying spaces BG is well known for the various commonly encountered groups G. Thus if a map $f : M \to BG$ is used to construct the bundle $E = f^*E_G$, then this same map f can be used to pull back elements of $H^*(BG)$ to $H^*(M)$. More precisely the *generators*, c_i say, of the cohomology ring $H^*(BG)$ are called universal characteristic classes, and their pullbacks f^*c_i are the characteristic classes which belong to the bundle $E = f^*E_G$. We have therefore

$$\begin{aligned} f^* : H^*(BG) &\longrightarrow H^*(M) \\ c_i &\longmapsto f^*c_i \end{aligned} \tag{1.50}$$

If one chooses the base space M to be a real or complex differentiable manifold, then the structure group G for a bundle E is naturally $U(n)$ or $O(n)$ respectively, where n is the real or complex dimension of M, whichever is appropriate. The corresponding characteristic classes $f^*c_i \in H^*(M; \mathbf{Z})$ are the Pontrjagin classes in the real case and the Chern classes in the complex case and they are determined ultimately by $H^*(BO(n); \mathbf{Z})$ or $H^*(BU(n); \mathbf{Z})$ respectively. In the real case one also considers cohomology classes with $\mathbf{Z}_2$ coefficients, these are the universal Stiefel–Whitney classes which belong to $H^*(BO(n); \mathbf{Z}_2)$; finally, in the case where the structure group reduces from $O(n)$ to $SO(n)$ with n *even*, there is an extra class—the celebrated Euler class.

A bundle E with fibre F, base M, and a projection π which preserves base points, is often represented by

$$F \xrightarrow{i} E \xrightarrow{\pi} M \tag{1.51}$$

where i denotes inclusion. From this follows the valuable fibre homotopy (exact) sequence

$$\cdots \longrightarrow \pi_n(F) \longrightarrow \pi_n(E) \longrightarrow \pi_n(M) \longrightarrow \pi_{n-1}(F) \longrightarrow \pi_{n-1}(E) \longrightarrow \cdots$$
$$\cdots \longrightarrow \pi_0(F) \longrightarrow \pi_0(E) \longrightarrow \pi_0(M) \longrightarrow 0$$
$$(1.52)$$

This sequence contains extremely useful information relating the homotopy groups of the total space, fibre and base—for a trivial bundle E one has $E = M \times F$ which implies immediately that $\pi_n(E) \simeq \pi_n(M) + \pi_n(F)$. In general this is false and one must fall back upon the fibre homotopy sequence. Unfortunately there is no simple relation between the *cohomology* groups of E, F and M; these latter have a much more complicated inter-relationship which is expressed by spectral sequences, cf. Bott and Tu [1].

An extremely fruitful generalisation of the idea of a fibre bundle is that of a *fibration*. The term fibration is applied to a map π of the form

$$\pi : E \longrightarrow M \qquad (1.53)$$

when π possesses the *homotopy lifting property*. The content of this is as follows: take another space F and a map $f : F \to E$, this map f can immediately be composed with π to give the map $g = \pi \circ f : F \to M$. This gives the situation

$$\begin{array}{ccc} F & \xrightarrow{f} & E \\ & {\scriptstyle g}\searrow & \downarrow \pi \\ & & M \end{array} \qquad (1.54)$$

and one refers to f as a *lift* of g. Now suppose that one extends g to a homotopy class of mappings g_t such that $g_0 = g$, and suppose also that this homotopy of g lifts to give a corresponding homotopy f_t of f, with $f_0 = f$, then $\pi : E \to M$ has the *homotopy lifting property*. We can summarise all this in the diagram below

$$\begin{array}{ccc} F & \xrightarrow{f_t} & E \\ & {\scriptstyle g_t}\searrow & \downarrow \pi \\ & & M \end{array} \qquad (1.55)$$

$$g_0 = g \quad f_0 = f \quad g_t = \pi \circ f_t$$

If E happens to already be a *fibre bundle* with projection π, then π is automatically a fibration, albeit a locally trivial one. Above each point $x \in M$ lies the fibre at x given by $\pi^{-1}(x)$—the fibre above the base point $*$ is known simply as the fibre F of the fibration. Thus one has $F = \pi^{-1}(*)$.

The various fibres $\pi^{-1}(x)$ of a fibration need not all be homeomorphic, as they are for a fibre bundle; they are, however, all of the same homotopy type.

Such fibrations $F \to E \to M$ also satisfy an exact fibre homotopy sequence identical to the one already given above. In addition the cohomology groups of E, F and M are also related via spectral sequences.

An important example of this latter kind of fibration is the *path fibration:* Take as total space E the based paths on M and denote this space by PM; the precise definition of PM is that if $I = [0, 1]$ then

$$PM = Map_0 [I, M] \qquad (1.56)$$

and the relevant base point information is that 0 is the base point on $[0, 1]$ and $*$ is the base point on M—thus an element of PM, or a based path, is a map from $[0, 1]$ to M which maps 0 to $*$. Each such path has an end point x, say, and this suggests at once that the fibration π be defined as the map which assigns to each path its end point. So far then we have the picture

$$
\begin{array}{c}
PM \\
\downarrow \pi \\
M
\end{array}
\qquad (1.57)
$$

The fibre $\pi^{-1}(x)$ above a point $x \in M$ is obviously all the paths from the base point $*$ to x. However, a fuller description is possible: if one chooses at the beginning one fixed path, $p_0(x)$ say, then one can describe any other path $p(x)$ in $\pi^{-1}(x)$ by saying how it differs from $p_0(x)$. But a pair such as $p_0(x)$, $p(x)$ join naturally to form a loop through $*$, and so they determine an element of the loop space ΩM. Finally, although one can see that one does not obtain all elements of ΩM in this way, it is now possible to prove that, up to homotopy, $\pi^{-1}(x) \simeq \Omega M$: this just requires two observations. First note that if we consider two different fibres $\pi^{-1}(x)$ and $\pi^{-1}(y)$ then one is homotopic to the other, i.e. $\pi^{-1}(x) \simeq \pi^{-1}(y)$; secondly notice that the fibre above the base point $*$ is simply the loop space ΩM, hence we have our result. Thus the fibre F above the base point is ΩM and we can complete the picture of the fibration to give

$$
\begin{array}{ccc}
\Omega M & \longrightarrow & PM \\
& & \downarrow \pi \\
& & M
\end{array}
\qquad (1.58)
$$

The map π can be easily seen to have the homotopy lifting property so that π is indeed an example of a fibration. This encoding of ΩM as the fibre of a fibration turns out to be a crucial step in calculating its cohomology via the technique of spectral sequences.

§ 6. Differentiable structures for manifolds

In §1 we distinguished three types of manifold M: M_{TOP}, M_{PL} and M_{DIFF}. It is possible to describe the more important differences between these three types using the topological techniques introduced in the previous sections.

Consider then a manifold[2] M of dimension n. The type of M is determined, as we saw in §1, by examining whether the transition functions[3] $g_{\alpha\beta}$ are homeomorphisms, piecewise-linear maps, or diffeomorphisms. Now, since the transition functions are maps from one subset of $\mathbf{R}^n$ to another, we introduce the groups TOP_n, PL_n and $DIFF_n$ which are all the homeomorphisms, piecewise-linear maps, and diffeomorphisms of $\mathbf{R}^n$ respectively. We can consider these three groups as functions of n, then as n increases we are naturally led to write down three sets of inclusions:

$$
\begin{array}{ccccccccc}
TOP_1 & \subset & TOP_2 & \subset & \cdots & \subset & TOP_n & \subset & \cdots \\
PL_1 & \subset & PL_2 & \subset & \cdots & \subset & PL_n & \subset & \cdots \\
DIFF_1 & \subset & DIFF_2 & \subset & \cdots & \subset & DIFF_n & \subset & \cdots
\end{array}
\qquad (1.59)
$$

For each of the three sets of inclusions we pass to the largest limiting space—this is known as taking the direct limit[4]—and construct the three limiting groups

$$TOP, \qquad PL, \qquad DIFF$$

With these three groups are associated the classifying spaces $BTOP$, BPL and $BDIFF$—these are just like the classifying spaces BG introduced in the previous section to classify fibre bundles over M with group G. In the case at hand the transition functions $g_{\alpha\beta}$ are those of the tangent bundle to M; and there are three possible tangent bundles depending on the type of M. We denote these tangent bundles by TM_{TOP}, TM_{PL} and TM_{DIFF} in an obvious notation. We can use the classifying spaces $BTOP$, BPL, and

---

[2] In this section M will always be assumed to be compact, connected and closed unless we indicate the contrary.

[3] Of course in the PL case we assume that a PL map has a PL inverse.

[4] The direct limit, or inductive limit, is a standard notion which is used in topology. It is defined for a sequence of objects O_i connected by maps ϕ_i which satisfy appropriate properties. It is denoted by $\varinjlim O_i$. By taking the dual of everything one obtains the corresponding inverse limit, or projective limit, which is written as $\varprojlim O_i$. for more details see, for example, Spanier [1].

$BDIFF$ to determine the tangent bundles TM_{TOP}, TM_{PL} and TM_{DIFF}: one simply selects an element of the homotopy classes

$$[M, BTOP] \qquad [M, BPL] \qquad \text{and} \qquad [M, BDIFF]$$

respectively.

It is now natural to ask when can one 'straighten out' a topological manifold to make it piecewise-linear; and also, when can one 'smooth' a piecewise-linear manifold to make it differentiable? Both these questions can be answered using fibrations.

Taking the first of them, so that we are comparing piecewise-linear and topological structures on M, consider BPL and $BTOP$. One can fibre the first space over the second, it is possible to check that the fibre over the base point $*$ is then simply TOP/PL. The fibration is therefore of the form

$$
\begin{array}{ccc}
TOP/PL & \longrightarrow & BPL \\
& & \downarrow \pi \\
& & BTOP
\end{array}
\qquad (1.60)
$$

A possible method for 'straightening out' a PL manifold is now apparent: select a topological manifold by choosing a map $\alpha : M \to BTOP$, now if we can *factorise* α through BPL then M acquires a PL structure. We show this below

$$
\alpha : M \xrightarrow{\beta} BPL \xrightarrow{\pi} BTOP \qquad \equiv \qquad
\begin{array}{ccc}
& & BPL \\
& {}^{\beta}\nearrow & \downarrow \pi \\
M & \xrightarrow{\alpha} & BTOP
\end{array}
$$
with $\quad \alpha = \pi \circ \beta$

In other words, the existence of the map $\beta : M \to BPL$ satisfying $\alpha = \pi \circ \beta$ provides M with a PL structure and is a lifting of the map α from the base $BTOP$ to the total space BPL. Actually this method does work *provided* $\dim M \geq 5$, cf. Kirby and Siebenmann [1].[5] This recasting of a factorisation problem as a lifting problem is standard in homotopy theory. The techniques for analysing it belong to what is called obstruction theory, for which two very good references are Steenrod [1] and Whitehead [1]. It is then standard that one can proceed as follows: If one has a fibration $F \to E \xrightarrow{\pi} B$ and one wishes to lift a map $\alpha : M \to B$ so that we have again

$$
\alpha : M \xrightarrow{\beta} E \xrightarrow{\pi} B \qquad \equiv \qquad
\begin{array}{ccc}
& & E \\
& {}^{\beta}\nearrow & \downarrow \pi \\
M & \xrightarrow{\alpha} & B
\end{array}
\qquad (1.61)
$$
with $\quad \alpha = \pi \circ \beta$

---

[5] We also assume that M is closed. If M has a boundary then we require $\dim M \geq 6$.

then the obstructions to the lifting are cohomology classes of the form

$$H^{n+1}(M; \pi_n(\mathcal{F})) = H^{n+1}(M; \pi_n(F)) \tag{1.62}$$

In 1.62 above $\pi_n(\mathcal{F})$ denotes cohomology with what are called local coefficients, but, as the equation indicates, the coefficients may be taken[6] to be just $\pi_n(F)$. Thus in our case the obstructions are in

$$H^{n+1}(M; \pi_n(TOP/PL)) \tag{1.63}$$

Fortunately, due to the work of Kirby and Siebenmann, it is known that only one of these cohomology classes is non-zero, for they found that

$$\pi_n(TOP/PL) = \begin{cases} \mathbf{Z}_2 & \text{if } n = 3 \\ 0 & \text{otherwise} \end{cases} \tag{1.64}$$

In other words, TOP/PL has only one non-trivial homotopy group and this group is known to be $\mathbf{Z}_2$. The actual cohomology class, $e(M)$ say, forming the obstruction to the lifting is therefore an element of[7]

$$H^4(M; \mathbf{Z}_2) \qquad \dim M \geq 5 \tag{1.65}$$

Now if the class $e(M)$ *vanishes* then the map β exists and furnishes M with a PL structure; and it is natural to go on to ask how many such β's exist? More precisely, since it is the homotopy class in $[M, BPL]$ that provides M with its PL structure, we ask how many distinct homotopy classes of such β's are there? This is also a standard question of obstruction theory and the answer is that the relevant homotopy classes are just the whole cohomology group

$$H^n(M; \pi_n(TOP/PL)) \tag{1.66}$$

However, in our case we have seen that the only non-trivial case occurs for $n = 3$. Thus we see that if a closed topological manifold M, with

---

[6] The point at issue is that the local coefficients $\pi_n(\mathcal{F})$ are really a set of copies of $\pi_n(F_b)$ with F_b the fibre above the point $b \in B$; these homotopy groups are all isomorphic since all fibres have the same homotopy type. However there is no natural isomorphism between them unless the fundamental group $\pi_1(B)$ acts trivially on them, for details cf. Whitehead [1] and Steenrod [1].

[7] One could also work with an element $d \in H^4(BTOP; \mathbf{Z}_2)$ where d is such that it is pulled back to $e(M)$ by the map α, i.e. $\alpha^* d = e(M)$, then d is referred to as the *universal* obstruction for the lifting problem.

$\dim M \geq 5$, acquires a PL structure by the lifting process just described, then the possible distinct PL structures are isomorphic to

$$H^3(M; \mathbf{Z}_2) \qquad (1.67)$$

which is not zero in general. The fact that there can be distinct PL structures on the same topological manifold M is sometimes referred to as saying that the Hauptvermutung is false. This result is of course very important: it means that one can construct distinct PL manifolds which, though they are all homeomorphic to one another, possess no invertible PL maps taking one into the next. Also if the Kirby–Siebenmann class $e(M) \in H^4(M; \mathbf{Z}_2)$ *does not vanish* then the topological manifold M cannot be given *any* PL structure.

If $\dim M < 5$ then all the interest lies in dimension 4. For $\dim M \leq 2$ Rado [1] proved that the notions PL and TOP coincide, while for $\dim M = 3$ this same fact was established by Moise [1]. Finally if $\dim M = 4$ then these notions do not coincide. However, before commenting further on this case we wish to give an account of the smoothing problem.

Similar ideas are used to address the question of smoothing a piecewise-linear manifold—however the results are different. Let us assume that M is a closed PL manifold with $\dim M \geq 5$. This time the fibration is

$$
\begin{array}{ccc}
PL/DIFF & \longrightarrow & BDIFF \\
& & \downarrow \pi \\
& & BPL
\end{array}
\qquad (1.68)
$$

The smoothing of a piecewise-linear M can be handled with obstruction theory and leads us immediately to the consideration of the homotopy groups $\pi_n(PL/DIFF)$. But the non-trivial homotopy groups of the fibre are much more numerous than in the piecewise-linear case. In fact one has

$$
\pi_n\,(PL/DIFF) = \begin{cases}
0 & \text{if } n = 5, 6 \\
\mathbf{Z}_{28} & \text{if } n = 7 \\
\mathbf{Z}_2 & \text{if } n = 8 \\
\vdots & \vdots \\
\mathbf{Z}_{992} & \text{if } n = 11 \\
\vdots & \vdots
\end{cases}
\qquad (1.69)
$$

From 1.69 above we see that if we choose $M = S^7$ then the first non-triviality occurs when $n = 7$ and so the obstruction to smoothing S^7 lies in

$$H^8(S^7; \pi_7(PL/DIFF)) \qquad (1.70)$$

which is of course zero for elementary dimensional reasons—this means that S^7 can be smoothed, a fact which we know from first principles. However, by the obstruction theory introduced above, the resulting smooth structures are isomorphic to

$$H^7(S^7; \pi_7(PL/DIFF)) = H^7(S^7; \mathbf{Z}_{28}) = \mathbf{Z}_{28} \qquad (1.71)$$

Hence we have the celebrated result (cf. Milnor [1] and Kervaire and Milnor [1]) that S^7 has 28 distinct differentiable structures 27 of which correspond to what are known as *exotic* spheres.

Now if $\dim M \leq 3$ then there is a bijection between PL and DIFF structures on M, cf. Kirby and Siebenmann [2] and Kuiper [1]—this leaves us with the case towards which we have been working in this section, namely $\dim M = 4$.

In four dimensions there are phenomena which have no counterpart in any other dimension. First of all there are topological 4-manifolds which have no smooth structure, though if they have a PL structure then they possess a unique smooth structure. Secondly the impediment to the existence of a smooth structure is of a completely different type to that met in the standard obstruction theory—it is not the pullback of an element in the cohomology of a classifying space, i.e. it is not a characteristic class. Also the 4-dimensional story is far from completely known. Nevertheless, there are some very striking results dating from the early 1980s onwards.

In Freedman [1] all, simply connected, *topological* 4-manifolds were classified by what is known as their intersection form q. We must define q and this can be achieved in more than one way, from which we choose the following: in brief q is a certain quadratic form constructed from the cohomology of M. Take two elements α and β of $H^2(M; \mathbf{Z})$ and form their cup product $\alpha \cup \beta \in H^4(M; \mathbf{Z})$; then we define $q(\alpha, \beta)$ by

$$q(\alpha, \beta) = (\alpha \cup \beta)[M] \qquad (1.72)$$

where $(\alpha \cup \beta)[M]$ denotes the integer obtained by evaluating $\alpha \cup \beta$ on the generating cycle $[M]$ of the top homology group $H_4(M; \mathbf{Z})$ of M. Poincaré duality ensures that such a form is always non-degenerate over $\mathbf{Z}$ and so has $det\, q = \mp 1$; q is then called unimodular. Also we refer to q, as *even* if all its diagonal entries are even, and as *odd* otherwise. The result of Freedman leads to the following:

Theorem (Freedman) *A simply connected 4-manifold M with even intersection form q belongs to a unique homeomorphism class, while if q is odd* *there are precisely two non-homeomorphic M with q as their intersection form.*

This is a very powerful result—the intersection form q very nearly determines the homeomorphism class of a simply connected M, and actually only fails to do so in the odd case where there are still just two possibilities. Further *every* unimodular quadratic form occurs as the intersection form of some manifold.

An illustration of the impressive nature of Freedman's work is rather easy to provide. Let us choose M to be the sphere S^4, since S^4 has trivial cohomology in two dimensions then its intersection form q is the zero quadratic form and is of course even; we write this as $q = \emptyset$. Now recollect that the Poincaré conjecture in four dimensions is the statement that any homotopy 4-sphere, S_h^4 say, is actually *homeomorphic* to S^4. Well, since a homotopy 4-sphere has the same homotopy type as S^4, and since a manifold with the same homotopy type as S^4 has the same cohomology, then it follows that any S_h^4 also has intersection form $q = \emptyset$. But Freedman's result says that for a simply connected M with even q there is only one homeomorphism class for M, therefore S_h^4 *is* homeomorphic to S^4 and we have established the conjecture. Incidentally this means that the Poincaré conjecture has now been proved for all n except $n = 3$—the case originally proposed by Poincaré.

There are also spectacular results for smoothability in four dimensions which we will now sketch—we will have cause to return to them in a later chapter. First, following Freed and Uhlenbeck [1], we point out that Freedman's result combined with a much earlier result of Rohlin [1] gives us an example of a non-smoothable 4-manifold: Rohlin's theorem asserts that given a smooth, simply connected, 4-manifold with even intersection form q, then the signature[8] $\sigma(q)$ of q is divisible by 16. Now if we set

$$q = \begin{pmatrix} 2 & -1 & 0 & 0 & 0 & 0 & 0 & 0 \\ -1 & 2 & -1 & 0 & 0 & 0 & 0 & 0 \\ 0 & -1 & 2 & -1 & 0 & 0 & 0 & 0 \\ 0 & 0 & -1 & 2 & -1 & 0 & 0 & 0 \\ 0 & 0 & 0 & -1 & 2 & -1 & 0 & -1 \\ 0 & 0 & 0 & 0 & -1 & 2 & -1 & 0 \\ 0 & 0 & 0 & 0 & 0 & -1 & 2 & 0 \\ 0 & 0 & 0 & 0 & -1 & 0 & 0 & 2 \end{pmatrix} = E_8 \qquad (1.73)$$

(E_8 is actually the Cartan matrix for the exceptional Lie algebra e_8) then, by inspection, q is even, and by calculation, it has signature 8. Thus by Freed-

---

[8] The signature of q is defined to be the difference between the number of positive and negative eigenvalues of q.

man's theorem there is a single, simply connected, 4-manifold with intersection form $q = E_8$. However, by Rohlin's theorem it cannot be smoothed
since its signature is 8.

The next breakthrough was due to Donaldson [1]. Donaldson's theorem is applicable to definite forms q, which by appropriate choice of orientation on M we can take to be *positive* definite. One has:

Theorem (Donaldson) *A simply connected, smooth 4-manifold, with positive definite intersection form q is always diagonalisable over the integers
to $q = diag(1, \ldots, 1)$*

Immediately one can go on to deduce that no, simply connected, 4-manifold
for which q is even and positive definite can be smoothed! For example
the manifold with $q = E_8 \oplus E_8$ has signature 16 and so gets by Rohlin's
theorem. But since E_8 is even then so is $E_8 \oplus E_8$ and so Donaldson's theorem
forbids such a manifold from existing smoothly. Before Donaldson's work
surgery techniques had been extensively used to try to construct smoothly
the manifold with intersection form $E_8 \oplus E_8$. We can now see that these
techniques were destined to fail.

In fact, in contrast to Freedman's theorem, which allows *all* unimodular quadratic forms to occur as the intersection form of some topological
manifold, Donaldson's theorem says that in the positive definite, smooth,
case only *one* quadratic form is allowed, namely I.

One of the most striking aspects of Donaldson's work is that his proof
uses the Yang–Mills equations. Since these equations will figure largely in
this book we shall only outline what is involved here; also the book Freed and
Uhlenbeck [1] is devoted to a detailed exposition of Donaldson's theorem.

In brief then let A be a connection on a principal $SU(2)$-bundle over
a simply connected 4-manifold M with positive definite intersection form.
If the curvature 2-form of A is F then F has an L^2 norm which is the
Euclidean Yang–Mills action S. One has

$$S = \|F\|^2 = -\int_M tr(F \wedge *F) \qquad (1.74)$$

where $*F$ is the usual dual 2-form to F. The minima of the action S are
given by those A, called instantons, which satisfy the famous self-duality
equations

$$F = *F \qquad (1.75)$$

Given one instanton A which minimises S one can perturb about A in an
attempt to find more instantons. This process is successful and the space of
all instantons can be fitted together to form a global moduli space of finite

dimension, cf. chapter 8. For the instanton which provides the absolute minimum of S, the moduli space $\mathcal{M}$ is a non-compact space of dimension 5.

We can now summarise the logic that is used to prove Donaldson's theorem: there are very strong relationships between M and the moduli space $\mathcal{M}$; for example, let q be regarded as an $n \times n$ matrix with precisely p unit eigenvalues (clearly $p \leq n$ and Donaldson's theorem is just the statement that $p = n$), then $\mathcal{M}$ has precisely p singularities which look like cones on the space CP^2. These combine to produce the result that the 4-manifold M has the same topological signature $Sign\,(M)$ as p copies of CP^2; now p copies of CP^2 have signature $a - b$ where a of the CP^2's are oriented in the usual fashion and b are given the opposite orientation. Thus we have

$$Sign\,(M) = a - b \qquad (1.76)$$

Now the definition of $Sign\,(M)$ is that it is the signature $\sigma(q)$ of the intersection form q of M. But since, by assumption, q is positive definite $n \times n$ then $\sigma(q) = n = Sign\,(M)$. So we can write

$$n = a - b \qquad (1.77)$$

However, $a + b = p$ and $p \leq n$ so we can assemble this information in the form

$$n = a - b, \qquad p = a + b \leq n \qquad (1.78)$$

but one always has $a + b \geq a - b$ so now we have

$$n \leq p \leq n \Rightarrow p = n \qquad (1.79)$$

and we have obtained Donaldson's theorem.

A final word about *non-compact* manifolds. There is a result due to Quinn [1] which says that non-compact manifolds are always smoothable. However, if we choose $M = \mathbf{R}^n$ then, for $n \neq 4$ there is only one smooth structure; but for $n = 4$ there is an exotic differentiable structure on $\mathbf{R}^4$, i.e. there exists a smooth $\mathbf{R}^4$ which is homeomorphic, but not diffeomorphic, to the standard $\mathbf{R}^4$. This is a consequence of Freedman's and Donaldson's work, cf. Gompf [1] and Freed and Uhlenbeck [1]. An exotic $\mathbf{R}^4$ is referred to as a *fake* $\mathbf{R}^4$. It is now known, Gompf [2], that there is a continuum of exotic differentiable structures that can be placed on $\mathbf{R}^4$. This is in sharp contrast to the finite discrete number of exotic structures that arise in the use of obstruction theory with compact manifolds. It is clear that this section shows that four dimensions is special in very many ways, and that, as yet, much remains to be understood; a useful reference is Kirby [1].

CHAPTER II

Elliptic Operators

§ 1. The meaning of ellipticity

In this chapter we shall discuss elliptic operators; and we shall see that these operators may be standard differential operators, or more generally, pseudo-differential operators.

In the next paragraph we meet our first elliptic operator but we begin with some introductory remarks. The property of ellipticity for an operator is of great importance in the theory of partial differential equations. This importance continually carries over into other branches of mathematics because of the appearance there of elliptic differential operators. For example, when these differential operators also carry some geometrical structure, e.g. the Laplacian on a compact Riemannian manifold M (which requires a metric for its definition), then both geometry and analysis come into play. Further, we can proceed from the geometrical to the topological because the Laplacian is also a source of topological information about M via Hodge–de Rham theory which provides us with the celebrated result (to which we return in § 3 of this chapter) that

$$\dim ker\, \Delta_p = \dim H^p(M; \mathbf{R}) \tag{2.1}$$

where Δ_p is the Laplacian on p-forms. This in its turn can be thought of as a point of departure for the immensely powerful Atiyah–Singer index theory for elliptic operators on compact manifolds which we shall discuss in chapter 4. Thus the theory of elliptic operators is central to differential topology.

We begin the process of defining ellipticity by taking some simple classical examples; we go on then to try and abstract from these examples what is essential to the notion of ellipticity. This motivates the formulation of a much more general definition which we shall need for this book. Begin then with the Laplacian Δ on the Euclidean space $\mathbf{R}^n$. To conform with common practice in differential geometry we multiply it by a minus sign so as to make Δ a positive operator. Thus we define

$$\Delta = -\frac{\partial^2}{\partial x_1^2} - \cdots - \frac{\partial^2}{\partial x_n^2} \tag{2.2}$$

Under Fourier transformation [1] this operator becomes simply multiplication by the expression

$$p_1^2 + \cdots + p_n^2 \tag{2.3}$$

where $p_1 \ldots p_n$ are the variables conjugate to $x_1 \ldots x_n$. Notice that 2.3 is simply a quadratic form, σ say, in the p_i and if we set σ equal to a constant we obtain a sphere. Now if we generalise the Laplacian slightly by making changes of scale on the x_i, we obtain the operator L given by

$$L = -\sum_i a_i \frac{\partial^2}{\partial x_i^2} \tag{2.4}$$

Under Fourier transformation L acts as multiplication by the quadratic form $\sigma(L)$ where

$$\sigma(L) = \sum_i a_i p_i^2 \tag{2.5}$$

and on setting $\sigma(L)$ equal to a constant c we obtain the equation

$$a_1 p_1^2 + \cdots + a_n p_n^2 = c \tag{2.6}$$

This represents an ellipsoid in $\mathbf{R}^n$ if the quadratic form $\sigma(L)$ has signature $+n$. If that is the case the operator L is called *elliptic*. We can now proceed to examine the beneficial properties that ellipticity bestows on L.

The most important property is that if f is a C^∞ function then the function u which is the solution to the equation

$$Lu = f \tag{2.7}$$

is automatically C^∞. It is instructive to follow an outline proof of this statement: if $G(x)$ is a Green's function (also referred to as a *fundamental solution* in the mathematical literature), then we have the solution formula

$$u(x) = G * f \qquad \text{where} \qquad LG(x) = \delta(x)$$
$$= \int_{\mathbf{R}^n} G(x - y) f(y) dy \tag{2.8}$$

---

[1] We define the Fourier transform $\mathcal{F}(f)$ of a function $f(x)$ by the formula $\mathcal{F}(f) = \frac{1}{(2\pi)^n} \int \exp(ipx) f(x)\, d^n x$.

Given this solution formula, then one knows that the presence of the Dirac delta function $\delta(x)$ in the equation for $G(x)$ means that $G(x)$ is not C^∞ at $x = 0$; however, provided $G(x)$ is C^∞ for $x \neq 0$, then it is easy to check, e.g. by attempting to differentiate the formula above, that $u(x)$ is indeed C^∞—it remains to establish that $G(x)$ is C^∞ for $x \neq 0$ and to do this we write

$$G(x) = \int_{\mathbf{R}^n} \frac{\exp[-ipx]}{\sum_i a_i p_i^2} dp \qquad (2.9)$$

Now the denominator of the above integral contains the quadratic form $\sum_i a_i p_i^2 = \sigma(L)$, and, by the assumption of ellipticity, $\sigma(L)$ only vanishes at the origin $p_i = 0$. Finally, by changing the scale of the variables p_i by a_i to $\tilde{p}_i$ say, and then using n-dimensional spherical polar coordinates, one sees that the only important integral is that over $\tilde{p} = \sqrt{\tilde{p}_1^2 + \cdots + \tilde{p}_n^2}$ which is done by appropriate use of Cauchy's theorem. Moreover, the fact that $\tilde{p}$ only vanishes at the origin (ellipticity again) means automatically that the Cauchy integral $G(x)$ is singularity free, and well behaved as $|\tilde{p}| \to \infty$ for $x \neq 0$. Thus $G(x)$ is C^∞ for $x \neq 0$ as required. Notice, though, that had the quadratic form $\sigma(L)$ vanished away from the origin then the whole argument would have failed. Let us move on to look at such cases.

Suppose then that the quadratic form $\sum_i a_i p_i^2 = \sigma(L)$ has signature $n-2$, i.e. it has $n-1$ positive eigenvalues and 1 negative eigenvalue, then the associated operator L is termed *hyperbolic*. For an example of a hyperbolic operator consider the wave equation—more precisely set $n = 2$ and set

$$L = \frac{\partial^2}{\partial x_1^2} - \frac{\partial^2}{\partial x_2^2}$$
$$\Rightarrow \sigma(L) = p_1^2 - p_2^2$$

Thus the so-called *level sets* or *level curves* of L, given by the equations $\sigma(L) = $ constant, are now hyperbolae rather than ellipses. We also expect that the solutions to $Lu = f$, with a smooth f, may contain singularities. We expect, too, the origin of these singularities to be traceable to singular behaviour of the Green's function $G(x)$.

All this is true and is rather well known: let us replace (x_1, x_2) by the more common variables (x, t) and examine the situation in some more detail. A Green's function for L is given by

$$G(x, t) = \frac{1}{2}\theta(x + t)\theta(x - t) \qquad \text{where} \quad \theta(x) = \begin{cases} 1 & \text{if } x \geq 0 \\ 0 & \text{if } x < 0 \end{cases} \qquad (2.10)$$

and $G(x, t)$ is thus singular everywhere *along the lines* $x = \mp t$. Also, if we make the simple C^∞ choice $f = 0$ for f, then among the solutions to

$Lu = f = 0$ is

$$u(x, t) = (x + t)^{5/2} \qquad (2.11)$$

which is also singular along the line $x = -t$ (e.g. its third derivative is infinite when $x = -t$). To define precisely the RHS of 2.11 we choose the principal branch of the fractional power which corresponds to $(x + t)^{5/2}$ positive, for $(x + t)$ positive, and to $(x + t)^{5/2}$ being pure imaginary with *positive* imaginary part for $(x + t)$ negative. Then $u(x, t)$ is only C^2 but it does obey the wave equation.

In fact the solving of the wave equation is a Cauchy problem, and it requires the specification of the usual Cauchy data for u and its first derivative at some initial time t_0. Consider then the general wave equation i.e.

$$Lu = f \qquad \text{with} \quad L = \frac{\partial^2}{\partial t^2} - \frac{\partial^2}{\partial x_1^2} - \cdots - \frac{\partial^2}{\partial x_n^2} \qquad (2.12)$$

the lines $x = \mp t$ become the familiar light-cones of an $(n + 1)$-dimensional Minkowski space. We can now characterise the circumstances under which $Lu = f$ can have a singular solution: let the Cauchy data be specified as just described and let this data be singular somewhere on its support[2]. Next consider the solution $u(x, t)$ at some *later* time t, and for (x, t) belonging to a set Ω say, then $u(x, t)$ will be singular in Ω if the light-cones passing through Ω extend back to the initial time t_0 *and* also pass through the *singular* part of the support of the Cauchy initial data. Extensions of this Cauchy problem discussion to include the non-singular support of u depend on the value of n. However, for $n = 3, 5, \ldots$ the support and the singular support of the Green's function $G(x, t)$ actually coincide, Treves [1], and are thus light-cones—since the Green's function is responsible for the propagation of the wave, such waves therefore do not diffuse and this fact is often referred to as Huygens' principle. For all other values of n the singular support of $G(x, t)$ is a cone, but the non-singular support is the whole interior of the cone.

Let us use the 1-dimensional wave equation to illustrate an important point. Because of the formal similarity between the Laplace equation and the wave equation in $\mathbf{R}^2$, one might replace t by iy in 2.11 and then go on to claim that the resultant function $u(x, y) = (x + iy)^{5/2}$ is a *singular* solution to the Laplace equation—thus contradicting its supposed ellipticity. This is not correct. The point is that the function $u(x, t)$ is single valued and C^2 in the whole of the (x, t) plane; however if $u(x, y) \equiv u(x + iy) = z^{5/2}$ then u is only single valued on the (x, y) plane *minus* some branch connecting zero to infinity. Hence $u = z^{5/2}$ fails to be a solution in any neighbourhood

---

[2] As usual the support of a function is the closure of the set on which it is non-zero.

intersecting the branch; indeed $z^{5/2}$ is not differentiable on the branch since it is discontinuous there.

As well as elliptic and hyperbolic equations there are also *parabolic* equations. An example of a parabolic equation is the Schrödinger equation

$$-i\frac{\partial u}{\partial t} - \frac{\partial^2 u}{\partial x^2} = Lu = 0 \qquad (2.13)$$

Under Fourier transform in $\mathbf{R}^2$ L becomes multiplication by $\sigma(L)$ where

$$\sigma(L) = -p_1 + p_2^2 \qquad (2.14)$$

and the level curves $\sigma(L) = c$ give the parabolas

$$p_2^2 - p_1 = c \qquad (2.15)$$

Despite the appearance of the equation for a parabola in 2.15 above, the actual definition of a parabolic equation does not require this: in fact the classification of L as parabolic is made by only examining the Fourier transform of the *highest order* derivatives in L—all derivatives were of the same order in the previous examples so this fact did not emerge until now.

It is now possible to give one definition for all three types of equation: Let L be the constant coefficient 2nd order operator

$$L = -\sum_{i,j} a_{ij}\frac{\partial^2}{\partial x_i \partial x_j} + \sum_i b_i \frac{\partial}{\partial x_i} + c \qquad (2.16)$$

Fourier transforming only the highest order derivatives gives the quadratic form

$$\sigma_2(L) = \sum_{i,j} a_{ij} p_i p_j \qquad (2.17)$$

Then we say that

$$L \text{ is } \begin{cases} elliptic & \text{if } \sigma_2(L) \text{ has signature } \mp n \\ hyperbolic & \text{if } det\,(a_{ij}) \neq 0 \text{ and } \sigma_2(L) \text{ has signature } \mp(n-2) \\ parabolic & \text{if } det\,(a_{ij}) = 0 \text{ i.e. } \sigma_2(L) \text{ has at least one zero} \\ & \text{eigenvalue} \end{cases}$$

$$(2.18)$$

We would like to pursue further the connections between the smoothness of f and the smoothness of u for an equation of the form $Lu = f$. This leads to two things: the enlargement of the notion of ellipticity to operators of arbitrary order with variable coefficients, and the introduction of the attractive property of *hypo-ellipticity*.

§ 2. Ellipticity and hypo-ellipticity

Let us now suppose that L is a linear differential operator on $\mathbf{R}^n$ of arbitrary order m with possibly variable coefficients $a_\alpha(x)$. Let us also employ the compact *multi-index* notation so that

$$L = \sum_{|\alpha| \leq m} a_\alpha(x) D^\alpha \qquad (2.19)$$

The explanation of the notation in 2.19 is quite straightforward: the symbol α is called a multi-index and is simply an n-tuple of integers $\alpha = (\alpha_1, \ldots, \alpha_n)$ and $\alpha_i \geq 0$ (notice that some of the α_i may be zero); $|\alpha|$ is the *length* of the multi-index α and is given by $|\alpha| = \sum \alpha_i$; also $p^\alpha = p_1^{\alpha_1} \cdots p_n^{\alpha_n}$; finally $D^\alpha = (-i)^{|\alpha|}(\partial/\partial x_1)^{\alpha_1} \cdots (\partial/\partial x_n)^{\alpha_n}$, and the inclusion of the factor $(-i)^{|\alpha|}$ is a convenience which removes it from expressions obtained later by Fourier transformation. Thus 2.19 is short for the expression

$$L = \sum_{|\alpha| \leq m} a_{(\alpha_1,\ldots,\alpha_n)}(x)(-i)^{|\alpha|} \frac{\partial^{\alpha_1}}{\partial x_1^{\alpha_1}} \cdots \frac{\partial^{\alpha_n}}{\partial x_n^{\alpha_n}} \qquad (2.20)$$

Using Fourier transformation we can represent the operation of L on $u(x)$ via the action induced on its Fourier transform $\mathcal{F}(u(x)) = \hat{u}(p)$. The result is that we can write

$$Lu(x) = \sum_{|\alpha| \leq m} a_\alpha(x) D^\alpha u(x) = \sum_{|\alpha| \leq m} a_\alpha(x) \int_{\mathbf{R}^n} \exp[-ipx] p^\alpha \hat{u}(p) \, dp$$

$$= \int_{\mathbf{R}^n} \sigma(x,p) \hat{u}(p) \exp[-ipx] \, dp$$

$$(2.21)$$

where the definition of $\sigma(x,p)$ is, as might be anticipated,

$$\sigma(x,p) = \sum_{|\alpha| \leq m} a_\alpha(x) p^\alpha \qquad (2.22)$$

The function $\sigma(x,p)$ defined by the summation in 2.22 is called the *symbol* of the differential operator L, and a vital part in the properties of L is played by the last term in the summation; this term is denoted by $\sigma_m(x,p)$ where

$$\sigma_m(x,p) = \sum_{|\alpha| = m} a_\alpha(x) p^\alpha \qquad (2.23)$$

Thus $\sigma_m(x,p)$ is built from only the highest order derivatives of L, i.e. those of order m, and $\sigma_m(x,p)$ is called the *leading symbol* of L (the terms *principal symbol* and *highest order symbol* are also used). Clearly, different operators of order m which have the same highest order terms but differ in lower orders will have the same leading symbol $\sigma_m(x,p)$, but they will have different (complete) symbols $\sigma(x,p)$. In any case it is the leading symbol which appears in the definition of ellipticity for a general operator L.

Definition (Ellipticity)   *A differential operator $L = \sum_{|\alpha|\leq m} a_\alpha(x)D^\alpha$ of order m is called elliptic if its leading symbol $\sigma_m(x,p)$ is always non-zero for non-zero p.*

One can think also of ellipticity for L as being characterised by the invertibility of the leading symbol of L, or the invertibility of L itself modulo lower order terms; this point will be developed further when we discuss vector bundles below.

We have seen in the previous section that the benefit of ellipticity is to give smooth solutions to $Lu = f$ given a smooth f. It is not necessary for an operator to be elliptic for this to be true—some parabolic equations possess this property, an example being the heat equation, see later in this section. This motivates the following definition of hypo-ellipticity:

Definition (Hypo-ellipticity)   *A differential operator L is hypo-elliptic if the condition Lu is C^∞ in an open set Ω implies that u is C^∞ in Ω.*

Thus when one solves $Lu = f$ for a hypo-elliptic L, f smooth implies the solution u is smooth. Clearly all elliptic operators are also hypo-elliptic but there are *non-elliptic*, hypo-elliptic operators: an example being the heat equation in $\mathbf{R}^{n+1}$ where we have

$$L = \frac{\partial}{\partial t} - \frac{\partial^2}{\partial x_1^2} - \cdots - \frac{\partial^2}{\partial x_n^2} \qquad (2.24)$$

The Green's function G for L is easily calculated by Fourier transformation to be

$$G \equiv G(x,t) = \frac{1}{(2\sqrt{\pi t})^n}\theta(t)\exp\left[-\frac{\sum x_i^2}{4t}\right] \qquad (2.25)$$

Now the formula 2.25 for $G(x,t)$ can be seen to give a singularity at $x = t = 0$, i.e. at the origin of $\mathbf{R}^{n+1}$. However, all other partial derivatives of $G(x,t)$ converge (strongly) for $x \neq 0$ and $t > 0$, and vanish (because of the $\theta(t)$) for $x \neq 0$ and $t < 0$, so $G(x,t)$ is C^∞ in $\mathbf{R}^{n+1}$ except at the origin—thus by the same proof as we sketched in § 1 for the elliptic operator $L = -\sum_i a_i \partial^2/\partial x_i^2$, the solutions to $Lu = f$ will be smooth given a smooth *f*. Hence L is hypo-elliptic even though the heat equation is parabolic rather than elliptic. Let us examine two more well known equations for hypo-ellipticity.

The two equations are the Schrödinger equation, which we have already introduced as an example of a parabolic equation, and the Cauchy–Riemann equation.

As regards the Schrödinger equation we can prove that it is not hypo-elliptic despite its similarity to the heat equation. As usual we take the formula for the Green's function, which, in the Schrödinger case in $\mathbf{R}^{n+1}$, is

$$G(x,t) = i\frac{1}{(2\sqrt{\pi it})^n}\theta(t)\exp\left[-\frac{\sum x_i^2}{4it}\right] \tag{2.26}$$

Now it is easy to see that because of the presence of the factor $\exp[-\sum x_i^2/4it]$ in 2.26 which behaves monstrously on the hyperplane $t = 0$, then $G(x,t)$ is not C^∞ anywhere on this hyperplane—the corresponding factor for the heat equation vanishes on this hyperplane, hence the difference between the two cases. In any event, because of the non-smooth behaviour of its Green's function the Schrödinger equation is definitely not hypo-elliptic. We move on to the Cauchy–Riemann equation.

The Cauchy–Riemann operator in the complex plane $\mathbf{C}$ is usually denoted by $\bar{\partial}$ and so we set

$$L = \bar{\partial} = \frac{\partial}{\partial\bar{z}} = \frac{1}{2}\left(\frac{\partial}{\partial x} + i\frac{\partial}{\partial y}\right) \tag{2.27}$$

where (x,y) are the coordinates of $\mathbf{C}$. The Green's function, $G(z) = G(x,y)$ say, of $\bar{\partial}$ obeys the equation

$$\frac{1}{2}\left(\frac{\partial}{\partial x} + i\frac{\partial}{\partial y}\right)G(x,y) = \delta(x,y) \tag{2.28}$$

which has a solution given by

$$G(x,y) = \frac{1}{\pi(x+iy)} = \frac{1}{\pi z} = G(z) \tag{2.29}$$

Notice that $G(x,y)$ is perfectly well behaved away from the origin in $\mathbf{C}$, and, as we have seen already, this is a property of elliptic operators. We can easily verify that the Cauchy-Riemann operator is elliptic for, using the definition of ellipticity introduced on p. 33, and denoting the Fourier transform variables by p_1 and p_2, we find that the leading symbol is

$$\sigma_1(z,p) = \frac{1}{2}(p_1 + ip_2) \tag{2.30}$$

and since $\sigma_1(z, p) \neq 0$ for $p \neq 0$ then $\bar{\partial}$ is elliptic, it is also, of course, hypo-elliptic. Indeed $G(z)$ is actually *real analytic* for $z \neq 0$; this means that $\bar{\partial}u = f$ will have analytic solutions u given an analytic f; such operators are called *analytic-hypo-elliptic*. It is not difficult to check that the Laplacian is another example of an analytic-hypo-elliptic operator but that the heat equation is hypo-elliptic but not analytic-hypo-elliptic [3]

We end this section with a short discussion of ellipticity for ordinary differential equations or ODEs for short. Roughly speaking, all ODEs are, either elliptic, or only depart in a rather obvious way from ellipticity. The same general definition of ellipticity given on p. 33 applies to ODEs. Suppose then that an nth order ODE $Lu = f$, with smooth coefficients $a_i(x)$ and a smooth f, is given by

$$a_n(x)\frac{d^n u(x)}{dx^n} + \cdots + a_1(x)\frac{du(x)}{dx} + a_0(x)u(x) = f(x) \tag{2.31}$$

Then the leading symbol of L is just

$$\sigma_n(x, p) = a_n(x)p^n \tag{2.32}$$

so that L is elliptic on $\mathbf{R}$ provided $a_n(x)$ is non-vanishing on $\mathbf{R}$. We see immediately that examples of elliptic ODEs are easy to obtain: all ODEs with constant coefficients are elliptic; further it is well known that such equations have smooth solutions. A typical non-elliptic ODE is Bessel's equation of order λ

$$x^2\frac{d^2 u(x)}{dx^2} + x\frac{du(x)}{dx} + (x^2 - \lambda^2)u(x) = 0 \qquad \lambda \in \mathbf{R} \tag{2.33}$$

The solutions to Bessel's equation are not, in general, C^∞ because they can misbehave at $x = 0$ where one has $u(x) \to x^{\mp\lambda}$ or $u(x) \to x^{\mp\lambda}\ln(x)$; Bessel's equation is then cited as an example of an equation with a regular singular point, this point being the origin. Of course the leading symbol is

$$\sigma_2(x, p) = x^2 p^2 \tag{2.34}$$

which vanishes at $x = 0$, hence the non-ellipticity.

---

[3] The relevant point for the heat equation is that, in $\mathbf{R}^{n+1} - \{0\}$, the Green's function $G(x, t) = \frac{1}{(2\sqrt{\pi t})^n}\theta(t)\exp\left[-\sum x_i^2/t\right]$ is C^∞ but not analytic—it is analytic, though, *off* the hyperplane in $\mathbf{R}^{n+1}$ given by $t = 0$.

Another non-elliptic example in one variable is Gauss's hypergeometric equation

$$x(1-x)\frac{d^2u(x)}{dx^2} + \{\gamma - (\alpha + \beta + 1)\}\frac{du(x)}{dx} - \alpha\beta u(x) = 0 \qquad (2.35)$$

whose leading symbol $\sigma_2(x, p)$ is given by

$$\sigma_2(x, p) = x(1-x)p^2 \qquad (2.36)$$

The zeros of $\sigma_2(x, p)$ at 0 and 1 render it non-elliptic; and the well known hypergeometric function $F(\alpha, \beta, \gamma; x)$, which is one of the solutions to the hypergeometric equation, is not smooth at $x = 1$ having a branch there, and there is a second solution singular at the origin.

In the case of one variable it is both easy and instructive to verify the smoothness of the solutions to elliptic equations. Suppose one takes the general ODE 2.31 and assumes it to be elliptic, then since we know that $a_n(x) \neq 0$ we may divide by it on both sides of 2.31 to obtain

$$\frac{d^n u(x)}{dx^n} + \cdots + \frac{a_1(x)}{a_n(x)}\frac{du(x)}{dx} + \frac{a_0(x)}{a_n(x)}u(x) = \frac{f(x)}{a_n(x)} \qquad (2.37)$$

Then the existence of 2.37 implies that $u(x)$ is C^n, but the non-vanishing of the denominator $a_n(x)$ and the smoothness of the $a_i(x)$ imply that one may differentiate both sides and infer that $u(x)$ is C^{n+1}; induction then establishes that $u(x)$ is C^∞ as required.

§ 3. Ellipticity and vector bundles

In this book the more common setting for the various differential operators that arise will be one in which vector bundles are present. More precisely, the general setting which we require is the following: Let U and V be vector bundles of rank q and p, respectively, over a compact manifold M. Denote by $\Gamma(M, U)$ and $\Gamma(M, V)$ the spaces of *smooth* sections of the respective bundles, then we wish to work with linear differential operators L of order m that map one space of sections into the other. Thus we can write

$$L : \Gamma(M, U) \longrightarrow \Gamma(M, V) \qquad (2.38)$$

and when no confusion can arise we shall abbreviate $\Gamma(M, U)$ and $\Gamma(M, V)$ to simply $\Gamma(U)$ and $\Gamma(V)$.

The motive for requiring that $\Gamma(M, U)$ and $\Gamma(M, V)$ should be spaces of *smooth* sections is simply that we are working within the field of differential topology. Further, given this choice of spaces, it is natural to expect that elliptic operators L will play a prominent part in their study; this is because we have seen that when L is elliptic the smoothness of its image implies the smoothness of its inverse image; thus since the image of L lies in the space $\Gamma(M, V)$, which is smooth by definition, then the inverse image of L is guaranteed to be in the other smooth space $\Gamma(M, U)$. Before going on it is clearly necessary for us to extend our definition of ellipticity to cover the case at hand. This can be accomplished in the following manner.

The space of smooth sections $\Gamma(M, U)$ can be thought of loosely as a space of vector valued functions on M; thus an operator L which acts on $\Gamma(M, U)$ can be thought of as a matrix valued operator. To make this more concrete in the case at hand we choose an open set $\Omega \subset M$ above which the bundle U is, of course, locally trivial and possessed of the local coordinates (x, u), where x is a local coordinate for Ω, and u a local coordinate for a q-dimensional vector belonging to the fibre $\mathbf{R}^q$ of U. Using these coordinates, a section in $\Gamma(M, U)$ is represented by the vector valued function $u(x)$ and an expression for Lu is provided by

$$(Lu(x))_i = \sum_{j=1}^{q} \sum_{|\alpha| \le m} a_\alpha^{ij}(x) D^\alpha u_j \qquad i = 1, \ldots p \qquad (2.39)$$

Referring to 2.39 we can confirm that the coefficients $a_\alpha(x)$ have become $p \times q$ matrices, and that $u(x)$ and $Lu(x)$ have become vectors of dimension q and p respectively. The symbol of L is also matrix valued and is given by $\sigma(x, k)$ where

$$\sigma(x, k) = \sum_{|\alpha| \le m} a_\alpha(x) k^\alpha \qquad (2.40)$$

and the k denotes the Fourier transform variable; actually the local coordinates (x, k) should be regarded as belonging to the cotangent bundle T^*M of M. As might be anticipated, the leading symbol $\sigma_m(x, k)$ of L is a $p \times q$ matrix or linear map between the fibres of U and V of the form $\sigma_m(x, k) : U_x \to V_x$ where $\sigma_m(x, k)$ is given by [4]

$$\sigma_m(x, k) = \sum_{|\alpha| = m} a_\alpha(x) k^\alpha \qquad (2.41)$$

---

[4] Strictly speaking, since the functions $a_\alpha^{ij}(x)$ clearly will change by a Jacobian-type transformation if we change local coordinates, we should write something like $a_{\alpha, \Omega}^{ij}(x)$ to display a dependence on Ω. This should also be done for the symbol which could more properly be written as $\sigma_m^\Omega(x, k)$; we omit all this to relieve the burden of too cumbersome a notation.

In the above we think of $\sigma_m(x, k)$ for each x and arbitrary k as a map between the appropriate fibres. It is also useful to work globally and avoid having to mention the local coordinates (x, k). To do this let $\pi : T^*M \to M$ be the projection for the cotangent bundle T^*M. Then, since U and V are bundles over M, the map π can be used to pull them back to bundles over T^*M itself. The resulting bundles are of course π^*U and π^*V respectively; since they are pullbacks they will have the same fibres as before but these fibres now sit over the base T^*M. Finally the map $\sigma_m(x, k)$ is equivalent to a map which we write simply as σ_m, where $\sigma_m : \pi^*U \to \pi^*V$ is a *bundle homomorphism* between π^*U and π^*V.

Let us just elucidate this last assertion. To say that a map between vector bundles is a bundle homomorphism means that, for each point p in the base manifold, it induces an ordinary vector space homomorphism between the fibres above p. Now in the case at hand the base is the cotangent bundle T^*M and a point p in the base is given by $p = (x, k)$; the fibre above such a p is U_x for π^*U and V_x for π^*V; thus a bundle homomorphism σ_m between π^*U and π^*V is given by supplying a linear map, which it is natural to denote by $\sigma_m(x, k)$, between U_x and V_x. However, this is precisely what we did when we first introduced $\sigma_m(x, k)$ and so our assertion is established.

We have seen that ellipticity is concerned with the invertibility of $\sigma(x, k)$ for $k \neq 0$. We can delete $k = 0$ from the picture by replacing the cotangent bundle T^*M with the bundle of *non-zero* cotangent vectors, i.e. we delete the subset of T^*M with local coordinates $(x, 0)$; since such a subset is what one obtains when one writes down the zero section of T^*M this is often called deleting the zero section.

With these facts in place we now have the ellipticity definition

Definition (Ellipticity, vector bundle case)   *The mth order linear differential operator L, defined as described above, is elliptic if, for any open set $\Omega \subset M$, its corresponding leading symbol $\sigma_m(x, k)$ is invertible for $k \neq 0$.*

Thus when L is elliptic we see at once that the bundles U and V must have the same rank, i.e. $p = q$—also in the coordinate free method of dealing with symbols we can equivalently define L as being elliptic as follows.

Definition (Ellipticity, more abstract definition)   *The mth order linear differential operator $L : \Gamma(M, U) \to \Gamma(M, V)$ is elliptic if*

$$\sigma_m : \pi^*U \to \pi^*V$$

is a bundle isomorphism off the zero section of T^*M. It is clear, too, that the definition of the ellipticity of L as being the invertibility of its leading symbol is the natural linear algebraic generalisation of the previous definition which simply required it to be non-vanishing.

To illustrate the vector bundle case we choose a specific example. Let M be a compact Riemannian manifold of dimension n with metric g_{ij}, and d be the usual exterior derivative, then the Laplacian Δ_p on p-forms $\Omega^p(M)$ is given in terms of d by

$$\Delta_p = (dd^* + d^*d)_p \tag{2.42}$$

where the suffix p on the RHS is simply used to denote restriction to p-forms, and, as usual, d^* denotes the adjoint of d relative to the standard inner product on p-forms, whose definition in turn requires use of the Hodge $*$. For future use it is convenient to summarise some of this information below

$$
\begin{aligned}
&< \omega, \nu > = \int_M \omega \wedge *\nu \quad \omega, \nu \in \Omega^p(M) \\
&< \omega, d\mu > = < d^*\omega, \mu > \quad \omega \in \Omega^p(M), \quad \mu \in \Omega^{p-1}(M) \\
&d : \Omega^p(M) \longrightarrow \Omega^{p+1}(M) \Rightarrow d^* = (-1)^{np+n+1} * d* \\
&* : \Omega^p(M) \longrightarrow \Omega^{n-p}(M) \quad *^2 = (-1)^{p(n-p)} \\
&\omega = \omega_{i_1 \ldots i_p} dx^{i_1} \wedge \ldots \wedge dx^{i_p} \\
&*\omega = \frac{\sqrt{g}}{(n-p)!} \epsilon^{i_1 \ldots i_p}{}_{i_{p+1} \ldots i_n} \omega_{i_1 \ldots i_p} dx^{i_{p+1}} \wedge \ldots \wedge dx^{i_n}
\end{aligned}
\tag{2.43}
$$

where the last two expressions in 2.43 are valid for local coordinates in some open set $\Omega \subset M$, also we have assumed that the metric is Riemannian positive definite—some signs would change if the metric was Lorentzian. Now, from the bundle standpoint, Nash and Sen [1], a p-form is a section of a bundle over M, the bundle being $\wedge^p T^*M$, i.e. the pth anti-symmetric power of the cotangent bundle T^*M. Thus over Ω we can pass to the appropriate sections and write

$$\Delta_p : \Gamma(M, \wedge^p T^*M) \longrightarrow \Gamma(M, \wedge^p T^*M) \tag{2.44}$$

An instructive way to calculate the leading symbol $\sigma_2(x, k)$ of Δ_p is to employ the following result: let P and Q be two differential operators of order i and j, with leading symbols denoted by $\sigma_i^P(x, k)$ and $\sigma_j^Q(x, k)$ respectively. Then their product $P \circ Q$ is of order $(i + j)$ and has leading symbol given by an appropriate product of the individual leading symbols. The reader can easily verify that one has

$$
\begin{aligned}
&P \circ Q : \Gamma(M, U) \xrightarrow{Q} \Gamma(M, V) \xrightarrow{P} \Gamma(M, W) \\
&\sigma_{i+j}^{P \circ Q}(x, k) = \sigma_i^P(x, k) \sigma_j^Q(x, k)
\end{aligned}
\tag{2.45}
$$

Thus if we choose $P = d$ and $Q = d^*$ then the symbol of the Laplacian obeys the equation

$$\sigma_2^{\Delta_p}(x, k) = \sigma_1^d(x, k)\sigma_1^{d^*}(x, k) + \sigma_1^{d^*}(x, k)\sigma_1^d(x, k) \qquad (2.46)$$

Another easily verifiable property of symbols is that the symbols of an mth order operator L and its adjoint L^* are related by

$$\sigma_m^{L^*}(x, k) = \left\{\sigma_m^L(x, k)\right\}^* \qquad (2.47)$$

Because we are working with differential forms we represent the Fourier transform variable k by $k = k_i dk^i$, with this done it is immediate that, if $v \in \wedge^p T^* M$, then we have

$$\sigma_1^d(x, k) : \wedge^p T_x^* M \longrightarrow \wedge^{p+1} T_x^* M$$
$$v \longmapsto k \wedge v \qquad (2.48)$$

On the other hand, since the action of $\sigma_1^d(x, k)$ is just wedge product with k, then we know that its adjoint is interior product with k, i.e. $v \mapsto k \lrcorner v$. An elementary calculation then gives the expected result for the symbol of the Laplacian:

$$\sigma_2^{\Delta_p}(x, k) = k^2 I \qquad (2.49)$$

where I represents a unit matrix in the vector space $\wedge^p T_x^* M$. We note that the symbol of Δ_p, though a matrix, is a multiple of the identity, however, the symbol $\sigma_1^{d^*}(x, k)$ is a non-square matrix. Various generalisations of the Laplacian can be formed by tensoring the bundle $\wedge^p T^* M$ with another vector bundle E, and by studying the equations of motion for Yang–Mills theories, we shall encounter these examples in later chapters.

The kernel of Δ_p plays a distinguished part in the theory of the Laplacian; the recognition that this is so and the subsequent development of the relevant results is due to Hodge. These results are particularly well known and are fortunately rather easy to describe. We shall now give a brief summary—for a proof the reader can consult Wells [1].

First we suppose that ω is a *harmonic p*-form on a compact manifold M so that $\Delta_p \omega = 0$. Thus we can make the following simple argument

$$< \omega, \Delta_p \omega > = < \omega, (dd^* + d^* d)\omega > = 0$$
$$\Rightarrow \quad < d^* \omega, d^* \omega > + < d\omega, d\omega > = 0$$
$$\Rightarrow \quad \|d^* \omega\|^2 + \|d\omega\|^2 = 0 \qquad (2.50)$$
$$\Rightarrow \quad d^* \omega = d\omega = 0$$

It is at once clear that

$$\omega \in ker\, \Delta_p \iff d^*\omega = d\omega = 0 \tag{2.51}$$

If we combine this with the construction of the de Rham cohomology groups (cf. p. 13 eq. 1.43) then we have

$$\omega \in ker\, \Delta_p \Rightarrow [\omega] \in \frac{ker\, d}{Im\, d} = H^p_{de\, Rham}(M) \tag{2.52}$$

The force of Hodge's result is that the two spaces in the equation above are actually isomorphic, giving us the powerful theorem

Theorem (Hodge) *On a compact closed Riemannian manifold there is precisely one harmonic p-form ω in each cohomology class $[\omega]$ of $H^p(M; \mathbf{R})$.*

This of course implies immediately a formula for calculating Betti numbers b_p as we quoted at the beginning of this chapter, i.e. we have

$$\dim ker\, \Delta_p = \dim H^p(M; \mathbf{R}) = b_p \tag{2.53}$$

Note that although the Laplacian depends on a choice of a Riemannian metric the theorem remains true regardless of which metric is used; indeed the Betti numbers are topological quantities which do not require a metric for their definition. The theorem of Hodge above generalises to include cases where the exterior derivative d is replaced by other differential operators. This generalisation and the whole question of kernels of elliptic operators is resumed in chapter 4 when we discuss index theory for elliptic operators.

§ 4. Pseudo-differential operators

We are now ready to introduce our most general kind of elliptic operator: this is what is known as an *elliptic pseudo-differential operator*. First we must define the term pseudo-differential operator.

We accomplish this by abstracting to a more general situation the main properties of the symbols of standard differential operators. If P is a general operator defined via its symbol $\sigma(x, p)$ by the equation

$$Pu(x) = \int_{\mathbf{R}^n} \sigma(x, p)\hat{u}(p)\exp[-ipx]\, dp \tag{2.54}$$

then P is said to be a *pseudo-differential operator of order m* if its symbol $\sigma(x, p)$ belongs to an appropriate space of symbols which we denote by $S^m(\Omega)$. It remains to define the space $S^m(\Omega)$; and to do this we simply try and imitate the standard situation where one thinks loosely of the symbol as being a polynomial of degree m in p with x-dependent coefficients. A definition that turns out to work is the following:

Definition (Symbol Space)   *Given an open set $\Omega \subset \mathbf{R}^n$, there is a symbol space $S^m(\Omega)$, with $m \in \mathbf{R}$, which consists of those functions $\sigma(x,p)$ which, for $(x,p) \in K \times \mathbf{R}^n$, with K a compact subset of Ω, satisfy the condition*

$$\left| D_x^\alpha D_p^\beta \sigma(x,p) \right| \le C_{\alpha,\beta}(K)(1+|p|)^{m-|\beta|} \tag{2.55}$$

where $C_{\alpha,\beta}(K)$ is a constant.[5] Before taking the next step and defining a pseudo-differential operator we have some observations.

First we point out that the intended definition of a pseudo-differential operator P has as its main idea that P is pseudo-differential of order m if its symbol $\sigma(x,p)$ belongs to $S^m(\Omega)$. Next notice that we have defined $S^m(\Omega)$ for *all real m*, thus the order m can be *non-integral or negative* etc; in fact we shall also encounter examples with infinite m. Thus we will be able to deal with operators of any real order, finite and infinite. Observe too that if $\sigma(x,p)$ belongs to $S^m(\Omega)$ it also belongs to $S^{m+1}(\Omega)$. Thus if an operator is of order m it is also of all higher orders—this suggests that the order be defined as the lowest possible such m: the infimum say. We do not do this because this infimum may not be attained, i.e it is possible to encounter cases where an operator is of order $m + \epsilon$ for all strictly positive ϵ, but we cannot pass to the limit where $\epsilon = 0$, cf. the example in 2.62 below.

We can now give our general definition of a pseudo-differential operator.

Definition (Pseudo-Differential operator of order m)   *Let M be a compact n-dimensional manifold, and U and V be vector bundles of rank s and r respectively over M. Denote the corresponding spaces of sections by $\Gamma(M,U)$ and $\Gamma(M,V)$ respectively. Then $P : \Gamma(M,U) \to \Gamma(M,V)$ is a pseudo-differential operator of order m if, for any choice of open set $\Omega \subset M$ with local coordinates (x,p), all the entries of the $r \times s$ matrix symbol $\sigma(x,p)$ belong to $S^m(\Omega)$.*

With this very general definition we also have to say what is meant by the leading symbol $\sigma_m(x,p)$ when $\sigma(x,p) \in S^m(\Omega)$. Having in mind the case of differential operators of order m where the leading symbol is

---

[5]   There is a larger symbol space than the one that we have just defined here—it is obtained by replacing the condition 2.55 by $\left| D_x^\alpha D_p^\beta \sigma(x,p) \right| \le C_{\alpha,\beta}(K)(1+|p|)^{m+\delta|\alpha|-\rho|\beta|}$, and this space is denoted by $S_{\rho,\delta}^m(\Omega)$, $0 < \rho$, $0 \le \delta$. Clearly our space $S^m(\Omega)$ is simply $S_{1,0}^m(\Omega)$.

homogeneous of order m as a function of p, we are led to make the following definition

Definition (m^{th} order symbol) *If P is a pseudo-differential operator of order m then its mth order symbol $\sigma_m(x, p)$ is given by the limit*

$$\sigma_m(x, p) = \lim_{\mu \to \infty} \frac{\sigma(x, \mu p)}{\mu^m} \qquad (2.56)$$

We wish to point out that in this definition μ is positive; thus $\sigma_m(x, p)$ satisfies $\sigma_m(x, \mu p) = \mu^m \sigma_m(x, p)$ for positive μ but not necessarily for negative μ. For example, $\sigma_m(x, p)$ could be the function $|p|^m$ and then P would be a pseudo-differential operator similar to the example considered in 2.61 below; however, if P is a *differential* operator then we will have $\sigma_m(x, \mu p) = \mu^m \sigma_m(x, p)$ regardless of the sign of μ. If necessary we shall refer to this homogeneity property possessed by the leading symbol of pseudo-differential operators as positive homogeneity.

It is also now possible to define ellipticity in the pseudo-differential case. No new idea is needed and so we proceed at once to the definition

Definition (Ellipticity, pseudo-differential case) *Suppose that in the above definition the vector bundles U and V have the same rank, r say, then the pseudo-differential operator $P : \Gamma(M, U) \to \Gamma(M, V)$ is elliptic of order m if, $\sigma(x, p) \in S^m(\Omega)$, and if, for any choice of open set $\Omega \subset M$ with local coordinates (x, p), its mth order symbol $\sigma_m(x, p)$ is an invertible $r \times r$ matrix for $p \neq 0$.*

Just as in the case of a differential operator we can give a more compact definition which we shall make use of in our subsequent treatment of index theory:

Definition (Ellipticity, pseudo-differential case, alternative definition) *The pseudo-differential operator $P : \Gamma(M, U) \to \Gamma(M, V)$ is elliptic of order m if*

$$\sigma_m : \pi^* U \to \pi^* V$$

is a bundle isomorphism off the zero section of $T^* M$.

Notice that ellipticity of order m *does* require the existence of the mth order symbol $\sigma_m(x, p)$; also, although an elliptic m^{th} order operator is also of order $m + 1$, it is not *elliptic* of order $m + 1$ since the hypothesis $\sigma(x, p) \in S^m(\Omega)$ implies that its $(m + 1)^{th}$ order symbol $\sigma_{m+1}(x, p)$ has to be zero and hence non-invertible.

Of course any differential operator of order m is automatically pseudo-differential of order m, this follows directly if one substitutes the relevant symbol expression 2.22 into the symbol space definition 2.55.

Examples of pseudo-differential operators which are not *differential operators* can arise when we try to solve integral equations of the form

$$Pu(x) = \int_{\mathbf{R}^n} K(x,y)u(y)\,dy = f(x) \tag{2.57}$$

where f and K are known. If we write this as

$$Pu(x) = \int_{\mathbf{R}^n} \sigma(x,p)\hat{u}(p)\exp[-ipx]\,dp = f(x)$$

$$\text{with} \quad \sigma(x,p) = \frac{1}{(2\pi)^n}\int_{\mathbf{R}^n}\exp[-ip(y-x)]K(x,y)dy \tag{2.58}$$

then P is pseudo-differential for suitable $\sigma(x,p)$. Let us be specific: choose $n = 1$ and let

$$\sigma(x,p) = (a^2 + p^2)^\lambda, \qquad a \neq 0 \quad -\frac{1}{2} < \lambda < 0 \tag{2.59}$$

Such a symbol clearly belongs to the symbol space $S^{2\lambda}$ so that P is pseudo-differential of order 2λ. The solution of the integral equation is easily given provided one can evaluate the appropriate Fourier transforms. Having carried out the transforms we can quote the equation and its solution together; this yields the rather complicated expressions

$$Pu(x) \equiv \frac{2^{1+\lambda}a^{\frac{1}{2}+\lambda}}{\sqrt{2\pi}\Gamma(-\lambda)}\int_{\mathbf{R}}\frac{K_{\frac{1}{2}+\lambda}(a(x-y))}{(x-y)^{\frac{1}{2}+\lambda}}u(y)\,dy = f(x)$$

$$u(x) = \frac{2^{1-\lambda}a^{\frac{1}{2}-\lambda}}{\sqrt{2\pi}\Gamma(\lambda)}\int_{\mathbf{R}}\frac{K_{\frac{1}{2}-\lambda}(a(x-y))}{(x-y)^{\frac{1}{2}-\lambda}}f(y)\,dy \tag{2.60}$$

$$\text{where} \quad K_\lambda(z) = \frac{\pi}{2\sin(\lambda\pi)}\left[e^{\frac{\lambda}{2}\pi i}J_{-\lambda}(iz) - e^{-\frac{\lambda}{2}\pi i}J_\lambda(iz)\right],$$

$J_\lambda(z)$ is the usual Bessel function of order λ, and we take an appropriate definition of the fractional powers. In fact we chose the above example because it has a particularly simple form and interpretation if $a = 0$. In this case the symbol $\sigma(x,p) = p^{2\lambda}$ and thus P is just *fractional integration of order* -2λ, and the solution can be thought of as the corresponding fractional *derivative* of order -2λ. The reason that we did not deal with this simpler form in the first place is just a technical one: when $a = 0$, P is not a pseudo-differential operator due to its singular derivatives at $p = 0$; however the integrals giving the equation and its solution are nevertheless still perfectly well defined and so we can write

$$Pu \equiv \frac{1}{\sqrt{2}\cos(\lambda\pi)\Gamma(-2\lambda)} \int_{\mathbf{R}} \frac{u(y)}{(x-y)^{1+2\lambda}} \, dy = f(x)$$

$$u(x) = \frac{1}{\sqrt{2}\cos(\lambda\pi)\Gamma(2\lambda)} \int_{\mathbf{R}} \frac{f(y)}{(x-y)^{1-2\lambda}} \, dy$$

(2.61)

Also, whether $a = 0$ or not, the operator P does possess one of the essential properties of a derivative: to see this denote P temporarily by P_λ, then it is easy to verify that we have $P_\lambda P_\mu = P_{\lambda+\mu}$. Moreover, even when λ is such that the integrals are not convergent P_λ can be defined in the distributional sense; cf. Gel'fand and Shilov [1] for the necessary distribution theory of these sort of integrals.

The next example is one where the operator has order $2\lambda + \epsilon$ for all positive ϵ but we may not set $\epsilon = 0$. Let P be given by the equation

$$Pu(x) = \int_{\mathbf{R}} (a^2 + p^2)^\lambda \ln(a^2 + p^2)\hat{u}(p) \, dp \tag{2.62}$$

with a and λ as before. The symbol $\sigma(x,p)$ of P is just $(a^2 + p^2)^\lambda \ln(a^2 + p^2)$, thus $\sigma(x,p) \in S^{2\lambda+\epsilon}$ provided $\epsilon > 0$; the presence of the logarithm prevents the passage to the limit $\epsilon = 0$. It is even possible to write P in convolution form should it be necessary; all one has to do is to differentiate the operator of preceding example with respect to λ, a procedure which is in fact permissible if one uses distribution theory. This gives the rather formidable looking result

$$Pu = \int_{\mathbf{R}} (a^2 + p^2)^\lambda \ln(a^2 + p^2)\hat{u}(p) \, dp$$

$$= \int_{\mathbf{R}} G(x - y)u(y) \, dy$$

with $\quad G(z) = \dfrac{2^{\lambda+1}a^{\frac{1}{2}+\lambda}}{\sqrt{2\pi}\Gamma(-\lambda)} \left\{ \ln(2a) - \dfrac{d}{d\lambda}\ln\Gamma(-\lambda) \right\} \dfrac{K_{\frac{1}{2}+\lambda}(az)}{z^{\frac{1}{2}+\lambda}}$

(2.63)

$$+ \frac{2^{\lambda+1}a^{\frac{1}{2}+\lambda}}{\sqrt{2\pi}\Gamma(-\lambda)z^{\frac{1}{2}+\lambda}} \left\{ \frac{d}{d\lambda}K_{\frac{1}{2}+\lambda}(az) - K_{\frac{1}{2}+\lambda}(az)\ln z \right\}$$

Next let us take the much simpler integral operator defined by the equation below

$$Pu(x) = \int_{\mathbf{R}} \exp\left[-\frac{(x-y)^2}{2} \right] u(y) \, dy \tag{2.64}$$

Because of the particularly simple Gaussian kernel the symbol of P is again Gaussian and is given by $\sigma(x,p) = \exp[-p^2/2]/\sqrt{2\pi}$. However, we now make the elementary observation that, for appropriate constants C_m, we have

$$\frac{1}{\sqrt{2\pi}} \exp\left[-\frac{p^2}{2}\right] \leq C_m(1+p)^m \qquad \forall m$$

$$\Rightarrow \sigma(x,p) \in S^m(\Omega) \qquad m = \ldots 2,1,0,-1,-2,\ldots \tag{2.65}$$

Thus P is in fact of order $-\infty$, or equivalently we write $\sigma(x,p) \in S^{-\infty}(\Omega)$. Such a P is called a *smoothing operator* since it has an effect on $u(x)$ like that of integrating it an infinite number of times: i.e. it produces the function Pu which is always C^∞ even though $u(x)$ may be not even be differentiable. To check that $Pu(x)$ is indeed C^∞, one only has to differentiate under the integral sign in 2.64 and note the convergence of the resulting integrals.

One could go on to construct more examples of smoothing operators with the Gaussian kernel replaced by a general kernel $K(x,y)$. Sufficient conditions on $K(x,y)$ to render P of order $-\infty$ are that it be smooth in x, and compactly supported in y—actually the Gaussian kernel does not have compact support so these conditions are clearly not necessary, in practice though we most frequently work with compact manifolds where the condition of compact support is trivially satisfied.

Our final example is a differential operator raised to a *complex* power. To describe this example we let Δ denote the Laplacian in $\mathbf{R}^n$ and a be a real constant, then the desired pseudo-differential operator P is given by

$$P = (a^2 + \Delta)^z, \quad z \text{ complex} \tag{2.66}$$

Thus P is clearly a complex power of the positive elliptic operator $(a^2 + \Delta)$. We can deduce immediately the symbol $\sigma(x,p)$ of P so that we have

$$\sigma(x,p) = (a^2 + p^2)^z \tag{2.67}$$

Of obvious interest is the order, m say, of P; m can be easily calculated from the definition of $S^m(\Omega)$ that we gave above, cf. 2.55. Applying this definition with $\alpha = \beta = 0$ and letting $z = u + iv$ we see that $\sigma(x,p)$ must obey

$$|\sigma(x,p)| \leq C(1+|p|)^m$$

$$\Rightarrow |(a^2 + p^2)^z| \leq C(1+|p|)^m \tag{2.68}$$

$$\Rightarrow |(a^2 + p^2)|^u \leq C(1+|p|)^m$$

where we have used the fact that $|(a^2 + p^2)^{iv}| = 1$. Thus the order of P is simply $2u$; also, since its symbol is non-vanishing, P is elliptic. Thus we can say that $(a^2 + \Delta)^z$, $z \in \mathbf{C}$ is an elliptic operator of order $2 Re\, z$.

Notice that, unlike differential operators, pseudo-differential operators are not, in general, *local* operators—by locality [6] of an operator P we mean that the support of Pu is contained in the support of u, or *supp Pu $\subset$ supp u* (this is often stated as: differential operators preserve supports). If we choose one of the examples above where the pseudo-differential operator P is an integral operator it is obvious that P is not local.

Instead of locality, pseudo-differential operators possess a property known as *pseudo-locality*. This is simply the property

$$sing\,supp\,Pu \subset sing\,supp\,u \qquad (2.69)$$

where, if f is any function, *sing supp f* stands for the subset of *supp f* on which f is singular [7] (one can say that pseudo-differential operators preserve singular supports). Now it may happen that *sing supp Pu* and *sing supp u* *coincide* so that we have

$$sing\,supp\,Pu = sing\,supp\,u \qquad (2.70)$$

Then it is clear that the property Pu is smooth (i.e. *sing supp Pu* $= \emptyset$) implies that u is smooth, i.e. P is hypo-elliptic; this can be thought of as a characterisation of hypo-ellipticity.

It is useful to have available an analytic description of the sort of spaces of functions that arise when studying pseudo-differential operators. Such a description can be constructed by using Sobolev spaces; we now give a short account of some of their more important properties.

§ 5. Pseudo-differential operators and Sobolev spaces

Sobolev spaces provide a natural framework within which to study differential or pseudo-differential operators. A Sobolev space is one for which a function and all its derivatives up to some order, k say, belong to a desired space. For example, if we were working in the Hilbert space $L^2(\mathbf{R}^n)$, then we might require, of a function $f : \mathbf{R} \to \mathbf{C}$, that all its derivatives up to order k belong to $L^2(\mathbf{R}^n)$—more formally we could demand that

$$\sum_{|\alpha|\leq k} \int_{\mathbf{R}^n} \frac{\partial^\alpha f}{\partial x} \frac{\partial^\alpha \bar{f}}{\partial x}\, dx = \sum_{|\alpha|\leq k} \int_{\mathbf{R}^n} \left|\frac{\partial^\alpha f}{\partial x}\right|^2 dx < \infty \qquad (2.71)$$

---

[6] This definition of locality can be seen to correspond to the usual intuitive idea of locality if one applies it to a few examples taken from the various operators considered in the previous sections of this chapter; locality can even be regarded as an abstract definition of a differential operator, cf. Peetre [1,2].

[7] More precisely *sing supp f* can be defined as the complement of the open set on which f is smooth.

If we replace $L^2(\mathbf{R}^n)$ by $L^p(\mathbf{R}^n)$ then we demand instead that

$$\sum_{|\alpha| \le k} \int_{\mathbf{R}^n} \left| \frac{\partial^\alpha f}{\partial x} \right|^p \, dx < \infty \tag{2.72}$$

A suitable definition which allows us to work with these sorts of functions is the following:

Definition (Sobolev Space for $k = 0, 1, \ldots$) *Let Ω be an open set in $\mathbf{R}^n$, then the Sobolev space $L_k^p(\Omega)$ consists of those functions $f : \Omega \to \mathbf{C}$ for which the norm $\|f\|_{k,p}$ is finite where*

$$\|f\|_{k,p} = \sum_{|\alpha| \le k} \left\{ \int_{\mathbf{R}^n} \left| \frac{\partial^\alpha f}{\partial x} \right|^p \, dx \right\}^{1/p} \tag{2.73}$$

with $k = 0, 1, 2, \ldots$ and $1 \le p < \infty$

Evidently the case when $p = 2$ is special for then we have a Hilbert space rather than just a Banach space. For this special value of p the notation conventionally used for the Sobolev space is $H^k(\Omega)$, i.e. we have

$$H^k(\Omega) = L_k^2(\Omega) \tag{2.74}$$

It is rather important to note that the derivatives must be understood in the distributional sense.

Notice, too, that if $k = 0$ then the space H^k becomes the ordinary space $L^2(\Omega)$ of square integrable functions on Ω—we then have the equalities $L_0^2(\Omega) = H^0(\Omega) = L^2(\Omega)$, but we shall usually use the notation H^0. Also, if M is an arbitrary manifold, then by carrying out the standard procedure of taking a partition of unity and its associated open covering, we can define Sobolev spaces of L_k^p-functions on M; we denote these spaces by $L_k^p(M)$ or $H^k(M)$ whichever is applicable. Still more generally if E is a vector bundle over M we can define Sobolev spaces of L_k^p-sections of E over M denoted by $L_k^p(M, E)$—these latter can be thought of as vector valued functions on M. When the context is sufficient to resolve any ambiguity the notation in these various cases may be abbreviated to just L_k^p or H^k.

Thus far we have only defined the Sobolev spaces L_k^p for *integral* k (and $1 \le p < \infty$), in fact they can be defined for all real k: to extend the definition the natural tool is the Fourier transform $\mathcal{F}$. Using $\mathcal{F}$ we formulate the following definition:

Definition (Sobolev space for general k) *Let Ω be an open set in $\mathbf{R}^n$, then the Sobolev space $L_k^p(\Omega)$ consists of those functions and distributions f for which the norm $\|f\|_{k,p}$ is finite where*

$$\|f\|_{k,p} = \left\{ \int_{\mathbf{R}^n} |\hat{f}(\xi)(1 + \xi^2)^{k/2}|^p \, d\xi \right\}^{1/p} \qquad \text{with } k \in \mathbf{R}, \text{ and } 1 \leq p < \infty$$

$$(2.75)$$

Notice that we have now allowed f to be a *distribution* as well as a function, we show that this is a natural thing to do in the next paragraph when we discuss H^k for *negative* k. First, though, it is necessary to check that, for non-negative integral k, this definition coincides with our first one. To this end let k be a non-negative integer so that we can make the finite expansion

$$(1 + \xi^2)^k = (1 + \xi_1^2 + \cdots + \xi_n^2)^k = \sum_{|\alpha| \leq 2k} a_\alpha \xi^\alpha \qquad (2.76)$$

for suitable positive constants a_α. Using this expression, and setting $p = 2$ so as to deal with the H^k, we can write

$$\|f\|_{k,2} = \left\{ \sum_{|\alpha| \leq 2k} a_\alpha \int_{\mathbf{R}^n} \xi^\alpha |\hat{f}(\xi)|^2 \, d\xi \right\}^{1/2}$$

$$= \left\{ \sum_{|\alpha| \leq k} a_{2\alpha} \int_{\mathbf{R}^n} |D^\alpha f(x)|^2 \, dx \right\}^{1/2} \qquad (2.77)$$

and the convergence of the integrals with respect to x implies the convergence of the previous defining expression 2.73 above—hence the two definitions do indeed coincide for non-negative integral k and $p = 2$; the straightforward extension to $p \neq 2$ we leave to the reader.

We shall deal mainly with the case $p = 2$ so that our spaces will be the Hilbert spaces H^k for various k. We wish to give some more details about the H^k for a general $k \in \mathbf{R}$, in particular we wish to discuss the case of negative k. It turns out that one passes between positive and negative k by the action of taking the dual—if $(H^k)^*$ denotes the dual of H^k then we now demonstrate that

$$H^{-k} = (H^k)^*$$

Recall that distributions can be regarded as linear functionals acting on appropriate spaces of functions, i.e. distributions belong to the dual of the function space. Now the standard definition of dual says that, for f and g

complex valued functions on $\mathbf{R}^n$, then g can be regarded as being that linear functional on f, which when evaluated on f gives the complex number $g(f)$ where

$$g(f) = \int_{\mathbf{R}^n} f(x)\bar{g}(x)\, dx =< f, g >$$

with the obvious definition for $< f, g >$. Suppose, then, that $k > 0$ and $f \in H^k$; then, setting $h^k = (1 + \xi^2)^{k/2}$, we can write

$$g(f) =< f, g >=< \hat{f}, \hat{g} >= \int_{\mathbf{R}^n} \hat{f}(\xi)\bar{\hat{g}}(\xi)\, d\xi$$

$$= \int_{\mathbf{R}^n} \hat{f}(\xi)(1+\xi^2)^{k/2}\bar{\hat{g}}(\xi)(1+\xi^2)^{-k/2}\, d\xi =< fh^k, gh^{-k} > \tag{2.78}$$

Next we use the Schwarz inequality on $< fh^k, gh^{-k} >$ and so obtain

$$|g(f)| = | < fh^k, gh^{-k} > | \leq < fh^k, fh^k >^{1/2} < gh^{-k}, gh^{-k} >^{1/2}$$

$$= \left\{ \int_{\mathbf{R}^n} |\hat{f}(\xi)(1+\xi^2)^{k/2}|^2 \right\}^{1/2} \left\{ \int_{\mathbf{R}^n} |\hat{g}(\xi)(1+\xi^2)^{-k/2}|^2 \right\}^{1/2}$$

$$= \|f\|_{k,2} \|g\|_{-k,2} \tag{2.79}$$

Thus, since by assumption $f \in H^k$, and this means that $\|f\|_{k,2} < \infty$, then $g(f) < \infty \iff \|g\|_{-k,2} < \infty$, i.e. $\iff g \in H^{-k}$ and so we have shown that $H^{-k} = (H^k)^*$.

The previous discussion was for $p = 2$. For a general p we simply quote the easily obtainable result: the dual of the Sobolev space L_k^p is given by, cf. Treves [1],

$$(L_k^p)^* = L_{-k}^{p/(p-1)}, \quad k \in \mathbf{R}, \ 1 < p < \infty \tag{2.80}$$

Having seen that the H^{-k} for negative k are naturally distributions, we make use of this opportunity to expand a little on the sorts of functions and distributions that can belong to the H^k for various k.

We assume for simplicity that everything is defined on $\mathbf{R}^n$—we have already indicated the formalism required to adapt properties valid on $\mathbf{R}^n$ to compact manifolds M and vector bundles over M.

There are three standard spaces of *functions* that occur, these are $\mathcal{S}$, $\mathcal{D}$ and $\mathcal{E}$ with the following definitions (α and β are arbitrary multi-indices)

$$f(x) \in \begin{cases} \mathcal{D} & \text{if } f(x) \text{ is } C^\infty \text{ with compact support} \\ \mathcal{S} & \text{if } f(x) \text{ is } C^\infty \text{ and } \lim_{|x|\to\infty} |x|^\alpha |D^\beta f(x)| \to 0 \\ \mathcal{E} & \text{if } f(x) \text{ is simply } C^\infty \end{cases} \tag{2.81}$$

The space $\mathcal{S}$, introduced originally by Schwartz, is often referred to as the space of smooth functions rapidly decreasing at infinity; the space $\mathcal{D}$ is also known as the space of *test functions;* finally the space $\mathcal{E}$ is also denoted by C^∞. It is also clear that each these spaces satisfy $\mathcal{D} \subset \mathcal{S} \subset \mathcal{E}$ so that, passing to distributions, their duals[8] $\mathcal{D}'$, $\mathcal{S}'$ and $\mathcal{E}'$, possess the same property with the order *reversed.* In other words we have

$$\mathcal{D} \subset \mathcal{S} \subset \mathcal{E}$$
$$\mathcal{E}' \subset \mathcal{S}' \subset \mathcal{D}' \tag{2.82}$$

Another point of nomenclature is that the distribution space $\mathcal{S}'$ is commonly known as the space of *tempered distributions.* Now the space $\mathcal{S}$ is certainly contained in $L^2 = H^0$ so we can write $\mathcal{S} \subset L^2$, and taking duals gives $(L^2)' \subset \mathcal{S}'$. However, as is well known, a Hilbert space is isomorphic with its own dual; if we use this isomorphism to identify L^2 and $(L^2)'$ we can then write

$$\mathcal{S} \subset L^2 \subset \mathcal{S}' \tag{2.83}$$

This identification is exactly what is commonly employed when a function is regarded as being a distribution.

Returning now to the H^k themselves we point out that the H^k *increase in size as k decreases,* and vice versa. Thus we have

$$\cdots \subset H^2 \subset H^1 \subset H^0 \subset H^{-1} \subset H^{-2} \subset \cdots \tag{2.84}$$

and in general we have $H^k \subset H^{k'}$ if $k > k'$ with $k, k' \in \mathbf{R}$. This being so it is of interest to identify both the largest and the smallest of the H^k. To do this we define the spaces H^∞ and $H^{-\infty}$ by

$$H^\infty = \bigcap_{k \in \mathbf{R}} H^k \quad \text{and} \quad H^{-\infty} = \bigcup_{k \in \mathbf{R}} H^k = (H^\infty)^* \tag{2.85}$$

We display the hierarchy of inclusions connecting these various spaces below

$$\mathcal{D} \subset \mathcal{S} \subset H^\infty \subset \cdots H^k \subset \cdots H^0 \subset H^{-1} \subset H^{-k} \subset \cdots H^{-\infty} \subset \mathcal{S}' \subset \mathcal{D}' \tag{2.86}$$

The definition of the space $\mathcal{S}$ and its dual $\mathcal{S}'$ requires us to work in $\mathbf{R}^n$; however, if we omit $\mathcal{S}$ and $\mathcal{S}'$ from 2.86 then 2.86 is still valid for the Sobolev spaces $H^k(\Omega)$ with Ω an open set in $\mathbf{R}^n$. Also, instead of functions on M,

---

[8] Usually we denote the dual of a vector space E by E^*; here we follow the common practice and replace the asterisk by a prime.

we can consider sections of a vector bundle E over a compact manifold M with its associated Sobolev space denoted by $H^k(M, E)$. Then 2.86, with S and S' omitted, will again be valid.

There is a very useful result which relates the L_k^p to the space C^l where C^l denotes the space of l-times differentiable functions with *continuous* derivatives. It is summarised in the following theorem

Theorem (Sobolev (i)) *If M is an n-dimensional manifold with $M = \mathbf{R}^n$ or M compact, the space L_k^p is contained in the space C^l if $k - n/p > l$ or*

$$L_k^p \subset C^l \text{ if } \left(k - \frac{n}{p} \right) > l \qquad (2.87)$$

Thus if k is big enough L_k^p sits inside the space C^l; in fact when this is so the inclusion map from L_k^p to C^l is a compact operator.[9] We can also compare two different Sobolev spaces and show that one can be embedded inside the other, the precise result is that

Theorem (Sobolev (ii)) *If M is an n-dimensional manifold with $M = \mathbf{R}^n$ or M compact, then the spaces L_k^p and L_l^q possess the property that*

$$L_k^p \subset L_l^q \quad \text{if} \quad k \geq l \quad \text{and} \quad \left(k - \frac{n}{p} \right) \geq \left(l - \frac{n}{q} \right) \qquad (2.88)$$

In this case, the natural injection from L_k^p to L_l^q is a compact operator if the inequalities are strict. These two results are usually known as the Sobolev embedding theorems. We can indicate a simple proof of Sobolev (i) for the case $p = 2$ without too much difficulty, cf. Wells [1]. Suppose that $f \in H_k$ and $f \in C^l$. Then we may represent $D^l f$ by the Fourier transform formula

$$D^l f = \int_{\mathbf{R}^n} \exp[-i\xi x]\xi^l \hat{f}(\xi) \, d\xi \qquad (2.89)$$

and the convergence of this integral implies that $D^l f$ both exists and is a continuous function. But with judicious use of Schwarz's inequality we can

---

[9] A compact operator T is one for which the image of the unit ball has a compact closure. It is perhaps helpful to note that an operator whose range $R(T)$ is finite dimensional is automatically compact. Further, though a compact operator can have an infinite dimensional range it can always be *approximated* by a sequence of operators with a finite dimensional range.

write

$$\left| \int_{\mathbf{R}^n} \exp[-i\xi x]\xi^l \hat{f}(\xi)\, d\xi \right| = \left| \int_{\mathbf{R}^n} \exp[-i\xi x]\hat{f}(\xi)(1+\xi^2)^{k/2} \frac{\xi^l}{(1+\xi^2)^{k/2}}\, d\xi \right|$$

$$\leq \left\{ \int_{\mathbf{R}^n} \left| \hat{f}(\xi)(1+\xi^2)^{k/2} \right|^2 d\xi \right\}^{1/2} \left\{ \int_{\mathbf{R}^n} \frac{\xi^{2l}}{(1+\xi^2)^k}\, d\xi \right\}^{1/2}$$

$$= \|f\|_{k,2} \left\{ \int_{\mathbf{R}^n} \frac{\xi^{2l}}{(1+\xi^2)^k}\, d\xi \right\}^{1/2}$$

(2.90)

However, since $f \in H^k$, then $\|f\|_{k,2} < \infty$, so we just require convergence of the explicit integral in 2.90; but power counting shows that the integral converges when $2l - 2k + n < 0$ or $(k - n/2) > l$ just as the theorem states. Thus we have proved Sobolev (i) for the case $M = \mathbf{R}^n$ and $p = 2$; it is easy enough to extend to the case where M is compact. For the case $p \neq 2$ it is easy to provide evidence that the condition of the theorem is a necessary one: choose a perfectly good C^l function such as $f(x) = (1+x_1^2+\cdots+x_n^2)^{l/2}$ and observe that $\|f\|_{k,p} < \infty$ requires $(k - n/p) > l$—for a full proof cf. Adams [1].

An instructive way to think about the result $H^k \subset C^l$ if $(k - n/2) > l$ of Sobolev (i) is to think of it as indicating the amount of smoothness possessed by the elements of H^k: if $\epsilon > 0$ and k is a non-negative integer we can restate the theorem as

$$H^{k+\epsilon} \subset C^{k-n/2}$$

(2.91)

This means that, rather than the elements of $H^{k+\epsilon}$ being C^k, they are less differentiable, being only $C^{k-n/2}$ (a similar formula applies to the L_k^p: one has $L_{k+\epsilon}^p \subset C^{k-n/p}$). Thus, in general, the Sobolev spaces L_k^p contain functions with a deficit in their smoothness measured by the number n/p, this being ultimately due to the distributional nature of the derivatives in the definition of the norm $\| \ \|_{k,p}$.

We are now able describe the connection between pseudo-differential operators and Sobolev spaces; we shall limit ourselves to remarks about the H^k. If P is a linear pseudo-differential operator of arbitrary real order m then P may be realised as a linear map between Sobolev spaces (which we also denote by P) of the form

$$P : H^k \longrightarrow H^{k-m}$$

(2.92)

for all $k \in \mathbf{R}$. If P is a differential operator then the order m will always be a non-negative integer, but we have given examples in §4 where m is fractional or negative, and we can see from 2.92 above that the use of a P with a fractional m requires the existence of a H^k with fractional k. Also we have encountered smoothing operators which have order $-\infty$. Suppose that P is a smoothing operator, then 2.92 becomes

$$P : H^k \longrightarrow H^\infty \subset C^\infty \tag{2.93}$$

Thus we see that, as we pointed out at the time, a smoothing operator maps any function onto a smooth function hence justifying its name.

If P is an elliptic m^{th} order differential operator on a compact manifold then P has a finite dimensional kernel, but P elliptic implies P^* elliptic (recall from § 3 that the leading symbol of P^* is the adjoint of that of P so the invertibility of the leading symbol of P implies the same thing for P^*) and so P^* also has a finite dimensional kernel. Thus we can say that P has the property of possessing a finite dimensional kernel and cokernel. When P is viewed as an operator between Sobolev spaces this remains true, but bounded linear operators between Hilbert spaces with this property are called Fredholm operators. Thus elliptic P are also Fredholm when realised on Sobolev spaces. It is of interest, though, to verify that such elliptic P are *bounded* operators on Sobolev space; this is because a typical elliptic P, such as Δ, has a spectrum tending to infinity, which might be thought to indicate unboundedness. We can show that this is not so in a simple example which also serves to illustrate the mechanism which operates to ensure boundedness. Let M be the circle S^1 with local coordinate θ, and P be the operator $-d^2/d\theta^2 = D_\theta^2$. Note that the spectrum of P is just $0, 1, 2^2, \ldots n^2, \ldots$ with eigenvectors $\exp[\mp in\theta] = e_n$ and so the spectrum tends to infinity. Now in general if L is some linear operator between Hilbert spaces $L : H \to H$ then the norm of L is defined by

$$\|L\| = \sup_{\substack{v \in H \\ v \neq 0}} \frac{\|Lv\|}{\|v\|} \tag{2.94}$$

Thus if L had a sequence of eigenvectors v_n whose eigenvalues λ_n tended to infinity it is evident that we would have $\|L\| = \sup_n\{\lambda_n\} = \infty$. However, our situation is a little different: since $P = D_\theta^2$ is of order 2 we have

$$P : H^k \longrightarrow H^{k-2} \Rightarrow \|P\| = \sup_{\substack{v \in H^k \\ v \neq 0}} \frac{\|Pv\|_{k-2,2}}{\|v\|_{k,2}} \tag{2.95}$$

Thus we can take the supremum over the eigenvectors e_n of P, and if we denote the ordinary L^2 norm by $\| \ \|$, then the norm of P is therefore given by the expression

$$\sup_n \frac{\|Pe_n\|_{k-2,2}}{\|e_n\|_{k,2}} = \sup_n \frac{\sum_{j=0}^{j=k-2} \|D_\theta^j(D_\theta^2 e_n)\|}{\sum_{j=0}^{j=k} \|D_\theta^j e_n\|} = \sup_n \frac{\sum_{j=0}^{j=k} \|D_\theta^j e_n\|}{\sum_{j=0}^{j=k} \|D_\theta^j e_n\|} = 1$$

(2.96)

Thus the norm of P is indeed finite as befits a Fredholm operator. A further well known property of Fredholm operators is that they are invertible modulo a compact operator. This is nothing other than the Sobolev space version of the property that elliptic operators possess of being invertible modulo a smoothing operator.

An inverse in this sense for an elliptic operator P is called a parametrix for P. More generally, if P is an elliptic pseudo-differential operator of order m realised as $P : H^k \to H^{k-m}$ then Q is a parametrix for P if there are compact operators K_1 and K_2 such that

$$P \circ Q_1 - K_1 = I_{H^k} \text{ and } Q \circ P - K_2 = I_{H^{k-m}} \qquad (2.97)$$

where I_{H^k} and $I_{H^{k-m}}$ denote the identities on H^k and H^{k-m} respectively.

We shall return to elliptic operators in chapter 4; in the next chapter we turn our attention to cohomology properties of vector bundles and introduce the very useful notion of a sheaf.

CHAPTER III

Cohomology of Sheaves and Bundles

§ 1. Sheaves

Sheaves are mathematical objects designed to deal with the problem of how to pass from local data on a space to global data on that space: it may sometimes be the case that something which is locally true is also globally true, for example if the dimension of an open set of a closed manifold is n then the manifold as a whole has dimension n; on the other hand the scalar curvature on a Riemannian manifold can be locally constant without being globally so.

A particular sheaf usually singles out some specific local property such as continuity, differentiability or holomorphicity and incorporates this into its definition; the non-triviality or otherwise of the passage from the local to the global is typically measured by what is called a sheaf cohomology group.

We turn now to the definition of a sheaf—the reader should not be unduly put off by its rather abstract appearance—we shall proceed quickly to some concrete examples.

Definition (Sheaf) *A sheaf $\mathcal{F}$ over a topological space M is an assignment to each open set $U \subset M$ of a group $\mathcal{F}(U)$, known as the sections of $\mathcal{F}$ over U, which possesses the following two properties*

  (i) Given two such open sets U and V, with $U \subset V$, there exist what are called restriction maps $r_U^V : \mathcal{F}(V) \to \mathcal{F}(U)$ which satisfy

$$r_U^U = identity \ \ and \ if \ U \subset V \subset W \ then \ r_U^W = r_U^V \circ r_V^W$$

  (If a section $s \in \mathcal{F}(V)$ then one should think of $r_U^V(s)$ as the restriction of s from V to its subset U. Thus the first property just says that the restriction of s from U to U leaves s unchanged, the second property says

that the restriction of s from W to U may be done via an intermediate subset V.)

(ii) Let U be expressed as a union of open sets according to $U = \bigcup U_i$ then

given two sections $s_1, s_2 \in \mathcal{F}(U)$ $r^U_{U_i}(s_1) = r^U_{U_i}(s_2)$ $\forall i, \Rightarrow s_1 = s_2$

if $r^{U_i}_{U_i \cap U_j}(s_i) = r^{U_j}_{U_i \cap U_j}(s_j)$, $\forall i, j$,

then there is a unique $s \in \mathcal{F}(U)$ such that $r^U_{U_i}(s) = s_i$

$$(3.1)$$

(The first property here asserts that if the restrictions of two sections always agree then the two sections are identical; the second one says that if sections s_i and s_j always agree on their overlap $U_i \cap U_j$ then one can assemble one global section s out of the local data given by the s_i.)

These two properties are somewhat complementary to one another. The latter shows that a section over U may be assembled from local data on subsets of U, it thus represents the passage from the local to the global. The former illustrates the fact that a section over U is determined by all its restrictions to subsets of U; it thus represents the passage from the global to the local.

Having in mind the remarks we made above contrasting the local and the global we see that, in general, the spaces of sections $\mathcal{F}(U)$ get smaller and smaller as U gets bigger. Conversely, as U shrinks down to a point the $\mathcal{F}(U)$ get bigger. This leads to the idea of a *stalk* $\mathcal{F}_x$ at x. Using a direct limit we set

$$\mathcal{F}_x = \varinjlim_{x \in U} \mathcal{F}(U) \qquad (3.2)$$

Stalks are used when we discuss sheaf cohomology. We are now ready to look at some examples.

Loosely speaking one can think of a sheaf as a kind of parametrised family of functions. Beginning with a differentiable manifold M of dimension n we have the sheaf of C^∞ functions on M which we denote by $\mathcal{E}$: in this example the sections $\mathcal{E}(U)$ over U are just the smooth functions defined on U, these are required by the definition of a sheaf to be a group and the (Abelian) group operation is simply that of addition of functions; finally the maps $r^V_U : \mathcal{E}(V) \to \mathcal{E}(U)$ are simply the ordinary restrictions of a smooth function from an open set V to an open subset U.

The verification of the definition for $\mathcal{E}$ is a simple task which we leave to the reader as an exercise; a further simple exercise for the reader is to check that this example gives rise to two more examples: we obtain the first by replacing the property of smoothness by that of real-analyticity, the resulting sheaf is denoted by $\mathcal{A}$; we obtain the second by requiring M to be a *complex* manifold of complex dimension n and replacing the property of real-analyticity by that of holomorphicity, and this sheaf is denoted by $\mathcal{O}$.

Sheaves in analysis are of considerable value. They can be used on non-compact manifolds and also on complex analytic spaces in which singularities are allowed. Because such spaces do not have good systems of local coordinates, there are problems using differential forms in such contexts. However the methods of partial differential equations, which are essentially local, are applicable; and combining these with sheaves gives a theory which can tackle problems which occur in this wider class of spaces.

Now that we have met some sheaves on M we give an example (cf. Wells [1]) of an object which *fails* to be a sheaf on M. Let M be the complex plane and try to define a sheaf $\mathcal{F}$ by defining $\mathcal{F}(\mathcal{U})$ to be all holomorphic, bounded functions on U, where U is an open set in M. Then $\mathcal{F}$ fails to satisfy the definition of a sheaf, in particular it fails to satisfy the very last property of the definition. To see this we observe that there are many *different* bounded holomorphic functions possible if one chooses varying open sets U in the complex plane. However the last part of the definition asserts that we can always find a global function on the complex plane which coincides locally with *all* the local versions. This we know to be impossible since Liouville's theorem tells us that there is only *one* bounded holomorphic function on the entire complex plane, namely the constant function. Hence $\mathcal{F}$ is not a sheaf. The reason for this failure should be quite clear: we have seen that a sheaf on M is a bearer of localised information about M; however, the boundedness requirement in $\mathcal{F}$ is not a sufficiently local property for the definition to work.

Having constructed $\mathcal{E}$, $\mathcal{A}$, and $\mathcal{O}$ one can easily manufacture more sheaves over M. To accomplish this all one has to realise is that a great many sheaves originate naturally as spaces of sections of vector bundles. To see this let E be a vector bundle over some manifold M. Then, if $\Gamma(M, E)$ denotes the sections of E over M, we simply define these to be the sections of a sheaf $\mathcal{F}$, with the sections $\mathcal{F}(U)$ over U being given in an obvious way by setting $\mathcal{F}(U) = \Gamma(U, E)$, and the restriction maps r_U^V being given by restriction of sections from V to U; the word restriction now being understood in its natural sense rather than an axiomatic one. We can immediately single out inside $\Gamma(M, E)$ the sheaves of continuous, differentiable, or, when M is complex, holomorphic sections over M.[1] Some of the most important commonly occurring sheaves have a geometric origin, i.e. the bundle E is

---

[1] The stalks $\mathcal{F}_x$ of $\mathcal{F}$ are often called the *germs* of $\mathcal{F}$ at x; consequently $\mathcal{F}$ is sometimes referred to as a sheaf of germs of differentiable or holomorphic sections over M.

chosen to be a bundle reflecting the geometry of M. Let us take E to be
the cotangent bundle T^*M of a smooth real manifold M. Thus

$$\Gamma(M, E) = \Gamma(M, T^*M)$$

and so the smooth sections of E comprise the sheaf of smooth 1-forms over
M; also, by taking the pth exterior power of T^*M, we obtain $\Gamma(M, \wedge^p T^*M)$
which are the sections of the corresponding sheaf of p-forms over M for which
we use the notation $\mathcal{E}^p$. Replacing M by a *complex* manifold, and setting E
equal to the *holomorphic* cotangent bundle T^*M, we obtain the sheaf Ω^p of
holomorphic p-forms over M—note that $\Omega^0 = \mathcal{O}$, the holomorphic functions
over M. Another useful example is the sheaf $\mathcal{O}^*$ of *non-zero* holomorphic
functions over M; obviously addition and subtraction of sections does not
preserve positivity, this means that $\mathcal{O}^*$ must be viewed as a multiplicative
sheaf, the group operation on the sections being multiplication.

We can also combine these geometric sheaves with an *additional* vector
bundle which we still denote by E. This process is called taking coefficients
in E. For example, the sheaf $\mathcal{E}^p(E)$ is known as the sheaf of smooth p-forms
with coefficients in E. More precisely $\mathcal{E}^p(E)$ is the sheaf whose sections are
given by the space

$$\Gamma(M, \wedge^p T^*M \otimes E) \tag{3.3}$$

In other words the coefficients are generated by tensoring with E; one can
think of the sections of $\mathcal{E}^p(E)$ as being vector valued p-forms—indeed for
an open set U over which the vector bundle E is locally trivial the sections
$\mathcal{E}^p(E)(U)$ *are* precisely vector valued p-forms. In an exactly similar way
one can construct the sheaf of holomorphic p-forms with coefficients in E,
which we write as $\Omega^p(E)$. Continuing in this fashion we can manufacture
new vector bundles by employing the operations of direct sum and tensor
product. Using the vector bundles E and F this results in new sheaves of
the form $\mathcal{E}(E \otimes F)$, $\mathcal{O}(E \oplus E \oplus \cdots)$ and so on.

§ 2. Sheaf cohomology

As we mentioned at the beginning of this chapter the non-triviality of passage from what is local to what is global is often measured by a sheaf
cohomology group. We can now have a look at sheaf cohomology. First
of all recall that in the ordinary cohomology theory of a manifold M one
computes $H^i(M; G)$ where G is some Abelian group such as $\mathbf{Z}$, $\mathbf{R}$, $\mathbf{Q}$, etc.
In sheaf cohomology the coefficients are taken, not in G, but in some sheaf
such as $\mathcal{E}$ or $\mathcal{O}$. Before being more precise about this it is useful to point
out that the two sorts of cohomology theories have some overlap. This is
because one can think of $\mathbf{R}$ as being the sheaf of constant functions on

a manifold M—the corresponding sheaf cohomology group is just the real cohomology group $H^i(M; \mathbf{R})$.

Since cohomology is a measure of how far some sequence is from being exact, then to define sheaf cohomology we introduce exact sequences of sheaves. Let $\mathcal{E}$, $\mathcal{F}$ and $\mathcal{G}$ be sheaves over M which are connected by two maps e and f as shown below

$$\mathcal{E} \xrightarrow{e} \mathcal{F} \xrightarrow{f} \mathcal{G} \tag{3.4}$$

Next we require the maps e and f to be *sheaf morphisms*—a map m between two sheaves $\mathcal{A}$ and $\mathcal{B}$ stands for a collection of maps between sections of the form $m_U : \mathcal{A}(U) \to \mathcal{B}(U)$ with m_U a homomorphism between the groups of sections $\mathcal{A}(U)$ and $\mathcal{B}(U)$. We say that m is a sheaf morphism if it renders the following diagram commutative

$$\begin{array}{ccc} \mathcal{A}(V) & \xrightarrow{m_V} & \mathcal{B}(V) \\ \downarrow r_U^V & & \downarrow r_U^V \\ \mathcal{A}(U) & \xrightarrow{m_U} & \mathcal{B}(U) \end{array} \quad U \subset V \subset M \tag{3.5}$$

Finally the sequence of maps

$$\mathcal{E} \xrightarrow{e} \mathcal{F} \xrightarrow{f} \mathcal{G} \tag{3.6}$$

is called *exact* if e and f are sheaf morphisms and they induce maps e_x and f_x which are exact on the stalks i.e. the following sequence is exact

$$\mathcal{E}_x \xrightarrow{e_x} \mathcal{F}_x \xrightarrow{f_x} \mathcal{G}_x, \qquad x \in M \tag{3.7}$$

An example of a sheaf exact sequence is easily found. Consider the sequence

$$0 \longrightarrow \mathbf{Z} \xrightarrow{i} \mathcal{O} \xrightarrow{\exp} \mathcal{O}^* \longrightarrow 0 \tag{3.8}$$

where i denotes inclusion and exp is a map which sends $f(z)$ to $\exp[2\pi i f(z)]$. Here $\mathbf{Z}$ is a constant sheaf and the exactness follows trivially from the elementary fact that, if $n \in \mathbf{Z}$, then $(\exp \circ i)(n) = \exp[2\pi i n] = 1$, where we also recall that $\mathcal{O}^*$, being a multiplicative sheaf, has the unit constant function as its 'zero'.

One should realise, though, that if one encounters a short exact sequence of sheaves over M

$$0 \longrightarrow \mathcal{E} \xrightarrow{e} \mathcal{F} \xrightarrow{f} \mathcal{G} \longrightarrow 0 \tag{3.9}$$

then the exactness of this sequence does *not* imply the exactness of the corresponding global sections over M; i.e. the sequence

$$0 \longrightarrow \mathcal{E}(M) \xrightarrow{e_M} \mathcal{F}(M) \xrightarrow{f_M} \mathcal{G}(M) \longrightarrow 0 \qquad (3.10)$$

need not be exact and sheaf cohomology can measure any obstruction to exactness. However, there is a circumstance which can render the sequence 3.10 exact. This can be seen in the following way. First of all it is easy to check that the sequence 3.10 can only fail to be exact at its right-hand end, i.e. at $\mathcal{G}$; in terms of maps we can say that the map f_M need not be surjective. Secondly suppose the sheaf $\mathcal{E}$ possesses the property that any section over a set $U \subset M$ can always be extended to a section over the whole of M; again in terms of maps this is the statement that the maps

$$r_U^M : \mathcal{E}(M) \longrightarrow \mathcal{E}(U) \qquad (3.11)$$

are surjective for all U. Then it is easy to show, cf. Wells [1], that the sequence 3.10 is now exact. In fact sheaves with the property 3.11 of always possessing global sections are called *soft sheaves*. As one might expect, the cohomology of soft sheaves is always trivial—we do not include a proof of this fact here but the proof is quite simple; see, for example, Wells [1]. In any case, the sheaf cohomology groups which we are about to construct are denoted by $H^q(M; \mathcal{F})$ and have the property that $\mathcal{F}$ soft implies $H^q(M; \mathcal{F}) = 0$, for $q > 0$. Standard examples of soft sheaves are $\mathcal{E}$ and the p-form sheaf $\mathcal{E}^p$. In fact these sheaves are an important *subclass* of soft sheaves which are called *fine*—a sheaf is called *fine* if it admits a locally finite partition of unity.

To construct sheaf cohomology we use a cochain complex and a coboundary operator. Let $\mathcal{F}$ be a sheaf over M and $O = \{O_\alpha\}$ be an open covering of M. Take a collection $(O_1, \ldots, O_{q+1}) = \Delta$ of $q + 1$ of the O_α and require them to have a non-empty intersection which we denote by $\bar{\Delta}$. Then we define the set of all q-cochains $C^q(O, \mathcal{F})$ by

$$C^q(O, \mathcal{F}) = \mathcal{F}(\bar{\Delta}) \qquad (3.12)$$

Next, let $f(\Delta) \in C^q(O, \mathcal{F})$, and define a coboundary operator d between the cochains by

$$df(\Delta) = \sum_{i=1}^{q+2} (-1)^i r_{\bar{\Delta}}^{\bar{\Delta}_i} f(\Delta_i) \qquad (3.13)$$

where $r_{\bar{\Delta}}^{\bar{\Delta}_i}$ is the usual sheaf restriction mapping and Δ_i is given by $\Delta_i = (O_1, \ldots, O_{i-1}, O_{i+1}, \ldots, O_{q+1})$. It is straightforward to verify that $d^2 = 0$

so that d is a true coboundary operator. This allows us to construct the complex

$$\cdots \xrightarrow{d_{q-2}} C^{q-1}(O,\mathcal{F}) \xrightarrow{d_{q-1}} C^q(O,\mathcal{F}) \xrightarrow{d_q} C^{q+1}(O,\mathcal{F}) \xrightarrow{d_{q+1}} \cdots \qquad (3.14)$$

We define the cohomology group for the complex $\{C^q(O,\mathcal{F}), d_q\}$ by the usual construction, i.e. we set

$$H^q(O;\mathcal{F}) = \frac{\ker d_q}{\operatorname{Im} d_{q-1}} \qquad (3.15)$$

Notice that we write $H^q(O;\mathcal{F})$ rather than $H^q(M;\mathcal{F})$. This is to draw attention to the fact that our construction could depend on the covering used. In fact, to prevent this, finer and finer coverings [2] are used and $H^q(M;\mathcal{F})$ is defined to be what is obtained as the limit of this process. Thus we define $H^q(M;\mathcal{F})$ by

$$H^q(M;\mathcal{F}) = \varinjlim_O H^q(O;\mathcal{F}) \qquad (3.16)$$

the limit being taken over the coverings O. Then $H^q(M;\mathcal{F})$ is referred to as the sheaf cohomology group of M of degree q with coefficients in $\mathcal{F}$. A useful fact which can be deduced directly from the cochain definitions above is that, for *any* sheaf $\mathcal{F}$ over M

$$H^0(M;\mathcal{F}) = \Gamma(M,\mathcal{F}) \qquad (3.17)$$

That is the zeroth cohomology group of M with coefficients in $\mathcal{F}$ is simply the global sections of $\mathcal{F}$ over M. Our definition of sheaf cohomology has used the techniques of Čech cohomology; there is also an alternative, more abstract, definition of sheaf cohomology using what are called sheaf resolutions, cf. Wells [1]. These two sorts of sheaf cohomology coincide if M is paracompact, something which is true for any manifold, cf. Spivak [1]—paracompactness for a space X is the property that any open cover $O = \{O_\alpha\}$ of X has a locally finite refinement; that is, any point $x \in X$ has a neighbourhood which intersects with only *finitely* many of the sets in the cover.

To calculate sheaf cohomology in practice requires a little more than just the definitions that we have presented above. One of the most fundamental calculational tools is the following: Given the short exact sheaf sequence

$$0 \longrightarrow \mathcal{E} \xrightarrow{e} \mathcal{F} \xrightarrow{f} \mathcal{G} \longrightarrow 0 \qquad (3.18)$$

---

[2] Given two coverings $O = \{O_\alpha\}_{\alpha \in I}$ and $O' = \{O'_\beta\}_{\beta \in I'}$ then O' is finer than, or is a refinement of, O if for each $\beta \in I'$ there is an $\alpha \in I$ such that $O'_\beta \subset O_\alpha$.

then the cohomology sequence below *is also exact*

$$
\begin{aligned}
0 &\longrightarrow H^0(M;\mathcal{E}) \longrightarrow H^0(M;\mathcal{F}) \longrightarrow H^0(M;\mathcal{G}) \\
&\longrightarrow H^1(M;\mathcal{E}) \longrightarrow H^1(M;\mathcal{F}) \longrightarrow H^1(M;\mathcal{G}) \longrightarrow \cdots \\
&\qquad\qquad\qquad\qquad \vdots \\
&\longrightarrow H^q(M;\mathcal{E}) \longrightarrow H^q(M;\mathcal{F}) \longrightarrow H^q(M;\mathcal{G}) \longrightarrow \cdots
\end{aligned}
\tag{3.19}
$$

Let us apply this result. Returning to the sheaf $\mathcal{E}^p$ of p-forms over M, we observed on p. 61 that $\mathcal{E}^p$ is a soft sheaf, thus its higher cohomology groups vanish so that $H^q(M;\mathcal{E}^p) = 0$ for $q > 0$. Denote by $\mathcal{Z}^p$ the sheaf of *closed* p-forms and by d the exterior derivative, and note, in passing, that $\mathcal{Z}^0 = \mathbf{R}$. Then we can immediately write down a series of short exact sequences of sheaves, and, to each, we can apply the result 3.19 above and the softness of $\mathcal{E}^p$. We record below the results.

$$
\begin{aligned}
0 &\longrightarrow \mathbf{R} \xrightarrow{\ d\ } \mathcal{E}^0 \xrightarrow{\ d\ } \mathcal{Z}^1 && \Rightarrow H^{q-1}(M;\mathcal{Z}^1) \simeq H^q(M;\mathbf{R}) \\
0 &\longrightarrow \mathcal{Z}^1 \xrightarrow{\ d\ } \mathcal{E}^1 \xrightarrow{\ d\ } \mathcal{Z}^2 && \Rightarrow H^{q-1}(M;\mathcal{Z}^2) \simeq H^q(M;\mathcal{Z}^1) \\
&\qquad\qquad \vdots && \qquad\qquad \vdots \\
0 &\longrightarrow \mathcal{Z}^p \xrightarrow{\ d\ } \mathcal{E}^p \xrightarrow{\ d\ } \mathcal{Z}^{p+1} && \Rightarrow H^{q-1}(M;\mathcal{Z}^{p+1}) \simeq H^q(M;\mathcal{Z}^p)
\end{aligned}
\tag{3.20}
$$

Applying a little induction to the cohomology results above yields the formula

$$
H^q(M;\mathbf{R}) \simeq H^{q-1}(M;\mathcal{Z}^1) \simeq H^{q-2}(M;\mathcal{Z}^2) \cdots \simeq H^1(M;\mathcal{Z}^{q-1}) \tag{3.21}
$$

But, since $H^0(M;\mathcal{E}^p) \neq 0$, we cannot conclude that $H^1(M;\mathcal{Z}^{q-1}) \neq H^0(M;\mathcal{Z}^p)$. Instead we have

$$
H^1(M;\mathcal{Z}^{q-1}) = \frac{H^0(M;\mathcal{Z}^p)}{Im\, dH^0(M;\mathcal{E}^{p-1})} \tag{3.22}
$$

However, the RHS of the last equation is just the closed p-forms on M modulo the exact ones, i.e. it is the qth de Rham cohomology group. Thus we have the well known result that

$$
H^q(M;\mathbf{R}) \simeq H^q_{de\,Rham}(M) \tag{3.23}
$$

There is a version of this result for complex manifolds which is also of some importance. To derive it we use an identical technique. We take a complex manifold M of complex dimension n and replace real differential forms by complex differential forms. In this latter connection we need to display the type of a complex differential form. To this end let ω be a differential form on M written in local coordinates $(z^i, \bar{z}^i)$ as

$$\omega = \omega_{i_1 \cdots i_p j_1 \cdots j_q} dz^{i_1} \wedge \cdots \wedge dz^{i_p} \wedge d\bar{z}^{j_1} \wedge \cdots \wedge d\bar{z}^{i_q} \tag{3.24}$$

Such an ω is obviously a $(p+q)$-form but is also known as a form of type (p, q). Having defined the type of a differential form it is appropriate to introduce the operators ∂ and $\bar{\partial}$ which are given in local coordinates by

$$\bar{\partial} = \sum_{i=1}^{n} \frac{\partial}{\partial \bar{z}^i} d\bar{z}^i$$

$$\partial = \sum_{i=1}^{n} \frac{\partial}{\partial z^i} dz^i \tag{3.25}$$

The familiar Cauchy–Riemann equations for a holomorphic function f are, of course, simply the statement $\bar{\partial} f = 0$. One should also note the various relations between the operators d, ∂ and $\bar{\partial}$, which are

$$d = \partial + \bar{\partial} \qquad \text{and} \qquad d^2 = \partial^2 = \partial \bar{\partial} + \bar{\partial} \partial = \bar{\partial}^2 = 0 \tag{3.26}$$

Using the notion of type we decompose the complex differential forms of degree n according to the sum

$$\bigoplus_{p+q=n} \mathcal{E}^{p,q} \tag{3.27}$$

where $\mathcal{E}^{p,q}$ is the sheaf of complex differential forms of type (p, q). It is also a *fine* sheaf.

Now we return to our cohomology calculation. Let $\mathcal{Z}^{p,q}$ be the sheaf of $\bar{\partial}$-*closed* differential forms of type (p, q). Notice that since *holomorphic p-forms* on M are necessarily of type $(0, p)$, then the sheaf Ω^p of holomorphic p-forms on M is the same as the sheaf $\mathcal{Z}^{p,0}$. We replace the exterior derivative d by the operator $\bar{\partial}$ and take the series of short sheaf exact sequences

$$0 \longrightarrow \Omega^p \xrightarrow{\bar{\partial}} \mathcal{E}^{p,0} \xrightarrow{\bar{\partial}} \mathcal{Z}^{p,1}$$

$$0 \longrightarrow \mathcal{Z}^{p,1} \xrightarrow{\bar{\partial}} \mathcal{E}^{p,1} \xrightarrow{\bar{\partial}} \mathcal{Z}^{p,2} \tag{3.28}$$

$$\vdots$$

$$0 \longrightarrow \mathcal{Z}^{p,q} \xrightarrow{\bar{\partial}} \mathcal{E}^{p,q} \xrightarrow{\bar{\partial}} \mathcal{Z}^{p,q+1}$$

Then, the vanishing of the higher cohomology of the fine sheaf $\mathcal{E}^{p,q}$, and an exactly similar inductive argument to that used in the de Rham case produces the result $H^q(M; \Omega^p) \simeq H^{q-1}(M; \mathcal{Z}^{p,1}) \cdots \simeq H^1(M; \mathcal{Z}^{p,q-1})$. Once again we stop before we get to $H^0(M; \mathcal{Z}^{p,q})$. But we can say instead that

$$H^1(M; \mathcal{Z}^{p,q-1}) \simeq \frac{H^0(M; \mathcal{Z}^{p,q})}{Im\,\overline{\partial}H^0(M; \mathcal{E}^{p,q-1})} \tag{3.29}$$

Also we know that the zeroth cohomology group was identified in 3.17 as the global sections of the sheaf in question. Applying this fact to the numerator and denominator of the preceding equation gives

$$\frac{H^0(M; \mathcal{Z}^{p,q})}{Im\,\overline{\partial}H^0(M; \mathcal{E}^{p,q-1})} = \frac{ker\,\left(\overline{\partial}: \mathcal{E}^{p,q} \longrightarrow \mathcal{E}^{p,q+1}\right)}{Im\,\left(\overline{\partial}: \mathcal{E}^{p,q-1} \longrightarrow \mathcal{E}^{p,q}\right)} \tag{3.30}$$

Following the usual practice we shall denote the RHS of 3.30 by $H^{p,q}_{\overline{\partial}}(M)$. Thus we have proved the following theorem.

Theorem (Dolbeault) *If M is a complex manifold then*

$$H^q(M; \Omega^p) \simeq H^{p,q}_{\overline{\partial}}(M) \tag{3.31}$$

$H^{p,q}_{\overline{\partial}}(M)$ is known as a *Dolbeault cohomology group;* as we have just seen, it is what the de Rham construction produces when one replaces, a real manifold by a complex one and the exterior derivative by the $\overline{\partial}$ operator. Dolbeault's theorem will be of use later on in this book.

§ 3. K-theory

In this section we would like to consider a cohomology theory for vector bundles known as K-theory. K-theory emphasises features which become prominent as the rank of the vector bundles become large. In this sense it is therefore a kind of linear algebra for large matrices. It is probably the most important example of a *generalised* cohomology theory. The word generalised refers to the fact that K-theory does not satisfy the axiomatic definition of a cohomology theory given by Eilenberg and Steenrod [1]; the particular axiom that is not satisfied being the dimension axiom which defines in advance the cohomology of a space consisting only of a point.

Let us consider real or complex vector bundles over a space X, which we may as well take to be a manifold M. We use I^j to denote a trivial bundle of rank j over M so that $I^j \simeq M \times \mathbf{R}^j$ in the real case and $I^j \simeq M \times \mathbf{C}^j$ in the complex case. K-theory considers all vector bundles over M and assembles them into two sorts of equivalence classes. We shall consider the first sort now: two bundles E and F are equivalent if the addition of a trivial bundle I^j to each of them renders them isomorphic. That is, one has

$$E \oplus I^j \simeq F \oplus I^k \tag{3.32}$$

where the relation $\simeq$ denotes isomorphism; and we do *not* assume that $j = k$ so that E and F are allowed to have *differing* ranks. It is easy to verify that 3.32 is an equivalence relation and when it is obeyed the two bundles E and F are called *stably equivalent*. We write this as

$$E \overset{s}{\sim} F \tag{3.33}$$

The corresponding equivalence classes are called *stable equivalence classes*.

Note that vector bundles have the property that

$$E \oplus G \simeq F \oplus G \nRightarrow E \simeq F \tag{3.34}$$

A well known counter example is used to illustrate this point. Let TS^n and $N(S^n)$ denote the tangent bundle and normal bundle to S^n, respectively. Notice that $N(S^n)$ is a line bundle and has an obvious global section given by an outward-pointing normal vector. A single non-vanishing global section is enough to trivialise a line bundle, cf. Nash and Sen [1], thus $N(S^n)$ is isomorphic to the trivial bundle I. Further, if we decompose any vector in $\mathbf{R}^{n+1}$ into a normal and transverse part, then we can record this elementary decomposition in a bundle-theoretic manner as the equation (valid on S^n)

$$T\mathbf{R}^{n+1} \simeq TS^n \oplus N(S^n) \tag{3.35}$$

Clearly $T\mathbf{R}^{n+1}$ is a trivial bundle so that $T\mathbf{R}^{n+1} \simeq I^{n+1}$. Thus we can write

$$TS^n \oplus N(S^n) \simeq I^n \oplus I \tag{3.36}$$

since both sides are the trivial bundle I^{n+1}. However, it does not follow that TS^n is trivial. Indeed TS^n is known to be trivial only for the case of parallelisable spheres which correspond to $n = 1, 3$, and 7; see, for example, Nash and Sen [1]. However, 3.36 also shows that TS^n is stably equivalent to the trivial bundle I^n. If a bundle is stably equivalent to a trivial bundle it is called stably trivial.

In fact it is generally true that if E, F and G are related as in 3.34 then E is stably equivalent to F. Thus we have the following theorem

Theorem (Stable equivalence) *If E, F and G are vector bundles over M satisfying $E \oplus G \simeq F \oplus G$ then E and F are stably equivalent, that is*

$$E \overset{s}{\sim} F \tag{3.37}$$

Proof: Let G' be a bundle such that,[3] for some n,

$$G \oplus G' \simeq I^n \tag{3.38}$$

Then observe that

$$E \oplus G \simeq F \oplus G \Rightarrow E \oplus G \oplus G' \simeq F \oplus G \oplus G'$$
$$\Rightarrow E \oplus I^n \simeq F \oplus I^n \Rightarrow E \overset{s}{\sim} F \tag{3.39}$$

K-theory associates to the vector bundles over M two groups, $K(M)$ and $\widetilde{K}(M)$. We are now ready to construct these groups. We shall see that $K(M)$ and $\widetilde{K}(M)$ provide us with a systematic way of handling the vector bundle properties that were developed in the preceding discussion.

Now let $Vect(M)$ denote all vector bundles over M and look again at 3.34, which stated that

$$E \oplus G \simeq F \oplus G \not\Rightarrow E \simeq F \tag{3.40}$$

One way of viewing this is to say that, although the direct sum operation $\oplus$ provides $Vect(M)$ with an addition, there is no subtraction defined for vector bundles. In group-theoretic language we say that $Vect(M)$ is an (Abelian) semi-group rather than a group. However an Abelian semi-group can always be completed to an Abelian group; this can also be done in such a way as to define a kind of subtraction for vector bundles. This group is easy to define. Suppose, for a moment, that S is any semi-group. Then form the Cartesian product $S \times S$ and quotient $S \times S$ by the equivalence relation $\sim$ which is defined by

$$(s_1, t_1) \sim (s_2, t_2) \iff \exists\, u \in S \text{ such that } s_1 + t_2 + u = s_2 + t_1 + u \tag{3.41}$$

The resulting group, $G(S)$ say, is given by $G(S) = (S \times S)/ \sim$ and, in $G(S)$, a pair (s, t) should be thought of as the (formal) difference $s - t$. An exactly similar method is often employed to construct the integers $\{\ldots, -1, 0, 1, \ldots\} = \mathbf{Z}$ from the natural numbers $\{1, 2, \ldots\} = \mathbf{N}$. Returning to the case at hand we have $S = Vect(M)$ and addition is given by $\oplus$ the direct sum. The equivalence relation becomes

$$(E, F) \sim (G, H) \iff \exists\, J \in Vect(M) \text{ such that } E \oplus H \oplus J \simeq G \oplus F \oplus J \tag{3.42}$$

---

[3] We assume M to be (compact) Hausdorff and then such a G always exists, cf. Atiyah [1].

Finally in the quotient $G(S) = (Vect(M) \times Vect(M))/ \sim$ we denote the equivalence class corresponding to a single bundle E by $[E]$; more generally we write an arbitrary element of $G(S)$ as $[E] - [F]$ and this corresponds to the formal difference $E - F$. We have now constructed the group $K(M)$ and so proceed to its formal definition

Definition $(K(M))$ *The group $K(M)$ of equivalence classes of formal differences of vector bundles over M is given by*

$$K(M) = (Vect(M) \times Vect(M))/ \sim \qquad (3.43)$$

This equivalence class of formal differences is the second set of equivalence classes whose existence we briefly alluded to on p. 65 at the beginning of the section. There is a close relationship between stable equivalence classes and equivalence classes of formal differences. We pass now to its discussion.

First notice that any element $[E] - [F]$ of $K(M)$ can be written in the form $[H] - [I^n]$ for some bundle H and integer n. To prove this we let F' be a bundle such that $F \oplus F' \simeq I^n$. Then we write

$$\begin{aligned}
[E] - [F] &= ([E] + [F']) - ([F] + [F']) = [E \oplus F'] - [F \oplus F'] \\
&= [E \oplus F'] - [I^n] = [H] - [I^n]
\end{aligned} \qquad (3.44)$$

Next we define the virtual dimension of $[E] - [F]$ by the difference of the ranks of E and F. Let $rk\,(E)$ denote the rank of a vector bundle E. Then we have

Definition (Virtual dimension) *The virtual dimension of $[E] - [F]$ is the integer $rk\,(E) - rk\,(F)$*

Clearly the virtual dimension can be positive, negative or zero. Note that, because of 3.44, we can express the virtual dimension by

$$rk\,(E) - rk\,(F) = rk\,(H) - n \qquad (3.45)$$

We shall now show that there is an isomorphism between the set of stable equivalence classes of $Vect(M)$, and, the subgroup of $K(M)$ whose elements have virtual dimension *zero*. The isomorphism is defined by assigning a vector bundle E of rank m the element $[E] - [I^m]$ of $K(M)$. To prove this, first suppose that E_m and F_n are two vector bundles, of ranks m and n respectively, which give rise under this isomorphism to the same element of $K(M)$, i.e. $[E_m] - [I^m] = [F_n] - [I^n]$. Then we must show that they are stably equivalent. We write,

$$\begin{aligned}
&[E_m] - [I^m] = [F_n] - [I^n] \\
\Rightarrow &E_m \oplus I^n \oplus K \simeq F_n \oplus I^m \oplus K \text{ for some } K \\
\Rightarrow &E_m \oplus I^n \overset{s}{\sim} F_n \oplus I^m \text{ by the theorem above} \\
\Rightarrow &E_m \overset{s}{\sim} F_n \text{ by definition}
\end{aligned} \qquad (3.46)$$

To prove the converse suppose that E_m and F_n are stably equivalent, then

$$E_m \oplus I^j \simeq F_n \oplus I^k$$

$$\Rightarrow [E_m \oplus I^j] - [I^{m+j}] = [F_n \oplus I^k] - [I^{n+k}]$$

But $\quad [E_m \oplus I^j] - [I^{m+j}] = [E_m] - [I^m] \qquad (3.47)$

and $\quad [F_n \oplus I^k] - [I^{n+k}] = [F_n] - [I^n]$

$$\Rightarrow [E_m] - [I^m] = [F_n] - [I^n]$$

and thus we have established that the isomorphism maps E_m and F_n onto the same element of $K(M)$, which has indeed vanishing virtual dimension.

We can now define what is sometimes called reduced K-theory.

Definition (Reduced K-theory) *The set of all stable equivalence classes of Vect(M) over M form the reduced K-theory of M which we write as*

$$\widetilde{K}(M) \qquad (3.48)$$

If we pursue further the notion of virtual dimension in $K(M)$ then we can define an obvious map $Rk : K(M) \to \mathbf{Z}$ by

$$
\begin{aligned}
Rk : \quad & K(M) \longrightarrow \mathbf{Z} \\
& ([E] - [F]) \longmapsto (rk\,(E) - rk\,(F))
\end{aligned}
\qquad (3.49)
$$

The kernel of this map, by what we have just proved, is clearly $\widetilde{K}(M)$. There is another way of looking at this which allows us to prove the very useful result that

$$K(M) = \widetilde{K}(M) \oplus \mathbf{Z} \qquad (3.50)$$

A short argument indicates how to prove this. Recall that if N and M are two manifolds connected by a continuous map $\alpha : N \to M$, then α induces the map α^* which pulls back cohomology on M to that on N, i.e. we have $\alpha^* : H^*(M; \mathbf{Z}) \to H^*(N; \mathbf{Z})$. Further, we know that if F is a bundle over M then α pulls back F to the bundle E over N where $E = \alpha^* F$ (note that, following conventional practice, we are using the *same* symbol α^* to denote the pulling back of both cohomology classes and bundles; the context should prevent any confusion). Since K-theory is a cohomology theory of vector bundles it is very easy to check that α induces a map, which we continue to denote by α^*, between the respective K-theories. Thus we write

$$\alpha^* : K(M) \longrightarrow K(N) \qquad (3.51)$$

Now let N be the very modest manifold consisting of just a single point which we can take to to be the base point m_0 of M. Having made this choice $N = \{m_0\}$ we make the natural choice for α, namely that α is the inclusion i of m_0 in M. We write then

$$i : \{m_0\} \longrightarrow M$$
$$i^* : K(M) \longrightarrow K(\{m_0\}) \tag{3.52}$$

Since the space $\{m_0\}$ consists of a single point all vector bundles over it are trivial and so are of the form I^k; and thus they are completely determined by this integer k which is their rank.[4] Thus

$$K(\{m_0\}) = \mathbf{Z} \tag{3.53}$$

By comparing with the definition of Rk above we can also verify that $ker\, i^* = \widetilde{K}(\{m_0\})$. There is a natural (right) inverse for i^* which we write as c^* where $c : M \longrightarrow \{m_0\}$ is the *collapsing map* which maps every point of M onto its base point m_0. So, if I denotes the inclusion of $\widetilde{K}(M)$ in $K(M)$, we can write down the exact sequence

$$0 \longrightarrow \widetilde{K}(M) \overset{I}{\longrightarrow} K(M) \overset{i^*}{\longrightarrow} K(\{m_0\}) \longrightarrow 0 \tag{3.54}$$

which splits, that is c^* is a right inverse for i^*. It is immediate that

$$K(M) = ker\, i^* \oplus K(\{m_0\}) = \widetilde{K}(M) \oplus \mathbf{Z} \tag{3.55}$$

and so we have the desired result.

So far we have shown that $K(M)$ and $\widetilde{K}(M)$ are additive groups. They also possess a multiplication which is distributive over the addition; thus they are actually *rings*. This multiplication is simply the operation induced in $K(M)$ by the tensor product $E \otimes F$ of two bundles. This existence of a multiplication is another sign that we are dealing with a cohomology theory, as such theories have a ring structure.

In § 5 of chapter 1 we used classifying spaces to discuss bundles. This can be done in K-theory too. The K-theory state of affairs is fairly simply described. Before giving this description we need to give some preliminary information on vector bundles and their classifying spaces. The reader will find more details in the references already cited in this chapter.

Let E_k be a real vector bundle of rank k over a real manifold M whose dimension is n. Then E_k has structure group $Gl(k, \mathbf{R})$ which is reducible

---

[4] Here we are assuming that M is connected

to the group $O(k)$ on choosing a metric on M. The classifying space for E_k is the Grassmannian $Gr(k, m, \mathbf{R})$ where

$$Gr(k, m, \mathbf{R}) = \frac{O(m)}{O(m-k) \times O(k)} \tag{3.56}$$

There is a universal bundle $Q(k, m, \mathbf{R})$ of rank k over this Grassmannian whose pullbacks generate bundles such as E_k. More precisely we say that

$$\begin{aligned} &\text{If} \quad f : M \longrightarrow Gr(k, m, \mathbf{R}) \quad \text{and if } m \geq (n + k + 1) \\ &\text{Then} \quad f^*Q(k, m, \mathbf{R}) \simeq E_k \quad \text{for some f} \end{aligned} \tag{3.57}$$

and if there is another map $g : M \to Gr(k, m, \mathbf{R})$ which is homotopic to f the bundles $g^*Q(k, m, \mathbf{R})$ and $f^*Q(k, m, \mathbf{R})$ are isomorphic—thus only homotopy classes in $[M, Gr(k, m, \mathbf{R})]$ are important. If we hold k fixed but increase m then observe that we have a sequence of inclusions for the Grassmannians

$$Gr(k, m, \mathbf{R}) \subset Gr(k, m+1, \mathbf{R}) \subset Gr(k, m+2, \mathbf{R}) \subset \cdots \tag{3.58}$$

and passing to the inductive limit we can define the infinite Grassmannian

$$Gr(k, \infty, \mathbf{R}) = \bigcup_m Gr(k, m, \mathbf{R}) \tag{3.59}$$

Now, whatever the dimension n of M, the bundle $Q(k, \infty, \mathbf{R})$ will be universal for E_k. We summarise this by writing

$$Vect_k(M) \simeq [M, Gr(k, \infty, \mathbf{R})] \tag{3.60}$$

where $Vect_k(M)$ stands for all real vector bundles of rank k over M. Referring back to the discussion on classifying spaces BG in chapter 1 we rewrite the above equation as

$$Vect_k(M) \simeq [M, BO(k)] \tag{3.61}$$

Thus we have identified $BO(k)$ the classifying space for bundles with group $O(k)$.

Now we return to the main discussion and consider the properties of vector bundles as the rank k is increased—what actually happens is that things simplify somewhat. To discover the details we must use the following vital theorem concerning $Vect_k(M)$.

Theorem (Stable Range) *If $E_k \in Vect_k(M)$ is a real vector bundle of rank k over the n-dimensional manifold M then, if $k > n$, E_k is isomorphic to $F_n \oplus I^{k-n}$ for some bundle F_n; and when $k > n$ such a bundle E_k is said to be in the stable range.*

The proof is by induction on k, cf. Husemoller [1]. In any case the importance of this theorem for us is that, in the language of K-theory, it says that a vector bundle E_k in the *stable range* is stably equivalent to some other bundle F_n of lower rank n where $n = \dim M$. It therefore belongs to the same stable equivalence class as F_n and corresponds to exactly the same element of $\widetilde{K}(M)$. In other words, as far as the K-theory is concerned, there is nothing to be gained by considering bundles of very high rank because, as soon as the stable range is reached, no new K-theory elements are obtained by increasing the rank. Notice, too, that, in the stable range, two isomorphic vector bundles are *necessarily* stably equivalent; the converse, though, is clearly false since two bundles of differing ranks can be stably equivalent. However, if we stick to bundles of the *same* rank, then, in the stable range, two bundles *are* stably equivalent if and only if they are isomorphic. We can present this more concisely by writing

$$\text{If } k > \dim M, \ \widetilde{K}(M) \simeq Vect_k(M)$$
$$\text{or } \ \widetilde{K}(M) = [M, Gr(k, \infty, \mathbf{R})] \tag{3.62}$$

This K-theory result can be thought of as expressing a kind of simplification which occurs in the linear algebra of 'large-dimensional' vector bundles.

We can put the preceding result 3.62 to immediate use in actual calculations. To see this we take M to be the sphere S^n. Therefore we have

$$\widetilde{K}(S^n) = [S^n, Gr(k, \infty, \mathbf{R})]$$
$$\simeq \pi_n\left(Gr(k, \infty, \mathbf{R})\right) \tag{3.63}$$

where we still demand that $k > n$. We also make the minor technical point that we have identified the homotopy group $\pi_n(Gr(k, \infty, \mathbf{R})$ with the *free* homotopy classes $[S^n, Gr(k, \infty, \mathbf{R})]$ rather than the more correct *based* homotopy classes $[S^n, Gr(k, \infty, \mathbf{R})]_0$. This amounts to choosing a fixed trivialisation over the base point of S^n.

Now to calculate the relevant homotopy groups we employ a useful technique (cf. for example Nash and Sen [1]) for classifying bundles over spheres: Let E be a vector bundle of rank k in the stable range over S^n. Cover S^n with two overlapping 'hemispheres' U and V. Then the contractibility of U and V implies that E is trivial over each hemisphere and so is determined by the single transition function g defined on the overlap $U \cap V$. But the overlap is homotopic to the equator S^{n-1} so g determines a map from S^{n-1} to $O(k)$, i.e an element of $\pi_{n-1}(O(k))$. It is this element of $\pi_{n-1}(O(k))$ which determines the bundle $E \in Vect_k(S^n)$ and, via 3.62, an element of $\tilde{K}(S^n)$—we digress briefly to point out that, for a general group G, this result amounts to the assertion $[S^n, BG] \simeq [S^{n-1}, G]$, but $[S^n, BG] = [SS^{n-1}, BG] = [S^{n-1}, \Omega BG]$ by 1.22; thus $[S^{n-1}, G] \simeq [S^{n-1}, \Omega BG]$ for each n and actually even more is true, in fact the space ΩBG is of the same homotopy type as G. In other words we can write $\Omega BG \simeq G$ and think of the loop space operand Ω as being a kind of homotopic inverse to the classifying space operand B. In any case we have found that

$$\tilde{K}(S^n) = \pi_{n-1}(O(k)), \quad k > n \tag{3.64}$$

Fortunately the homotopy groups of classical Lie groups such as $O(k)$ have been extensively studied. Also, though $\pi_{n-1}(O(k))$ is not known for all n and k, it is precisely in the stable range $k > n$ that we have complete information. Referring, for example, to Husemoller [1] we find that $\pi_{n-1}(O(k)) = \pi_{n-1}(O(k+1)) \cdots$ for $k > n$. If we want to find $\tilde{K}(S^n)$ for $n = 0, 1, \ldots, 4$, say, then we are immediately able to do so. By thinking of S^0 as two points it is clear that $\tilde{K}(S^0) = \mathbf{Z}$. For the other S^n we must use the homotopy properties of $O(k)$. Displaying the information in a list we obtain

$$\tilde{K}(S^0) = \mathbf{Z}$$
$$\pi_0(O(k)) = \mathbf{Z}_2, \ k \geq 1 \quad \Rightarrow \quad \tilde{K}(S^1) = \mathbf{Z}_2$$
$$\pi_1(O(k)) = \mathbf{Z}_2, \ k \geq 3 \quad \Rightarrow \quad \tilde{K}(S^2) = \mathbf{Z}_2 \tag{3.65}$$
$$\pi_2(O(k)) = 0, \ k \geq 4 \quad \Rightarrow \quad \tilde{K}(S^3) = 0$$
$$\pi_3(O(k)) = \mathbf{Z}, \ k \geq 5 \quad \Rightarrow \quad \tilde{K}(S^4) = \mathbf{Z}$$

So far we have only given explicit details for the K-theory of *real* vector bundles. We would like to remedy this omission now. As well as the real bundles there are complex and quaternionic vector bundles. In these latter two cases, for a vector bundle of rank k, the fibre $\mathbf{R}^k$ of the real case is replaced by $\mathbf{C}^k$ and $\mathbf{H}^k$ respectively ($\mathbf{H}$ stands for the quaternions). The place where the discussion needs to be altered is only from 3.56 onwards. First of all, the real Grassmannian must give way to a complex or quaternionic one with a corresponding change in classifying space. Secondly, the theorem above about the stable range undergoes a little change.

For complex and quaternionic vector bundles we use the Grassmannians

$Gr(k, m, \mathbf{C})$ and $Gr(k, m, \mathbf{H})$ which are given by

$$Gr(k, m, \mathbf{C}) = \frac{U(m)}{U(m-k) \times U(k)}, \quad \text{and} \quad Gr(k, m, \mathbf{H}) = \frac{Sp(m)}{Sp(m-k) \times Sp(k)}$$

$$(3.66)$$

where $U(k)$ and $Sp(k)$ are the usual unitary and symplectic groups. Now let $\mathbf{F}$ stand for $\mathbf{R}$, $\mathbf{C}$ or $\mathbf{H}$ so that we can introduce the convenient notation $Vect_k(M, \mathbf{F})$ to denote real, complex or quaternionic vector bundles of rank k. Proceeding in analogy to the real case we have the following

$$Vect_k(M, \mathbf{C}) \simeq [M, Gr(k, \infty, \mathbf{C})]$$
$$Vect_k(M, \mathbf{H}) \simeq [M, Gr(k, \infty, \mathbf{H})]$$

$$(3.67)$$

In the classifying space language this amounts to saying that $BU(k) = Gr(k, \infty, \mathbf{C})$ and $BSp(k) = Gr(k, \infty, \mathbf{H})$. Next we have the theorem that describes the stable range for the complex and quaternionic cases.

Theorem (Stable range: complex and quaternionic case) *Let E_k^c and E_k^h be complex and quaternionic vector bundles of rank k respectively over a manifold M of real dimension n. Let $[m]$ denote the integral part of m (i.e. the largest integer not greater than m), and let $k(c) = [(n+1)/2]$ and $k(h) = [(n+1)/4]$. Then, if E_k^c has $k > k(c)$, E_k^c is isomorphic to $F_{k(c)}^c \oplus I^{k-k(c)}$ for some bundle $F_{k(c)}^c$; and if E_k^h has $k > k(h)$, E_k^h is isomorphic to $F_{k(h)}^c \oplus I^{k-k(h)}$ for some bundle $F_{k(h)}^c$.*

Thus to be in the stable ranges in the complex case and quaternionic cases it is sufficient to require $k > n/2$ and $k > n/4$ respectively. To distinguish the K-theory in the three possible cases we introduce the notation $\widetilde{K}O$ for the real case, and $\widetilde{K}U$ and $\widetilde{K}Sp$ for the complex and quaternionic cases respectively. Using the above theorem in exactly the same manner as we used its counterpart in the real case we obtain

$$Vect_k(M, \mathbf{C}) \simeq \widetilde{K}U(M), \quad \text{for } k > \frac{n}{2}$$
$$Vect_k(M, \mathbf{H}) \simeq \widetilde{K}Sp(M), \quad \text{for } k > \frac{n}{4}$$

$$(3.68)$$

We can also carry out the calculations of $\widetilde{K}U(M)$ and $\widetilde{K}Sp(M)$ when $M = S^n$. As should be expected, all that one needs are the relevant homotopy groups of the unitary and symplectic groups. The formulae that we need are

$$\widetilde{K}U(S^n) = \pi_{n-1}(U(k)) \quad k > \frac{n}{2}$$
$$\widetilde{K}Sp(S^n) = \pi_{n-1}(Sp(k)) \quad k > \frac{n}{4}$$

$$(3.69)$$

Inserting the appropriate homotopy groups we find, cf. Husemoller [1], that

$$
\begin{array}{ll}
\widetilde{K}U(S^0) = \mathbf{Z} & \widetilde{K}Sp(S^0) = \mathbf{Z} \\
\widetilde{K}U(S^1) = 0 & \widetilde{K}Sp(S^1) = 0 \\
\widetilde{K}U(S^2) = \mathbf{Z} & \widetilde{K}Sp(S^2) = 0 \\
\widetilde{K}U(S^3) = 0 & \widetilde{K}Sp(S^3) = 0 \\
\widetilde{K}U(S^4) = \mathbf{Z} & \widetilde{K}Sp(S^4) = \mathbf{Z}
\end{array}
\tag{3.70}
$$

We point out that the $\widetilde{K}U(S^n)$ exhibit a periodicity of 2 in n in the above list. This periodicity is genuine and persists for all n. It is an expression of the property known as Bott periodicity to which we now turn our attention.

§ 4. Bott periodicity

The word periodicity here refers to a periodicity in the homotopy of the classical Lie groups. First we shall describe what is meant by this statement. Then we shall go on to show how to obtain a K-theoretic version of Bott periodicity.

For convenience we shall temporarily restrict ourselves to *complex* vector bundles. At the end of the discussion we shall describe the analogous situations for the real and quaternionic cases. Drawing on the previous section we have the fact that $\widetilde{K}U(S^n) = \pi_{n-1}(U(k))$ for $k > n/2$. Passing to the limit $k \to \infty$ we have

$$
\widetilde{K}U(S^n) = \pi_{n-1}(U(\infty))
\tag{3.71}
$$

where $U(\infty) = \cup_k U(k)$ is the infinite unitary group. However, the periodicity theorem of Bott [1] says that the homotopy groups of $U(\infty)$ are periodic with period 2, i.e.

$$
\pi_n(U(\infty)) = \pi_{n+2}(U(\infty))
\tag{3.72}
$$

Therefore this implies immediately that

$$
\widetilde{K}U(S^n) = \widetilde{K}U(S^{n+2})
\tag{3.73}
$$

as we claimed above. One way of thinking about this result is via the properties of the loop space ΩX of a space X. Recall our discussion of homotopy groups in chapter 1 where we deduced that

$$
\begin{aligned}
\pi_n(X) &= \pi_{n-1}(\Omega X) \\
\Rightarrow \pi_n(X) &= \pi_{n-2}(\Omega^2 X)
\end{aligned}
\tag{3.74}
$$

Applying this to the case in hand means setting $X = U(\infty)$. The periodicity theorem is then the statement that $\Omega^2 U(\infty)$ has the same homotopy groups as $U(\infty)$ itself. This is indeed the case and can be established by proving that $\Omega^2 U(\infty)$ has the *same homotopy type* as $U(\infty)$, cf. Bott [1].

We begin our account of the K-theoretic formulation of the periodicity theorem by introducing, for each space X, the spaces SX, $S^2 X$, ... where SX denotes the reduced suspension of X that was defined in chapter 1. Next we define the group $\tilde{K}U^{-n}(X)$ by

$$\tilde{K}U^{-n}(X) = \tilde{K}U(S^n X), \quad n = 0, 1, 2, \ldots \tag{3.75}$$

Remembering that if X is a sphere S^m we have $SS^m \simeq S^{m+1}$, we deduce that

$$\tilde{K}U^{-n}(S^m) = \tilde{K}U(S^{m+n}) \tag{3.76}$$

We have referred already to the existence of a product in $KU(X)$, this product being induced by the tensor product of vector bundles. This product can be viewed as a map [5] $\beta : KU(X) \otimes KU(X) \to KU(X)$. There is also another product called the *external product* which involves two spaces X and Y rather than one space X. Thus we have a map α, say, of the form

$$\alpha : KU(X) \otimes KU(Y) \longrightarrow KU(X \times Y) \tag{3.77}$$

We shall now define the map α: Let $a, b \in KU(X)$ and write their product, computed using β, as $a \bullet b$ so that $a \bullet b \in KU(X)$. Next consider the natural projections from $X \times Y$ onto each of its factors. These are the maps $\pi_X : X \times Y \to X$ and $\pi_Y : X \times Y \to Y$ under which (x, y) is mapped onto x or y respectively. These projections induce maps between the K-rings over X, Y and $X \times Y$: in the notation of 3.51 we have

$$\pi_X^* : KU(X) \longrightarrow KU(X \times Y) \text{ and } \pi_Y^* : KU(Y) \longrightarrow KU(X \times Y) \tag{3.78}$$

With these preliminaries out of the way we define α by choosing $c \in KU(X)$ and $d \in KU(Y)$ and writing

$$\begin{aligned} \alpha : KU(X) \otimes KU(Y) &\longrightarrow KU(X \times Y) \\ c \otimes d &\longmapsto \pi_X^*(c) \bullet \pi_Y^*(d) \end{aligned} \tag{3.79}$$

---

[5] Usually one thinks of a product on an arbitrary space S as being a map $S \times S \to S$ rather than a map of the form $S \times S \to S$. The appearance of the tensor product in the present case is because $KU(X)$ is a ring.

We shall see that the relevance of this to the periodicity theorem is that if we choose $Y = S^2$ then α is an *isomorphism*, cf. Husemoller [1]. In other words

$$KU(X) \otimes KU(S^2) \simeq KU(X \times S^2) \tag{3.80}$$

To show that this is related to our earlier statement of periodicity requires us to rewrite this isomorphism in terms of $\widetilde{K}U$. This is simply done: Use $KU(X) = \widetilde{K}U(X) \oplus KU(\{x_0\})$ and a similar equation for Y. Then note that

$$
\begin{aligned}
KU(X) \otimes KU(Y) &= (\widetilde{K}U(X) \oplus KU(\{x_0\})) \otimes (\widetilde{K}U(Y) \oplus KU(\{y_0\})) \\
&= (\widetilde{K}U(X) \otimes \widetilde{K}U(Y)) \oplus \widetilde{K}U(X) \oplus \widetilde{K}U(Y) \oplus \mathbf{Z}
\end{aligned}
\tag{3.81}
$$

where we have used some simple properties of the product which the reader may verify for interest. What we have just done relates the LHS of the map α to $\widetilde{K}U$. We now do the same for the RHS. The reduced join $X \vee Y$ defined in chapter 1 can be viewed as a subset of the Cartesian product $X \times Y$. This inclusion $X \vee Y \subset X \times Y$ suggests that we write down the natural set of maps (i denotes inclusion)

$$X \vee Y \xrightarrow{i} X \times Y \longrightarrow \frac{X \times Y}{X \vee Y} = X \wedge Y \tag{3.82}$$

This induces a sequence of maps in the K-theory which go in the reverse direction:

$$0 \longrightarrow \widetilde{K}U(X \wedge Y) \longrightarrow \widetilde{K}U(X \times Y) \longrightarrow \widetilde{K}U(X \vee Y) \longrightarrow 0 \tag{3.83}$$

Moreover, this sequence is exact, cf. Husemoller [1] for a proof. Using the exactness of the sequence, and assuming the isomorphism $\widetilde{K}U(X \vee Y) = \widetilde{K}U(X) \oplus \widetilde{K}U(Y)$ (cf. Husemoller [1]), we can say that

$$
\begin{aligned}
\widetilde{K}U(X \times Y) &= \widetilde{K}U(X \wedge Y) \oplus \widetilde{K}U(X \vee Y) \\
&= \widetilde{K}U(X \wedge Y) \oplus \widetilde{K}U(X) \oplus K(Y)
\end{aligned}
\tag{3.84}
$$

This completes our work on the LHS and RHS of α and on substitution we can rewrite α as the map

$$\alpha : (\widetilde{K}U(X) \otimes \widetilde{K}U(Y)) \oplus R \longrightarrow \widetilde{K}U(X \wedge Y) \oplus R \tag{3.85}$$

where, for shorthand, we have written $R = \widetilde{K}U(X) \oplus \widetilde{K}(Y) \oplus \mathbf{Z}$. Now the point is that since R occurs on both sides we can eliminate it by an appropriate restriction. This provides us with the simpler looking map $\widetilde{\alpha}$

$$\widetilde{\alpha} : \widetilde{K}U(X) \otimes \widetilde{K}U(Y) \longrightarrow \widetilde{K}U(X \wedge Y) \tag{3.86}$$

Then, as before, if we make the choice $Y = S^2$ Bott periodicity tells us that the resulting map is an isomorphism. We make this choice and also replace X by $S^n X$, i.e. we have

$$\tilde{K}U(S^n X) \otimes \tilde{K}U(S^2) \simeq \tilde{K}U(S^n X \wedge S^2)$$
$$\Rightarrow \tilde{K}U^{-n}(X) \otimes \tilde{K}U(S^2) \simeq \tilde{K}U(S^{n+2} \wedge X) \qquad (3.87)$$
$$\Rightarrow \tilde{K}U^{-n}(X) \otimes \tilde{K}U(S^2) \simeq \tilde{K}U^{-(n+2)}(X)$$

As a final step we define the map $\gamma : \tilde{K}U^{-n}(X) \to \tilde{K}U^{-(n+2)}(X)$ as follows: we know already that $\tilde{K}U(S^2) = \mathbf{Z}$ and that any element of $\tilde{K}U(S^2)$ is of the form $[E] - [I^{rk(E)}]$. If we take E to be the line bundle η associated to the famous Hopf fibration of S^3 over S^2 (η is the canonical line bundle over $\mathbf{CP}^1$ and this fibration is also the one that gives the Dirac monopole) then $[E] - [I^{rk(E)}] = [\eta] - [I]$, and $[\eta] - [I]$ actually *generates* this $\mathbf{Z}$. Now let $a \in \tilde{K}U^{-n}(X)$ and denote $[\eta] - [I]$ by b, then the map γ is defined by

$$\gamma : \tilde{K}U^{-n}(X) \longrightarrow \tilde{K}U^{-(n+2)}(X)$$
$$a \longmapsto \tilde{\alpha}(a \otimes b) \qquad (3.88)$$

and γ is an isomorphism, i.e.

$$\tilde{K}U^{-n}(X) = \tilde{K}U^{-(n+2)}(X) \qquad (3.89)$$

This is the K-theoretic statement of periodicity that we have been after—if we let $X = S^0$ we regain the original periodicity result that we had in 3.73.

    We simply quote the corresponding facts for the real and the quaternionic cases. Both theories possess periodicities and these can be traced back to the stable homotopy of the infinite orthogonal and infinite symplectic groups respectively. In summary we have

$$\pi_n(O(\infty)) = \pi_{n+8}(O(\infty)) \quad \text{and} \quad \tilde{K}O^{-n}(X) = \tilde{K}O^{-n-8}(X)$$
$$\pi_n(Sp(\infty)) = \pi_{n+4}(O(\infty)) \quad \text{and} \quad \tilde{K}Sp^{-n}(X) = \tilde{K}O^{-n-4}(X) \qquad (3.90)$$

There is now enough information to calculate any of $\tilde{K}O(S^n)$, $\tilde{K}U(S^n)$ or $\tilde{K}Sp(S^n)$ for all n.

§ 5. Some characteristic classes

In this section we shall discuss characteristic classes beginning with the Chern character and the Todd class. The Chern character is an extremely valuable tool in K-theory. Before we discuss its connection with K-theory we briefly review its basic properties when viewed purely as a characteristic class—further background can be found in Bott and Tu [1] and Nash and Sen [1]. The Chern character can be used to provide a link between the ring $K(X)$ and the cohomology ring $H^*(X)$. We shall see in chapter 4 that this link is vital in providing a practical formula for calculating the index of elliptic operators. The Todd class also turns up in this index formula and so it is convenient to discuss it here as well.

To introduce the Chern character we consider $Vect(X, \mathbf{C})$ the set of complex vector bundles over X. A bundle $E \in Vect(X, \mathbf{C})$ possesses Chern classes $c_i(E)$ which belong to the cohomology group $H^{2i}(X; \mathbf{Z})$. If we suppose that E has rank k and thus group $U(k)$, then these $c_i(E)$ are the pullbacks by the map $f : X \to BU(k)$ of the cohomology of the classifying space $BU(k)$ for $U(k)$ bundles. Using the notation of chapter 1 we have that $E_{U(k)}$ is the universal bundle over $BU(k)$ and

$$f : X \longrightarrow BU(k), \quad E = f^* E_{U(k)}$$

$$H^*(BU(k); \mathbf{Z}) \quad \text{has the even dimensional generators } c_i(E_{U(k)})$$

where $c_i(E_{U(k)}) \in H^{2i}(BU(k); \mathbf{Z})$

so that $c_i(E) = f^*(c_i(E_{U(k)})), \, c_i(E) \in H^{2i}(X; \mathbf{Z}), \, i = 1, \ldots, k$

$$(3.91)$$

A convenient device when working with Chern classes is the total Chern class $c(E)$ defined by

$$c(E) = 1 + c_1(E) + c_2(E) + \cdots + c_k(E) \tag{3.92}$$

A key property of $c(E)$ is its behaviour with respect to Whitney sums. This is that if $E, F \in Vect(X, \mathbf{C})$ we have

$$c(E \oplus F) = c(E)c(F)$$

This property can be proved using a technique known as the *splitting principle* to which we now devote some remarks.

The splitting principle uses the fact that in order to prove an identity for characteristic classes it is sufficient to prove it only for bundles E which decompose into a sum of line bundles, i.e. if E has rank k and $L_1, L_2, \ldots L_k$ are line bundles then E is expressible as

$$E = L_1 \oplus L_2 \oplus \cdots \oplus L_k$$

The validity of identities true for such E follows from the fact that if F is any rank k bundle over X, and F is not necessarily a sum of line bundles, then F can always be regarded as the pullback of another rank k bundle E over Y which is a sum of line bundles. For a detailed account see Bott and Tu [1].

If we calculate the total Chern class of E in 3.93 then, since L_i is a line bundle, $c(L_i) = 1 + c_1(L_i)$ and we find that

$$
\begin{aligned}
c(E) &= c(L_1 \oplus L_2 \oplus \cdots \oplus L_k) \\
&= \prod_{i=1}^{k} c(L_i) = \prod_{i=1}^{k}(1 + x_i), \quad \text{where } x_i = c_1(L_i)
\end{aligned}
\tag{3.93}
$$

If we now compare the two formulae 3.92 and 3.93 we obtain expressions for the Chern classes $c_i(E)$ in terms of the x_i. These are

$$
\begin{aligned}
c_1(E) &= \sum_i x_i, \quad c_2(E) = \sum_{i<j} x_i x_j, \quad \cdots \\
c_j(E) &= \sum_{i_1 < i_2 < \cdots < i_j} x_{i_1} x_{i_2} \cdots x_{i_j}, \quad \cdots \quad c_k(E) = x_1 x_2 \cdots x_k
\end{aligned}
\tag{3.94}
$$

and we see that the $c_i(E)$ are the elementary symmetric functions in the x_i. The formula 3.93 above can be used in general for doing algebra with Chern classes. The definition of the Chern character of a rank k bundle E is that $ch\,(E)$ is given in terms of the x_i by

$$
ch\,(E) = \sum_{i=1}^{k} e^{x_i}
\tag{3.95}
$$

One can view this expression for $ch\,(E)$ as a kind of generating function. An important fact about $ch\,(E)$ is that it is a *rational* linear combination of the x_i with coefficients $1/i!$. The fact that the coefficients are rational rather than integral means that $ch\,(E)$ is only a rational cohomology class instead of an integral one; as a consequence $ch\,(E)$ cannot detect any torsion in $H^*(X)$. Thus we write

$$
ch\,(E) \in H^*(X; \mathbf{Q})
\tag{3.96}
$$

If we use the relation above between the Chern classes and the x_i we can write the Chern character in terms of the $c_i(E)$ and we obtain thereby

$$
ch\,(E) = k + c_1(E) + \frac{1}{2}(c_1^2(E) - 2c_2(E)) + \cdots
\tag{3.97}
$$

The Chern character is well behaved with respect to both sums and products. It is straightforward to check that it satisfies

$$ch\,(E \oplus F) = ch\,(E) + ch\,(F)$$
$$ch\,(E \otimes F) = ch\,(E)\,ch\,(F) \tag{3.98}$$

We now return to K-theory. Next we use the Chern character to define a map between $K(X)$ and $H^*(X)$ (in this section $K(X)$ denotes complex K-theory). We denote the map by ch and define it as follows. Let $[E] - [F]$ be an element of $K(X)$ where E and F are complex vector bundles. Then define the map ch by

$$ch : K(X) \longrightarrow H^*(X; \mathbf{Q})$$
$$[E] - [F] \longmapsto ch\,(E) - ch\,(F) \tag{3.99}$$

It is necessary to check that ch is well defined. This amounts to verifying that

$$[E] - [F] = [G] - [H] \Rightarrow ch\,(E) - ch\,(F) = ch\,(G) - ch\,(H) \tag{3.100}$$

This follows rather easily since using 3.42 we have

$$[E] - [F] = [G] - [H] \Rightarrow E \oplus H \oplus J \simeq G \oplus F \oplus J$$
$$\Rightarrow ch\,(E \oplus H \oplus J) = ch\,(G \oplus F \oplus J)$$
$$\Rightarrow ch\,(E) + ch\,(H) + ch\,(J) = ch\,(G) + ch\,(F) + ch\,(J) \tag{3.101}$$
$$\Rightarrow ch\,(E) - ch\,(F) = ch\,(G) - ch\,(H)$$

Not only is the map ch well defined but, because of its good behaviour 3.98 for sums and products, the ring structure of $K(X)$ is mapped into that of $H^*(X)$. That is to say ch is a ring homomorphism and we can write

$$ch : K(X) \longrightarrow H^*(X; \mathbf{Q}) \tag{3.102}$$

Actually, since the Chern classes $c_i(E)$ are all even dimensional, we can refine this slightly by writing

$$ch : K(X) \longrightarrow \bigoplus_{i \geq 0} H^{2i}(X; \mathbf{Q}) \tag{3.103}$$

A point to bear in mind when using the Chern character is that, when evaluating the rational cohomology class $ch\,(E)$ on some cycle of X, the answer need only be rational rather than integral.

We now come to our second class: The Todd class. The Todd class is defined by another generating function formula. Let E be a complex vector bundle of rank k, then the *total Todd class* of E is written as $td(E)$ where

$$td(E) = \prod_{i=1}^{k} \frac{x_i}{1 - e^{-x_i}} \tag{3.104}$$

For Whitney sums the splitting principle shows that the Todd class behaves multiplicatively

$$td(E \oplus F) = td(E)\, td(F) \tag{3.105}$$

If we wish we can expand $td(E)$ in terms of the Chern classes $c_i(E)$ and obtain

$$td(E) = 1 + \frac{1}{2}c_1(E) + \frac{1}{12}(c_1^2(E) + c_2(E)) + \frac{1}{24}c_1(E)c_2(E) + \cdots \tag{3.106}$$

It is clear, too, that $td(E)$ is a *rational* cohomology class.

We move on to give a brief summary of a few useful properties of real and complex vector bundles and their characteristic classes.

Firstly let E be a complex vector bundle of rank k. The bundle E possesses a dual E^*. We remind the reader that E^* is the bundle obtained by replacing the transition functions g, say, of E by their inverse transpose, i.e. by $(g^{-1})^T$. Next consider the conjugate bundle $\overline{E}$ of E. The conjugate $\overline{E}$ is obtained by applying complex conjugation to the coordinates of the fibres $\mathbf{C}^k$ of E; in terms of transition functions this is equivalent to replacing the transition functions g by their complex conjugates $\bar{g}$ ($\bar{g}$ is the matrix whose entries are the complex conjugates of those of g). However, for a complex vector bundle of rank k we have $g \in U(k)$ so that g, being unitary, satisfies $\bar{g} = (g^{-1})^T$. Thus the transition functions of the dual and the conjugate bundles coincide or

$$E^* \simeq \overline{E}$$

We will normally make use of this isomorphism to identify[6] such bundles.

Let us consider only complex line bundles rather than vector bundles of any higher rank. Line bundles are rather special because the tensor

---

[6] This identification is made possible by the choosing of a Hermitian inner product on the fibres of E. It is with respect to this choice of metric or inner product that the transition functions are unitary; this same choice is what permits the reduction of the structure group of the bundle from $Gl(k, \mathbf{C})$ to $U(k)$.

product of two line bundles is another a line bundle. This means that they form an Abelian group with the group multiplication being tensor product; the inverse of a line bundle L is clearly its dual L^*; this is because the transition function of one is the inverse of the other. One can emphasise this group property by writing L^{-1} instead of L^*, or L^n (and even nL) instead of $\otimes^n L$.

Now let V be a *real* vector bundle of rank k over X. The choice of a Riemannian metric reduces the structure group from $Gl(n, \mathbf{R})$ to $O(n)$. The dual of V^* of V is a bundle with transition function $(g^{-1})^T$ where g is the transition function of V. However, since $g \in O(n)$, the transition functions of V and V^* coincide. Hence we shall almost always identify V and V^* when V is a real vector bundle; also, because V has real fibres, it is automatic that conjugation has no effect on V, i.e. $\bar{V} = V$. However, though V and V^* coincide topologically it can be important to distinguish them analytically. For example, let $V = TX$ be the tangent bundle to a *real* manifold X, then V^* is the cotangent bundle T^*X and we usually wish to differentiate TX from T^*X in this case.

It is useful to be able to construct a complex vector bundle from V by the process known as complexification. The complexification of V is the bundle over X formed from V by replacing its fibres $\mathbf{R}^k$ by $\mathbf{R}^k \otimes_{\mathbf{R}} \mathbf{C} = \mathbf{C}^k$; the transition functions g are still real valued matrices but now they act on $\mathbf{C}^k$ instead of $\mathbf{R}^k$; more concisely, if g is a map $g : \mathbf{R}^k \to \mathbf{R}^k$, the complexification process changes g into the map $g \otimes \mathbf{C} : \mathbf{C}^k \to \mathbf{C}^k$. We denote the complexification of a real vector bundle V by $V \otimes_{\mathbf{R}} \mathbf{C}$. If we construct the dual or conjugate of $V \otimes_{\mathbf{R}} \mathbf{C}$ then we obtain nothing new. To see this think of $\mathbf{C}$ as $\mathbf{R} \oplus i\mathbf{R}$, then $V \otimes_{\mathbf{R}} \mathbf{C}$ and its conjugate are given by

$$V \otimes_{\mathbf{R}} \mathbf{C} = V \otimes (\mathbf{R} \oplus i\mathbf{R}), \quad \text{and} \quad \overline{V \otimes_{\mathbf{R}} \mathbf{C}} = V \otimes (\mathbf{R} \oplus -i\mathbf{R}) \quad (3.107)$$

Also, since the transition functions g of $V \otimes_{\mathbf{R}} \mathbf{C}$ are still real valued, then they remain the transition functions of the conjugate. Thus the conjugation of $V \otimes_{\mathbf{R}} \mathbf{C}$ just amounts to the complex conjugation on the fibres; but this a linear map transforming each fibre into itself and so is a bundle isomorphism. In sum we have that if V is a real vector bundle then

$$V \otimes_{\mathbf{R}} \mathbf{C} \simeq \overline{V \otimes_{\mathbf{R}} \mathbf{C}} \simeq (V \otimes_{\mathbf{R}} \mathbf{C})^* \quad (3.108)$$

Complex vector bundles can also be complexified by first converting them into real vector bundles and then complexifying the result. To see how this works let E be a complex vector bundle of rank k over X. Denote the underlying real vector bundle (of real rank 2k) by E_r. Then complexify the result to obtain

$$E_r \otimes_{\mathbf{R}} \mathbf{C}$$

This unwieldy looking bundle has a rather simple structure, which we can easily discover: use again the decomposition $\mathbf{C} = \mathbf{R} \oplus i\mathbf{R}$ so that we get

$$
\begin{aligned}
E_r \otimes_{\mathbf{R}} \mathbf{C} &= E_r \otimes (\mathbf{R} \oplus i\mathbf{R}) \\
&= (E_r \otimes \mathbf{R}) \oplus (E_r \otimes i\mathbf{R}) \simeq E \oplus \overline{E}
\end{aligned}
\tag{3.109}
$$

where in the last step we restore the complex structure to the two summands. The isomorphism that we have uncovered is thus the following: if E is a complex vector bundle of rank k then the complexification of its underlying real bundle E_r is given by

$$
E_r \otimes_{\mathbf{R}} \mathbf{C} = E \oplus \overline{E}
\tag{3.110}
$$

and is thus another complex vector bundle of twice the rank of E.

Having discussed complexification of real vector bundles we can introduce the Pontrjagin classes of a real vector bundle. Let V be a real vector bundle of real rank k over X. Form the complexification $V \otimes_{\mathbf{R}} \mathbf{C}$ which we denote by $V_{\mathbf{C}}$. Then the Pontrjagin classes of V are defined in terms of the Chern classes of $V_{\mathbf{C}}$. More precisely the ith Pontrjagin class $p_i(V)$ of V is defined by

$$
p_i(V) = (-1)^i c_{2i}(V_{\mathbf{C}})
\tag{3.111}
$$

Note that since the Chern classes c_i are even dimensional cohomology classes then the $p_i(V)$ are cohomology classes in dimension 4, 8, etc. That is $p_i(V) \in H^{4i}(X; \mathbf{Z})$. The total Pontrjagin class $p(V)$ is defined in the usual way by

$$
p(V) = 1 + p_1(v) + \cdots + p_{[k/2]}(V)
\tag{3.112}
$$

where $[k/2]$ denotes the largest integer not greater than k—the reason that the top Pontrjagin class is $p_{[k/2]}(V)$ is that the top Chern class for $V_{\mathbf{C}}$ (since $V_{\mathbf{C}}$ has rank k) is $c_k(V_{\mathbf{C}})$ and we have $p_i(V) = (-1)^i c_{2i}(V_{\mathbf{C}})$. Of course, in general, both the top Chern class and the top Pontrjagin class might vanish for elementary dimensional reasons; this would happen if the dimension of X were too small to give non-zero cohomology in the dimensions in which these characteristic classes live. The Pontrjagin classes are not quite as well behaved as the Chern classes under Whitney sums: if V and W are two real vector bundles over X then we have

$$
p(V \oplus W) = p(V) p(W) + r
\tag{3.113}
$$

where r is an element of *order 2* in $H^*(X; \mathbf{Z})$ so that $2r = 0$. Clearly, if the cohomology of X has no torsion then such elements r do not exist and the Pontrjagin class behaves well under Whitney sums; moreover, if one just uses de Rham cohomology to calculate characteristic classes then the presence of such an r cannot be detected and again we have a well-behaved Pontrjagin class. Summarising, we write

$$p(V \oplus W) = p(V)p(W) \text{ modulo elements of order 2 in } H^*(X; \mathbf{Z}) \quad (3.114)$$

Example *The Pontrjagin classes of spheres*
A simple illustration of K-theory and characteristic class theory is provided by calculating the Pontrjagin classes of TS^n. Firstly, since the cohomology of S^n is torsion free, then real vector bundles over S^n satisfy $p(V \oplus W) = p(V)p(W)$. Secondly, the Pontrjagin classes $p_i(TS^n)$ (conventionally often written just as $p_i(S^n)$) all vanish. This triviality of the Pontrjagin classes follows from the stably trivial nature of TS^n. In 3.36 we had

$$TS^n \oplus N(S^n) \simeq I^n \oplus I \simeq I^{n+1}$$

Thus

$$p(TS^n \oplus N(S^n)) \simeq p(I^n \oplus I) = p(I^{n+1}) = 1 \quad (3.115)$$

But the LHS of 3.115 is equal to $p(TS^n)p(I) = p(TS^n)$, therefore $p(TS^n) \equiv p(S^n) = 1$ so that $p_i(S^n) = 0$ for $i > 0$.

Though Pontrjagin classes are defined at the outset for *real* vector bundles their definitions are always extended to cover *complex* vector bundles. We explain this now dealing first with the Pontrjagin case. If E is a complex vector bundle of rank k over X then the total Pontrjagin class of E is defined to be the Pontrjagin class of the underlying real bundle of (real) rank $2k$. This means that

$$p(E) = p(E_r)$$

Applying the definition of the p_i to real bundle E_r gives

$$p_i(E) \equiv p_i(E_r) = (-1)^i c_{2i}(E_r \otimes_{\mathbf{R}} \mathbf{C}) = (-1)^i c_{2i}(E \oplus \overline{E}) \quad (3.116)$$

But since $c(E \oplus \overline{E}) = c(E)c(\overline{E})$ we get

$$c(E \oplus \overline{E}) = (1 + c_1(E) + \cdots + c_k(E))(1 - c_1(E) + \cdots + (-1)^k c_k(E)) \quad (3.117)$$

If we multiply out the above expression only the even Chern classes survive and we can pick off the expressions for the Pontrjagin classes in terms of the Chern classes. In this way we get

$$p_1(E) = c_1^2(E) - 2c_2(E), \qquad p_2(E) = c_2^2(E) - 2c_1(E)c_3(E) + 2c_4(E),$$

and $p_i(E) = c_i^2(E) - 2c_{i-1}(E)c_{i+1}(E) + \cdots + (-1)^i 2c_{2i}(E)$

$$= \sum_{k=0}^{2i} (-1)^{k-i} c_k(E) c_{2i-k}(E)$$

$$(3.118)$$

Alternatively if we use the splitting principle to calculate $c(E)$ and $c(\overline{E})$ then we obtain the equivalent formula

$$c(E_r \otimes_{\mathbf{R}} \mathbf{C}) = c(E \oplus \overline{E}) = c(E)\, c(\overline{E}) = \prod_{i=1}^{k}(1+x_i)(1-x_i) = \prod_{i=1}^{k}(1-x_i^2)$$

$$\Rightarrow c_{2j}(E_r \otimes_{\mathbf{R}} \mathbf{C}) = (-1)^j \sum_{i_1,\ldots,i_j} x_{i_1}^2 x_{i_2}^2 \cdots x_{i_j}^2$$

$$\Rightarrow p_1(E) = \sum_i x_i^2, \qquad p_2(E) = \sum_{i,j} x_i^2 x_j^2, \qquad \ldots \quad , p_k(E) = x_1^2 x_2^2 \cdots x_k^2$$

$$\tag{3.119}$$

If we consider real vector bundles V of *even rank* over an *orientable* space X then there is an extra characteristic class that can exist; this is the Euler class. The origin of this extra class is that if X is orientable then a choice of orientation reduces the structure group of V from $O(k)$ to $SO(k)$; if, in addition, k is even then the transition functions of V are even dimensional matrices and so have the property that their determinants possess an $SO(k)$-invariant square root known as the Pfaffian (cf. for example Nash and Sen [1]). We shall write the Euler class of V as $e(V)$. For a more detailed account of $e(V)$ see also Bott and Tu [1]. Here we shall only define $e(V)$ up to a sign; that is we just use the fact that the square of $e(V)$ is the highest Pontrjagin class $p_{k/2}(V)$ of V. In other words $e(V)$ obeys

$$e^2(V) = p_{k/2}(V) \tag{3.120}$$

Thus $e(V)$ can be thought of loosely as the 'square root' of $p_{k/2}(V)$. Because $p_k(V) \in H^{4k}(X;\mathbf{Z})$ then for the Euler class we have

$$e(V) \in H^k(X;\mathbf{Z}) \text{ where } V \text{ has rank } k \tag{3.121}$$

The Euler class behaves well under Whitney sums. One has

$$e(V \oplus W) = e(V)\, e(W) \tag{3.122}$$

where V and W are two real vector bundles over X.

We can also define the Euler class of a complex vector bundle E: we repeat the procedure used in the Pontrjagin case: that is the Euler class of E is defined to be the Euler class of the underlying real bundle E_r. In other words

$$e(E) = e(E_r) \tag{3.123}$$

Let E have rank k; then we can show that the top Chern class $c_k(E)$ and the Euler class are equal. We just use the fact that $e^2(E) = p_k(E)$ and refer to 3.119, from which we obtain

$$c_k(E) = x_1 \cdots x_k, \qquad p_k(E) = x_1^2 \cdots x_k^2$$
$$\Rightarrow e(E) = c_k(E) \tag{3.124}$$

where the choice of sign implicit in the taking of the square root can be justified—cf. Bott and Tu [1] and Milnor and Stasheff [1].

In conclusion we recall that characteristic classes are often calculated using curvature methods; we shall often use these methods in subsequent chapters. Further background can be found in Nash and Sen [1] but the main idea is the following. Let F be a curvature form on a rank k vector bundle with group G, G being $U(k)$, $O(k)$ or $SO(k)$ whichever is appropriate. Such an F is a 2-form with values in the Lie algebra of G. Consider the invariant polynomial $P(t)$ given by

$$P(t) = det\left(tI + \frac{iF}{2\pi}\right) \tag{3.125}$$

Evidently $P(t)$ is a polynomial of degree k in the variable t whose coefficients are differential forms. The important point is that these differential forms are closed and so determine elements of the de Rham cohomology of the base X. These forms are just the characteristic classes of E; moreover, $P(t)$ provides a formula for these classes in terms of polynomials in the curvature. If E is a complex vector bundle then $P(t)$ determines the Chern forms $c_i(F)$ according to

$$P(t) = det\left(tI + \frac{iF}{2\pi}\right) = \sum_j t^j c_{k-j}(F) \tag{3.126}$$

Any expression obtained using splitting methods has an analogue written in terms of curvature: for example, the splitting principle gives the Chern character of E as

$$ch(E) = \sum_{i=1}^{k} e^{x_i}$$

Assume that the curvature form F, viewed as a $k \times k$ matrix in the Lie algebra $u(n)$, is diagonal, i.e.

$$F = \begin{pmatrix} F_1 & 0 & \cdots & 0 \\ 0 & F_2 & \cdots & 0 \\ \vdots & \vdots & \ddots & \vdots \\ 0 & 0 & \cdots & F_k \end{pmatrix}$$

Then the curvature version of the Chern character is $ch\,(F)$ where

$$ch\,(F) = tr\,\exp\frac{iF}{2\pi} \qquad (3.127)$$

More generally, suppose that $f(x_1,\ldots,x_k)$ is some expression obtained using the splitting principle. To produce its curvature counterpart one simply replaces the variables x_i by the curvature variables $iF_i/2\pi$. The validity of deriving properties of such functions $f(x_1,\ldots,x_k)$ using only diagonal matrices (Eguchi, Gilkey and Hanson [1]) is that such results can be viewed as continuous functions on the space of matrices: but, since diagonalisable matrices can be used to approximate any matrix, the diagonal matrices are dense in the space of all matrices; and if a property of a continuous function is verified on a dense subspace then it extends to the whole space. Thus we have a kind of linear algebraic analogue of the validity of the splitting principle.

§ 6. Fredholm operators and $K(X)$

We end this chapter with a description of $K(X)$ which can be made using Fredholm operators. Let $\mathcal{F}$ denote the space of all Fredholm operators then consider the space of homotopy classes $[X,\mathcal{F}]$. Now since the product of two Fredholm operators is also Fredholm, this product induces a product in $[X,\mathcal{F}]$ so that $[X,\mathcal{F}]$ is a semi-group. Actually it can be shown that $[X,\mathcal{F}]$ is isomorphic to the group $K(X)$. Then in the trivial case where X is a point, and thus $K(X) = \mathbf{Z}$, this isomorphism is just the assignment of a Fredholm operator F to its index, that is to the integer

$$\dim ker\,F - \dim coker\,F \qquad (3.128)$$

Since elliptic operators are Fredholm operators when viewed as operators in Hilbert space, one might expect $K(X)$ to turn up in a discussion of the index of elliptic operators. We shall see that this does happen in our next chapter, which is concerned with index theory.

CHAPTER IV

Index Theory for Elliptic Operators

§ 1. The index of an elliptic operator

We have already discussed elliptic operators in chapter 2 where we saw that such operators have finite dimensional kernels and cokernels. This being so one can define the index of an elliptic operator O by the usual expression

$$index\, O = \dim ker\, O - \dim coker\, O \qquad (4.1)$$

so that $index\, O$ is always a well-defined integer. Let this operator O act on a topological space X—in practice X will usually be a compact real or complex manifold. For the moment we require X to have no boundary; later we shall come to situations in which X has a boundary.

We would like to describe roughly what the Atiyah–Singer index theorem says and the strategy employed in its proof. Firstly the index theorem says that if O is an elliptic differential or pseudo-differential operator then the integer $index\, O$, as well being calculable by computing $\dim ker\, O$ and $\dim coker\, O$, can also be calculated using a purely topological formula consisting of characteristic classes. The index is also stable under perturbations of the operator O by a compact operator. The strategy of the proof is to give two apparently different definitions of an 'index' for O—these definitions are formulated so that the task of proving the theorem is equivalent to showing that these two definitions coincide.

These two definitions are referred to as the analytical index and the topological index. Before giving an account of them we need a little preliminary discussion designed to allow us to exploit some K-theory. In our earlier definition of K-theory the essential step was to point out that the vector bundles $Vect(X)$ over a space X form an Abelian semi-group under the operation of direct sum. This semi-group can then be completed to a group which is thereafter denoted by $K(X)$—in this chapter we shall use the notation $K(X)$ rather than $KU(X)$, $KO(X)$, or $KSp(X)$; the context should prevent any confusion.

For index theory we require an alternative definition, Segal [1], in which direct sums of vector bundles are replaced by direct sums of complexes. To this end consider the complex E below

$$0 \longrightarrow E^0 \xrightarrow{\alpha_0} E^1 \xrightarrow{\alpha_1} \cdots \xrightarrow{\alpha_n} E^{n+1} \longrightarrow 0$$
$$\alpha_i \circ \alpha_{i-1} = 0 \tag{4.2}$$

where, in the above, the E_i's are vector bundles over X and the α_i's are homomorphisms between them. We want to consider all such complexes over X. However we do not want to distinguish complexes E and F which are homotopic to one another. Homotopy between complexes is defined in an obvious way: two complexes E and F are homotopic if there exists a complex G over $X \times [0,1]$ such that $G|X \times \{0\} \simeq E$ and $G|X \times \{1\} \simeq F$. We denote all the homotopy classes of complexes over X by $C(X)$. Then the direct sum operation $\oplus$ of vector bundles induces an addition between complexes which renders $C(X)$ an Abelian semi-group. To complete this semi-group to a group we define the *support* of a complex E to be the set of $x \in X$ for which the sequence at x

$$0 \longrightarrow E_x^0 \xrightarrow{\alpha_0} E_x^1 \xrightarrow{\alpha_1} \cdots \xrightarrow{\alpha_n} E_x^{n+1} \longrightarrow 0 \tag{4.3}$$

fails to be exact.[1] Now if we consider all complexes in $C(X)$ with *empty* support, i.e. support ϕ, then these also form a semi-group $C_\phi(X)$, say, and clearly $C_\phi(X) \subset C(X)$. It transpires, Segal [1], that forming the natural quotient provides us with a group which is precisely $K(X)$. That is we have

$$K(X) = \frac{C(X)}{C_\phi(X)} \tag{4.4}$$

In imitation of our previous K-theory notation, if E is a complex over X, we denote the corresponding element of $K(X)$ by $[E]$. Also when X is compact there is an important formula which gives $[E]$ in terms of the elements $[E^i]$ of $K(X)$; it is the alternating sum formula

$$[E] = \sum_i (-1)^i [E^i] \tag{4.5}$$

---

[1] Should X be non-compact we restrict ourselves to complexes with compact support.

The relevance of the preceding discussion to index theory is that K-theory arises when one considers certain complexes which are natural to the elliptic theory. If D is an elliptic operator on a (compact closed) manifold X then we shall see that it is the K-theory of T^*X, i.e. $K(T^*X)$, rather than that of X itself, that we encounter in index theory.[2] Returning to the occurrence of complexes in the theory of elliptic operators, consider the bundles E^i again. Then form a complex $\Gamma(X, E)$ by letting a sequence of differential or pseudo-differential operators d_i connect their respective spaces of sections $\Gamma(X, E^i)$ so that we have

$$0 \longrightarrow \Gamma(X, E^0) \xrightarrow{d_0} \Gamma(X, E^1) \xrightarrow{d_1} \cdots \xrightarrow{d_n} \Gamma(X, E^{n+1}) \longrightarrow 0 \tag{4.6}$$
$$d_i \circ d_{i-1} = 0$$

We also require the d_i's to be operators of the *same* order m—one can allow operators of different orders, Atiyah and Bott [1], but we do not need to do that here. Granted this state of affairs we can use the results on p. 38 of chapter 2 to construct the leading symbols $\sigma_m(d_i)$ of the d_i's and write down the associated symbol complex, which we denote by $\sigma(E)$. This is the complex

$$0 \longrightarrow \pi^* E^0 \xrightarrow{\sigma_m(d_0)} \pi^* E^1 \xrightarrow{\sigma_m(d_1)} \cdots \xrightarrow{\sigma_m(d_n)} \pi^* E^{n+1} \longrightarrow 0 \tag{4.7}$$

Now, though the complex $\Gamma(X, E)$ has no particular reason to be exact, we shall *require* its associated symbol complex $\sigma(E)$ to be exact. Since complexes of this kind are important we describe them in a definition

Definition (Elliptic complex) *A complex $\Gamma(X, E)$ of sections of vector bundles over X of the type shown in 4.6 is called elliptic if its associated symbol complex $\sigma(E)$ over T^*X is exact off the zero section of T^*X.*

Notice that if this elliptic complex had only two non-trivial terms, so that there was only *one* differential operator d_0, then this definition would just be that of ellipticity for d_0. In general this is not so but, nevertheless, underlying each elliptic complex E is a *single* elliptic operator D. To see this consider the two vector bundle direct sums

$$E^+ = E^0 \oplus E^2 \oplus \cdots$$
$$E^- = E^1 \oplus E^3 \oplus \cdots \tag{4.8}$$

---

[2] The bundle T^*X is non-compact because its fibres are vector spaces and so we take $K(T^*X)$ to mean K-theory with compact supports. What we mean by this is that for a non-compact space S, say, one calculates the reduced K-theory of its one-point compactification $\bar{S}$, i.e. $K(S)$ denotes $\widetilde{K}(\bar{S})$; alternatively one uses 4.4 and complexes with compact support.

The next step is to introduce metrics on the bundles E^i so that the adjoints of the d_i's can be defined: using d_i and d_i^* we construct the single operator D by writing

$$D : \Gamma(X, E^+) \longrightarrow \Gamma(X, E^-) \qquad \text{where } D = \sum_i (d_{2i} + d_{2i+1}^*)$$

(4.9)

that is $D(e_0, e_2, e_4, \ldots) = (d_0 e_0 + d_1^* e_2, d_2 e_2 + d_3^* e_4, \ldots)$

To prove that D is elliptic we write down the complex connecting E^+ to E^- together with its associated symbol complex. This yields

$$0 \longrightarrow \Gamma(X, E^+) \xrightarrow{D} \Gamma(X, E^-) \longrightarrow 0$$
$$0 \longrightarrow \pi^* E^+ \xrightarrow{\sigma_m(D)} \pi^* E^- \longrightarrow 0$$

(4.10)

Now it follows at once from the fact that $d_i \circ d_{i-1} = 0$ that $d_{i-1}^* \circ d_i^* = 0$ and so $D^2 = 0$ as it should. More to the point, though, is the fact that the exactness of the original symbol complex $\sigma(E)$ of 4.7 implies the exactness of the present symbol complex for D. But, since this complex is of length one, exactness is simply the statement that $\sigma_m(D)$ is an isomorphism (off the zero section of $T^* X$). This is precisely the definition of ellipticity so that D is elliptic as claimed.

Having obtained an elliptic operator D from our elliptic complex E we are ready to examine its index. To do this we employ the standard device of introducing the Laplacians $\Delta_i(E)$ given by $\Delta_i(E) = d_i^* \circ d_i + d_{i-1} \circ d_{i-1}^*$. Just as in chapter 2 the kernels of these Laplacians are characterised by

$$\Delta_i(E) e_i = 0 \iff d_i e_i = d_{i-1}^* e_i = 0$$

It is also natural to construct the operators $\Delta^+(E)$ and $\Delta^-(E)$ where

$$\Delta^+(E) = D^* D = \Delta_0(E) \oplus \Delta_2(E) \oplus \cdots$$
$$\Delta^-(E) = DD^* = \Delta_1(E) \oplus \Delta_3(E) \oplus \cdots$$

(4.11)

The kernels of $\Delta^+(E)$ and $\Delta^-(E)$ are now determined and we have

$$ker \, \Delta^+(E) = \bigoplus_{i \geq 0} ker \, \Delta_{2i}, \qquad ker \, \Delta^-(E) = \bigoplus_{i \geq 0} ker \, \Delta_{2i+1}$$

(4.12)

But if $\mathbf{e} \in ker \, \Delta^+(E)$ then $D^* D \mathbf{e} = 0$ so that

$$< \mathbf{e}, D^* D \mathbf{e} > = 0 \iff < D \mathbf{e}, D \mathbf{e} > = 0 \iff D \mathbf{e} = 0$$

(4.13)

Thus $ker\, D = ker\, \Delta^+(E)$; similarly $ker\, D^* = ker\, \Delta^-(E)$. We have therefore reduced the computation of the index of D to counting harmonic sections over the E^i. More precisely we can write that

$$index\, D = \dim ker\, D - \dim ker\, D^* = \dim ker\, \Delta^+(E) - \dim ker\, \Delta^-(E)$$
$$= \sum_i \dim ker\, \Delta_{2i} - \sum_i \dim ker\, \Delta_{2i+1}$$
$$= \sum_i (-1)^i \dim ker\, \Delta_i$$

(4.14)

However, we already know from the Hodge theory of chapter 2 that the de Rham complex is elliptic and that its cohomology is given by the harmonic sections. It is now time to quote the generalisation of this result to any elliptic complex. The theorem we need (cf. Wells [1]) is

Theorem (Hodge, general case) *Let E be an elliptic complex over a compact closed manifold X. Define the standard cohomology $H^i(E)$ of the complex by $H^i(E) = ker\, d_i / Im\, d_{i-1}$. Then the $H^i(E)$ are isomorphic to the spaces of harmonic sections, that is*

$$H^i(E) \simeq ker\, \Delta_i \qquad (4.15)$$

Thus we have an alternative way of expressing the index of D. In summary we have $\dim H^i(E) = \dim ker\, \Delta_i$. So if we define the generalised Euler characteristic of the complex by $\chi(E) = \sum(-1)^i \dim H^i(E)$ the index formula becomes

$$index\, D = \chi(E) \qquad (4.16)$$

Having reached this point we can put together the definition of K-theory via complexes of vector bundles and that of an elliptic complex to obtain the definition of the analytical index of an operator. Given an elliptic complex the symbol complex $\sigma(E)$ provides us straightaway with a complex of vector bundles over T^*X:

$$0 \longrightarrow \pi^*E^+ \xrightarrow{\sigma_m(D)} \pi^*E^- \longrightarrow 0 \qquad (4.17)$$

This means that the leading symbol $\sigma_m(D)$ of D determines an element $[\sigma_m(D)]$, say, of $K(T^*X)$. It is also known that two elliptic D with the same leading symbol have the same index. Thus we can think of the process of calculating the integer $index\, D = \dim ker\, D - \dim ker\, D^*$ as being a map from $K(T^*X)$ to $\mathbf{Z}$. This map is none other than the *analytical index*. Turning to its definition we have

Definition (Analytical index) *For an elliptic complex E, with associated elliptic, m^{th} order, pseudo-differential operator D, the analytical index ind_a is the map*

$$ind_a : K(T^*X) \longrightarrow \mathbf{Z}$$
$$[\sigma_m(D)] \longmapsto index\ D \qquad\qquad (4.18)$$

It should be mentioned that it is tacit in the above definition that all elements of $K(T^*X)$ arise as $[\sigma_m(D)]$ for some D. It turns out, though, cf. Atiyah and Singer [2], that we can represent all of $K(T^*X)$ by working only with the symbol complexes of elliptic differential (*and pseudo-differential*) operators. Thus the analytical index ind_a is well defined.

Our next task is to introduce the topological index. This requires two things: the idea of a tubular neighbourhood and the homomorphism known as the Thom homomorphism. We shall explain them briefly, much fuller accounts can be found in Bott and Tu [1] and Atiyah and Singer [2].

Consider a curved length of string S in $\mathbf{R}^3$. Imagine that this string is embedded in a thick tube U. This tube U is called a tubular neighbourhood of S in $\mathbf{R}^3$. One can think of U as being made up of a union of cross-sectional discs each of which is perpendicular to the string at its centre. This amounts to saying that such a union of normal discs is homotopic to the normal bundle $N(S)$ of S in $\mathbf{R}^3$. More generally one has the result that if S is a sub-manifold of M, then there is an tubular neighbourhood U of S in M and U is identifiable with the normal bundle $N(S)$ of S in M.

To describe the Thom homomorphism we start with a complex vector bundle F over X. The Thom homomorphism is a certain map ϕ between the K-theory of X and that of F itself. It is thus of the form

$$\phi : K(X) \longrightarrow K(F) \qquad\qquad (4.19)$$

Now to define ϕ we take the exterior algebra $\wedge^* F = \oplus_i \wedge^i F$ of F and use it to define a complex. Since $\wedge^i F$ is mapped into $\wedge^{i+1} F$ by the action of the wedge product then we have the following complex over F

$$0 \longrightarrow \wedge^0 F \xrightarrow{\wedge} \wedge^1 F \xrightarrow{\wedge} \cdots \xrightarrow{\wedge} \wedge^n F \longrightarrow 0 \qquad (4.20)$$

Further, this complex is exact (off the zero section)—this follows at once from the linear algebraic fact that if v is an i-form, and $k = k_i dk^i \neq 0$, then $k \wedge v = 0 \Rightarrow v = k \wedge w$, where w is an $(i-1)$-form. Let us call this complex the exterior complex and denote it by $\wedge(F)$. The complex $\wedge(F)$ is a complex over F, it therefore determines an element of $K(F)$, an element which we denote by λ_F. Now $\lambda_F \in K(F)$ can act by multiplication on

$K(X)$ giving another element of $K(F)$. This multiplicative action viewed as a map from $K(X)$ to $K(F)$ is the Thom homomorphism ϕ

$$\phi : K(X) \longrightarrow K(F) \qquad (4.21)$$

The definition of ϕ can also be extended (Atiyah and Singer [2]) to the case when X is non-compact; we shall actually use this fact below because we shall use ϕ with X replaced by T^*X.

Actually the map ϕ is an *isomorphism* and, in the case where $X = \{x_0\}$ and V is the trivial bundle $V = \mathbf{C}^n$, this statement is just the Bott periodicity result in the form $\widetilde{K}(S^{2n}) = \mathbf{Z}$. To see this recall that $K(\{x_0\}) = \mathbf{Z}$ and $K(\mathbf{C}^n) \simeq K(\mathbf{R}^{2n}) = \widetilde{K}(S^{2n}) = \mathbf{Z}$; note that in calculating $K(\mathbf{C}^n)$ we followed the instructions of the footnote on p. 91, that is we first compactified $\mathbf{C}^n$ using $\mathbf{C}^n \cup \{\infty\} \simeq \mathbf{R}^{2n} \cup \{\infty\} \simeq S^{2n}$ and then used reduced K-theory.

We also need a simple property of inclusions of an open sub-manifold S in another manifold M. Let $i : S \to M$ be such an inclusion. Consider the de Rham cohomology of S and choose a p-form $\omega \in \Omega^p(S)$. This p-form ω can be extended to the whole of M by the simple device of letting it be zero outside S. In this way we obtain a map from $H^p_{de\ Rham}(S)$ to $H^p_{de\ Rham}(M)$; consideration of all values of p provides a map between the cohomology rings $H^*_{de\ Rham}(S)$ and $H^*_{de\ Rham}(M)$. Finally the same idea applied to a generalised cohomology such as K-theory provides a map between $K(S)$ and $K(M)$. This natural map we write as i_*. In summary, then, we have the two maps

$$i : S \longrightarrow M \quad \text{and} \quad i_* : K(S) \longrightarrow K(M) \qquad (4.22)$$

We are now ready to apply the maps ϕ and i_* to construct the topological index. Let D be an elliptic operator on X. Embed the manifold X in $\mathbf{R}^n$ for an appropriate n. Construct an open tubular neighbourhood $N(X)$ of X in $\mathbf{R}^n$. Notice that embedding X in $\mathbf{R}^n$ also induces an embedding of T^*X in $T^*\mathbf{R}^n$. The tubular neighbourhood $N(T^*X)$ of this latter embedding is just $T^*N(X)$ and can be thought of as two copies of $N(X)$ sitting over T^*X—that is, they are the pullback under $\pi : T^*X \to X$ of $N(X) \oplus N(X)$ from X to T^*X. We wish to render this pullback a complex vector bundle and so we impose the identification of $N(X) \oplus N(X)$ with $N(X) \oplus iN(X)$. But $N(X) \oplus iN(X) \simeq N(X) \otimes_{\mathbf{R}} \mathbf{C}$. Thus the complex vector bundle which is the tubular neighbourhood of T^*X in $\mathbf{R}^n$ is actually the bundle $\pi^*(N(X) \otimes_{\mathbf{R}} \mathbf{C})$ over T^*X. Next we use the Thom homomorphism with $F = \pi^*(N(X) \otimes_{\mathbf{R}} \mathbf{C})$. This is therefore a map

$$\phi : K(T^*X) \longrightarrow K(F)), \qquad (F = \pi^*(N(X) \otimes_{\mathbf{R}} \mathbf{C})) \qquad (4.23)$$

Having introduced ϕ we recall that F is the tubular neighbourhood of T^*X in $T^*\mathbf{R}^n$, therefore it is contained in $T^*\mathbf{R}^n$. So using this simple inclusion we have our second map

$$i_* : K(F) \longrightarrow K(T^*\mathbf{R}^n) \tag{4.24}$$

Putting these two maps ϕ and i_* together we get a homomorphism from $K(T^*X)$ to $K(T^*\mathbf{R}^n)$ which we denote by $i_!$. This yields

$$i_! : K(T^*X) \xrightarrow{\phi} K(F) \xrightarrow{i_*} K(T^*\mathbf{R}^n), \quad i_! = i_* \circ \phi, \quad (F = \pi^*(N(X) \otimes_{\mathbf{R}} \mathbf{C})) \tag{4.25}$$

To construct the topological index we need one further map $j_!$ which we now introduce. Let P stand for the space $\{x_0\}$ consisting of a single point which we take to be the origin of the space $\mathbf{R}^n$. Consider the inclusion of P in $\mathbf{R}^n$ and denote this second inclusion by j so that we have

$$j : P \to \mathbf{R}^n \tag{4.26}$$

Associated to j is its Thom homomorphism which we denote by $j_!$.

Displaying $i_!$ and $j_!$ together gives us the diagram

$$K(T^*X) \xrightarrow{i_!} K(T^*\mathbf{R}^n) \xleftarrow{j_!} K(T^*P) \tag{4.27}$$

The *topological index* is the map

$$(j_!)^{-1} \circ i_! : K(T^*X) \longrightarrow K(T^*P) \tag{4.28}$$

Before commenting further on the topological index we wish to calculate $K(T^*P)$ and $K(T^*\mathbf{R}^n)$. In fact we shall find that $K(T^*P) = K(T^*\mathbf{R}^n) = \mathbf{Z}$. We begin with $K(T^*P)$. Since P is a point, $T^*P \simeq P$ and so $K(T^*P) = K(P) = \mathbf{Z}$ from chapter 3. Moving on to $K(T^*\mathbf{R}^n)$ we can use the following argument. The bundle $T^*\mathbf{R}^n$, having a contractible base, is trivial; thus $T^*\mathbf{R}^n = \mathbf{R}^n \times \mathbf{R}^n = \mathbf{R}^{2n}$. But, as we saw on p. 95 $K(\mathbf{R}^{2n}) = \mathbf{Z}$. In other words we have shown that

$$K(T^*\mathbf{R}^n) = \mathbf{Z} \tag{4.29}$$

as claimed. We finish by defining the topological index.

Definition (Topological index) *Let E be an elliptic complex and D its associated elliptic operator as already introduced above. Then the topological index ind_t is the map $(j_!)^{-1} \circ i_!$, that is*

$$ind_t : K(T^*X) \longrightarrow \mathbf{Z}$$
$$[\sigma_m(D)] \longmapsto ((j_!)^{-1} \circ i_!)([\sigma_m(D)]) \tag{4.30}$$

Since the construction of the topological index uses embeddings and a tubular neighbourhood it is necessary to show, Atiyah and Singer [2], that ind_t is independent of the particular embedding and neighbourhood chosen. Actually the map $j_!$ is a ring isomorphism from $\mathbf{Z}$ to $\mathbf{Z}$ and it is an elementary fact that it is therefore the identity; nevertheless, we include $j_!$ in our definition of ind_t because it is needed in the more general case of the index theory for elliptic operators invariant under a compact Lie group G, cf. § 6 below.

With the two vital index definitions in place we can give the Atiyah–Singer index theorem itself. As we have already explained the theorem is now the assertion:

Theorem (Atiyah–Singer) *The topological and the analytical index are equal.*

The proof of this theorem is carried out in an axiomatic fashion whose logical structure is as follows. The topological index is defined and shown to be the *unique* map which satisfies a certain set of index axioms. The analytical index is then introduced and, after considerable amount of work, it is proved to obey these same axioms. The uniqueness then establishes the theorem. Slightly more concretely one can draw on an earlier proof (Atiyah and Singer [1]) of the index theorem and view the task of proving it in the following way. Firstly we should think of the index theorem as providing a topological formula for the index of a *general* elliptic operator D on a *general* manifold X. This is a formidable task as both the operator and the manifold are unrestricted. The way round these two obstacles is to show that, as far as the calculation of *index D* is concerned, both D and X may be replaced by much simpler counterparts. For X this is achieved by the inclusion $i : X \rightarrow \mathbf{R}^n$; X's counterpart being $\mathbf{R}^n$ which gets compactified to a sphere in the details of the calculation. For D we now only have to work on a sphere and one shows, Atiyah and Singer [1], that any D is somehow equivalent to a classical operator known as a generalised signature operator (of which more in § 3). The remaining work then is to show by explicit calculation that the index theorem is true for classical operators on spheres.

§ 2. Some examples

When calculating *index D* in practice one can make very profitable use of a certain index formula. This formula expresses the index in terms of characteristic classes and so is cohomological in content. Thus far we have a K-theoretic formulation of the index. A key part in producing the index formula is played by the Chern character. What is required is a mechanism for passing from K-theory to cohomology. But we saw in chapter 3 that precisely this property is possessed by the Chern character ch and it was used there to construct a homomorphism $ch : K(X) \rightarrow H^*(X; \mathbf{Q})$.

To obtain this cohomological formula for the index is a slightly lengthy but routine exercise in the translation of a K-theory formula into a cohomological formula. The details are in Atiyah and Singer [3] and they show that, if $[TX]$ denotes the fundamental homology class of TX, the integer $index\, D$ is given by

$$index\, D = (-1)^n \left\{ ch\left([\sigma_m(D)]\right) \cdot td\left(TX_\mathbf{c}\right)\right\} [TX] \qquad (4.31)$$

where D is an elliptic operator of order m, n is the dimension of X, $TX_\mathbf{c} = TX \otimes_\mathbf{R} \mathbf{C}$ denotes the complexification of the tangent bundle TX, and ch and td denote Chern character and Todd class respectively. A similar formula with the same combination of the Chern character and Todd class already occurs in earlier work on the Riemann–Roch theorem cf. Hirzebruch [1].

Notice that the cohomological formula 4.31 expresses the far from obvious fact that the RHS $(-1)^n \left\{ ch\left([\sigma_m(D)]\right) \cdot td\left(TX_\mathbf{c}\right)\right\} [TX]$ is an integer: since it is an expression in rational cohomology it might have been expected only to be rational.

It is useful to note that, for X a real manifold, $TX_\mathbf{c}$ is the complexification of a real vector bundle, and we saw in 3.108 of chapter 3 that this means that $TX_\mathbf{c}$ is self-dual. Thus its odd Chern classes vanish, and the expansion 3.106 of the Todd polynomial converts into a useful expansion in terms of the Pontrjagin classes of TX via their definition as $p_i(TX) = (-1)^i c_{2i}(TX_\mathbf{c})$. This expansion is easily found to be

$$td\left(TX_\mathbf{c}\right) = 1 - \frac{1}{12}p_1(TX) + \frac{1}{720}(3p_1^2 - p_2)(TX) + \cdots \qquad (4.32)$$

We can now put the index formula to use in some specific examples. The first two of these are special cases where geometry plays no particular rôle. We begin by dealing with the calculation of the index when the manifold X is odd dimensional.

Example *The index when* $\dim X$ *is odd*

If X is odd dimensional and D is elliptic then it turns out that $index\, D = 0$ when D is a genuine *differential* operator. However the same is not necessarily true if D is only a *pseudo-differential* operator.

The idea behind the calculation is to reverse the orientation on the cotangent bundle T^*X. Then, in the differential case, this will imply that

$index\, D = -index\, D$ thus forcing the index to vanish. Suppose, to begin with, that D is a differential operator of integral order m with leading symbol $\sigma_m(D)$. We know that locally $\sigma_m(D)$ is described by a function of the form

$$\sigma_m^D(x, p) = \sum_{|\alpha|=m} a_\alpha(x) p^\alpha$$

so that $\sigma_m^D(x, p)$ has the homogeneity property $\sigma_m^D(x, \lambda p) = \lambda^m \sigma_m^D(x, p)$. Now subject T^*X to the antipodal map R under which $p \mapsto -p$. Note that R reverses the orientation on T^*X because its volume element is given locally by $dx^1 \wedge dx^2 \ldots \wedge dx^n \wedge dp^1 \wedge dp^2 \ldots \wedge dp^n$ and a change of sign in all the p^i produces the factor $(-1)^n$, which is of course a change of sign for n odd. A change of orientation produces a change of sign of the corresponding fundamental class so that $[TX]$ becomes $-[TX]$. The effect of R on $\sigma_m^D(x, p)$ is simply that we have

$$\sigma_m^D(x, -p) = (-1)^m \sigma_m^D(x, p) \tag{4.33}$$

Now at the K-theory level this means that the element $[\sigma_m(D)] \in K(T^*X)$ is replaced by $[(-1)^m \sigma_m(D)] \in K(T^*X)$ but these elements are the same since multiplication by the constant $(-1)^m$ does not change the equivalence class of $[\sigma_m(D)]$ in $K(T^*X)$. The net effect of R on the index is therefore the following

$$index\, D = (-1)^n \left\{ ch\left([\sigma_m(D)]\right) \cdot td\left(TX_{\mathbf{c}}\right) \right\} [TX]$$
$$\text{use of } R \;\Rightarrow\; index\, D = (-1)^n \left\{ ch\left([(-1)^m \sigma_m(D)]\right) \cdot td\left(TX_{\mathbf{c}}\right) \right\} \{-[TX]\}$$
$$= -(-1)^n \left\{ ch\left([\sigma_m(D)]\right) \cdot td\left(TX_{\mathbf{c}}\right) \right\} [TX]$$
$$= -index\, D$$
$$\tag{4.34}$$

Thus we have established that

$$index\, D = 0$$

Suppose now that D is a pseudo-differential operator of order m—where m may or may not be integral—then the above argument fails. This is because under the antipodal map the symbol need not obey a symmetry relation such as 4.33. For example, let us choose X to be the circle S^1, with local coordinate θ, then there is an elliptic pseudo-differential operator of order 1 given by

$$D = \exp[-i\theta] \left(-i\partial_\theta + \sqrt{-\partial_\theta^2} \right) - \left(\partial_\theta + \sqrt{-\partial_\theta^2} \right) \tag{4.35}$$
$$\Rightarrow \sigma_1^D(\theta, p) = \exp[-i\theta](p + |p|) + (p - |p|)$$

The symbol $\sigma_1^D(\theta, p)$ is clearly that of an elliptic operator since it is non-zero away from $p = 0$. Applying the antipodal map R we obtain no simple relation between $\sigma_1^D(\theta, p)$ and $\sigma_1^D(\theta, -p)$; in fact D has indeed a non-zero index, cf. Eguchi, Gilkey and Hanson [1].

Example *The index when D acts on functions*

For this example we consider an mth-order elliptic (differential or pseudo-differential) operator $D : \Gamma(X, E) \rightarrow \Gamma(X, F)$ in the case where the vector bundles E and F are both *trivial*. This means that D is just acting on a system of *functions*. This system of functions has a dimension equal to the rank of E (this is of course the same as the rank of F since D is elliptic) which we take to be k. The leading symbol of $\sigma_m(D)$ of D is now literally a matrix valued function of x and p. In local coordinates we write $\sigma_m(D)$ as $\sigma_m^D(x, p)$ and, by definition, $\sigma_m^D(x, p)$ is positively homogeneous of degree m in p; this means that its p-dependence is determined by its values on the unit sphere in the variable p—in bundle theoretic language we say that $\sigma_m(D)$ is determined by its restriction to the unit sphere bundle $S(X)$ of T^*X. In any case we can now think of $\sigma_m(D)$ as a map,

$$\sigma_m(D) : S(X) \longrightarrow Gl(k, \mathbf{C}) \qquad (4.36)$$

The topological content of $\sigma_m(D)$ resides in its cohomological properties which are contained in the pullback

$$\sigma_m^*(D) : H^*(Gl(k, \mathbf{C}); \mathbf{Z}) \longrightarrow H^*(S(X); \mathbf{Z}) \qquad (4.37)$$

However, $U(k)$ is a deformation retract of $Gl(k, \mathbf{C})$ and so $H^*(Gl(k, \mathbf{C}); \mathbf{Z})$ can be replaced by $H^*(U(k); \mathbf{Z})$. Next let us simplify matters considerably by specialising to the case $k = 1$; this corresponds to the operator D acting just on functions rather than a system of functions. Having done this $U(k)$ becomes $U(1) = S^1$ and so $H^*(U(k); \mathbf{Z}) = H^*(S^1; \mathbf{Z})$ has a single 1-dimensional generator which we denote by h. We can now argue informally that, with one exception, *index* $D = 0$. We proceed as follows. The index formula can now be evaluated on $S(X)$ rather than TX and we have

$$index\ D = (-1)^n \{ch\left([\sigma_m(D)]\right) \cdot td\left(TX_\mathbf{c}\right)\}[S(X)] \qquad (4.38)$$

But the 1-dimensionality of $H^*(S^1; \mathbf{Z})$ means that $ch\left([\sigma_m(D)]\right)$ must be proportional to the 1-dimensional pullback $\sigma_m^*(D)h \in H^1(S(X); \mathbf{Z})$; also, using 4.32, the Todd class can be expanded in terms of Pontrjagin classes which are all in dimensions divisible by 4; note too that the sphere bundle $S(X)$ has dimension one less than TX, i.e. it is of dimension $2n - 1$.

Applying all of these dimensional facts to the formula for $index\,D$ means that the cohomology class on the RHS has dimension $4i + 1$, for some i, and that evaluating this class on $S(X)$ requires

$$4i + 1 = 2n - 1 \Rightarrow n = 2i + 1 \tag{4.39}$$

Thus X is odd dimensional, but we have just seen that differential operators have zero index for $\dim X$ odd, hence we infer that $index\,D = 0$ when D is a differential operator. Hence we have established that *the index of an elliptic differential operator acting on functions is always zero.*

If D is pseudo-differential rather than differential we know from our first example that the result is false when $\dim X = 1$; however, this is the only exception to the vanishing of $index\,D$. To see this we only have to observe that the Pontrjagin classes of dimension $4i$ occuring in the expansion of $td\,(TX_{\mathbf{c}})$ have a maximum possible dimension of n. Thus we can replace 4.39 by the stronger statement that

$$0 \leq 4i \leq n, \qquad n = 2i + 1 \tag{4.40}$$

and this forces $n = 1$. Thus we have the result that *the index of a pseudo-differential operator acting on functions is always zero if* $\dim X > 1$.

If D acts on a $k \times k$ system of functions with $k > 1$ then these results no longer hold—however, there are still vanishing statements for $index\,D$ for a certain range of k and n (Atiyah and Singer [3]).

For the remainder of this section we shall impose a slight restriction on the manifolds and the elliptic operators that we consider. On the manifold side we shall insist that X be *orientable;* on the operator side we shall only allow *differential* operators.

A benefit that derives immediately from the orientability is that the fundamental cycle $[X]$ of X now exists; we can use this fact to transfer the evaluation of the index from $[TX]$ to $[X]$.

We proceed then to the variant of the index formula that results when we transfer the evaluation to $[X]$. The standard map employed to pass between cohomology on a bundle E and that on its base X is the Thom isomorphism for cohomology. We denote this map by ψ to distinguish it from the corresponding map in K-theory for which we use the symbol ϕ. In the case at hand the bundle E is the tangent bundle TX with projection $\pi : TX \to X$ so that ψ is a map of the form $\psi : H^*(X; \mathbf{Z}) \to H^*(TX; \mathbf{Z})$. More precisely, let u denote the generator of $H^*(TX; \mathbf{Z})$, then the pullback π^* and the generator u combine to give the usual definition of ψ. This is that ψ is the composition of π^* with the action of cup product with u:

$$\psi : H^*(X; \mathbf{Z}) \xrightarrow{\pi^*} H^*(TX; \mathbf{Z}) \xrightarrow{\cdot u} H^*(TX; \mathbf{Z})$$
$$\psi(a) = \pi^*(a) \cdot u, \qquad a \in H^*(X; \mathbf{Z}) \tag{4.41}$$

Actually it is clear that we must apply not ψ but ψ^{-1} to achieve our end. On applying ψ^{-1} to the RHS of the index formula we obtain

$$
\begin{aligned}
\psi^{-1}\left[(-1)^n\left\{ch\left([\sigma_m(D)]\right)\cdot td\left(TX_{\mathbf{c}}\right)\right\}[TX]\right] &\\
= (-1)^{n(n+1)/2}\psi^{-1}\left\{ch\left([\sigma_m(D)]\right)\cdot td\left(TX_{\mathbf{c}}\right)\right\}[X] &\qquad (4.42)\\
= (-1)^{n(n+1)/2}\psi^{-1}\left\{ch\left([\sigma_m(D)]\right)\right\}\cdot td\left(TX_{\mathbf{c}}\right)[X] &
\end{aligned}
$$

The central object to calculate in the above formula is the part that comes from the elliptic operator D. In other words we need to compute $[\sigma_m(D)]$, which belongs to $K(T^*X)$. To accomplish this we recall the expression 4.5 for K-theory elements in terms of complexes $0 \to E^0 \to E^1 \to \cdots$ which is

$$
[E] = \sum_i (-1)^i[E^i] \qquad (4.43)
$$

For an elliptic operator we have

$$
0 \longrightarrow \Gamma(X, E^0) \xrightarrow{d_0} \Gamma(X, E^1) \xrightarrow{d_1} \cdots \xrightarrow{d_n} \Gamma(X, E^{n+1}) \longrightarrow 0
$$

and the all important symbol complex

$$
0 \longrightarrow \pi^* E^0 \xrightarrow{\sigma_m(d_0)} \pi^* E^1 \xrightarrow{\sigma_m(d_1)} \cdots \xrightarrow{\sigma_m(d_n)} \pi^* E^{n+1} \longrightarrow 0 \qquad (4.44)
$$

We know that it is the symbol complex which determines the K-theory element $[\sigma_m(D)]$; and comparing this complex with 4.43 shows that, in the symbol complex, the E^i of 4.43 is replaced by $\pi^* E^i$. Thus we have

$$
\begin{aligned}
[\sigma_m(D)] &= \sum_p (-1)^p[\pi^* E^p]\\
\Rightarrow\; ch\left([\sigma_m(D)]\right) &= ch\left(\sum_p (-1)^p[\pi^* E^p]\right) \qquad (4.45)\\
&= \pi^* ch\left(\sum_p (-1)^p[E^p]\right)
\end{aligned}
$$

Inserting this into our index calculation gives

$$
\begin{aligned}
(-1)^{n(n+1)/2}\psi^{-1}\left\{ch\left([\sigma_m(D)]\right)\right\}\cdot td\left(TX_{\mathbf{c}}\right)[X] &\\
= (-1)^{n(n+1)/2}\psi^{-1}\left\{\pi^* ch\left(\sum_p (-1)^p[E^p]\right)\right\}\cdot td\left(TX_{\mathbf{c}}\right)[X] &\qquad (4.46)
\end{aligned}
$$

Using the definition of ψ it is then possible to verify, Atiyah and Singer [3], that

$$\psi^{-1}\{\pi^* u\} = \frac{u}{e(X)}, \quad u \in H^*(TX; \mathbf{Z})$$

$$\Rightarrow \psi^{-1}\{\pi^* ch\left(\sum_p (-1)^p[E^p]\right)\} = \frac{ch\left(\sum_p (-1)^p[E^p]\right)}{e(X)} \tag{4.47}$$

where $e(X)$ (also more accurately written as $e(TX)$) is the Euler class of the tangent bundle TX of X—since $e(X)$ is non-zero only if X is even dimensional we assume this fact whenever using the above expression. In any case we have already seen that, for differential operators, the index vanishes when X is odd dimensional. Thus, since we have just restricted ourselves to differential operators, we only have a need for 4.47 when X is even dimensional. Putting these various results together provides us with our second index formula which is

$$index\, D = (-1)^{n/2} \frac{ch\left(\sum_p (-1)^p[E^p]\right)}{e(X)} \cdot td\,(TX_{\mathbf{c}})[X] \tag{4.48}$$

which is applicable to elliptic operators on *even dimensional, orientable* manifolds. We now move on to consider examples in which we use this new formula 4.48.

These remaining examples are four classical elliptic complexes where both geometry and topology play an essential part. The simplest of them is the de Rham complex.

Example *The de Rham complex*

We consider the usual de Rham complex with the single change that we take the n-forms to be *complex valued* instead of real valued. The reason for this is that the index theorem uses the Chern character, for which we need complex vector bundles. The manifold X is not assumed complex, however. We do not alter our notation so that the (complex valued) de Rham complex is

$$\cdots \xrightarrow{d_{p-2}} \Omega^{p-1}(X) \xrightarrow{d_{p-1}} \Omega^p(X) \xrightarrow{d_p} \Omega^{p+1}(X) \xrightarrow{d_{p+1}} \cdots \tag{4.49}$$

with d_p denoting the exterior derivative on the p-forms $\Omega^p(X)$. In the language of sections we can express the space $\Omega^p(X)$ of complex valued forms as $\Omega^p(X) = \Gamma(X, \wedge^p T^* X_{\mathbf{c}})$, where $T^* X_{\mathbf{c}} = T^* X \otimes_{\mathbf{R}} \mathbf{C}$, that is to say the complexification of the cotangent bundle. It follows that the associated symbol complex is given by

$$\cdots \xrightarrow{\sigma_1(d_{p-1})} \pi^*(\wedge^p T^* X_{\mathbf{c}}) \xrightarrow{\sigma_1(d_p)} \pi^*(\wedge^{p+1} T^* X_{\mathbf{c}}) \xrightarrow{\sigma_1(d_{p+1})} \cdots \qquad (4.50)$$

and this complex is exact by the same argument that we gave for 4.20. We know, therefore, that the de Rham complex is an example of an elliptic complex. Its underlying elliptic operator is the map from the even forms to the odd forms given by restricting $d + d^*$ to the even forms—cf. the general explanation in 4.9. For convenience we shall denote it by d and so we have

$$d : (\Omega^0(X) \oplus \Omega^2(X) \oplus \cdots) \longrightarrow (\Omega^1(X) \oplus \Omega^3(X) \oplus \cdots) \qquad (4.51)$$

and since $H^i(E) = H^i(\wedge^* T^* X_{\mathbf{c}}) = H^i_{de\ Rham}(X)$

$$index\ d = \sum (-1)^i \dim_{\mathbf{c}} H^i_{de\ Rham}(X) = \sum (-1)^i b_i = \chi(X) \qquad (4.52)$$

where b_i are the Betti numbers [3] and $\chi(X)$ is the usual Euler–Poincaré characteristic of X.

On the other hand we shall now show that if we use the index formula 4.48 to calculate $index\ d$ we get

$$index\ d = \int_X e(X) \qquad (4.53)$$

where $e(X)$ is the Euler class of X. Thus the index theorem applied to the de Rham complex gives the famous Gauss–Bonnet theorem

$$\int_X e(X) = \chi(X) \qquad (4.54)$$

To obtain this result is an instructive exercise in manipulating characteristic classes. Turning to the index formula 4.48 and the symbol complex it is evident that we have to compute

$$ch\left([\sigma_1(d)]\right) = ch\left(\sum_p (-1)^p [E^p]\right), = \sum_p (-1)^p ch\left(E^p\right), \text{ with } E^p = \wedge^p T^* X_{\mathbf{c}}$$

$$(4.55)$$

---

[3] Because we used complex valued forms we have $H^i_{de\ Rham}(X) \simeq H^i(X; \mathbf{C})$; as a consequence the Betti number b_i denotes $\dim_{\mathbf{c}} H^i(X; \mathbf{C})$ the *complex* dimension of $H^i(X; \mathbf{C})$, we still agree with the usual definition since it is clear that we can write $b_i = \dim H^i(X; \mathbf{R}) = \dim_{\mathbf{c}} H^i(X; \mathbf{C})$.

To compute $ch\left(\wedge^p T^* X_{\mathbf{c}}\right)$ we employ the splitting principle. Suppose, to begin with, that a vector bundle F is a sum of n line bundles $L_1, L_2, \ldots L_n$. If we now wish to compute $\wedge^p F$ then it is a standard linear algebraic fact that

$$F = L_1 \oplus \cdots \oplus L_n \Rightarrow \wedge^p F = \bigoplus_{1 \leq i_1 < i_2 < \cdots < i_p \leq n} \left(L_{i_1} \otimes \cdots \otimes L_{i_p}\right) \quad (4.56)$$

Applying the Chern character to $\wedge^p F$ and using its properties 3.98 under direct sum and tensor product gives

$$ch\left(\wedge^p F\right) = \sum_{1 \leq i_1 < i_2 < \cdots < i_p \leq n} ch\left(L_{i_1}\right) ch\left(L_{i_2}\right) \cdots ch\left(L_{i_p}\right) \quad (4.57)$$

But for a line bundle L_i we have $ch\left(L_i\right) = e^{x_i}$, where $x_i = c_1(L_i)$ so that we obtain

$$\begin{aligned} ch\left(\wedge^p F\right) &= \sum_{1 \leq i_1 < i_2 < \cdots < i_p \leq n} e^{x_{i_1}} e^{x_{i_2}} \cdots e^{x_{i_p}} \\ &= \sum_{1 \leq i_1 < i_2 < \cdots < i_p \leq n} e^{\left(x_{i_1} + \cdots + x_{i_p}\right)} \end{aligned} \quad (4.58)$$

Finally we need an expression for $\sum_p (-1)^p ch\left(\wedge^p F\right)$, which is given by

$$\begin{aligned} \sum_{p=0}^n (-1)^p ch\left(\wedge^p F\right) &= \sum_{p=0}^n (-1)^p \sum_{1 \leq i_1 < i_2 < \cdots < i_p \leq n} e^{\left(x_{i_1} + \cdots + x_{i_p}\right)} \\ &= \prod_{i=1}^n \left(1 - e^{x_i}\right) \end{aligned} \quad (4.59)$$

But we should now notice that the RHS of 4.59 is reminiscent of the expression that defines the Todd class. Let us recall that the Todd class of a bundle E of rank k was defined in 3.104 by

$$td\left(E\right) = \prod_{i=1}^k \frac{x_i}{1 - e^{-x_i}} \Rightarrow td(E^*) = (-1)^k \prod_{i=1}^k \frac{x_i}{1 - e^{x_i}}$$

where we take advantage of the occasion to quote $td\left(E^*\right)$ as well—the expression for $td\left(E^*\right)$ follows at once from the splitting principle because taking the dual simply changes the sign of the x_i. If we use this definition, together with the fact that the splitting principle says that the top

Chern class $c_n(F)$ of F is given by $c_n(F) = x_1 x_2 \cdots x_n$, meaning also that $c_n(F^*) = (-1)^n x_1 x_2 \cdots x_n$, then we find that

$$\sum_{p=0}^{n} (-1)^p ch\left(\wedge^p F\right) = \prod_{i=1}^{n} (1 - e^{x_i})$$

$$= (-1)^n x_1 x_2 \cdots x_n (-1)^n \prod_{i=1}^{n} \frac{(1 - e^{x_i})}{x_i} \qquad (4.60)$$

$$\Rightarrow \sum_{p=0}^{n} (-1)^p ch\left(\wedge^p F\right) = \frac{c_n(F^*)}{td(F^*)}$$

This completes our characteristic class manipulations. All that remains is for us to return to the de Rham calculation and apply 4.60 and 4.48 with $F = T^* X_{\mathbf{c}}$—the distinction between F and F^* will not matter now since F is the complexification of a real bundle and we saw in 3.108 of chapter 3 that this implies $F = F^*$. When we do this we discover that

$$index\ d = (-1)^{n/2} \frac{c_n(TX_{\mathbf{c}})}{e(X)\, td(TX_{\mathbf{c}})}\, td(TX_{\mathbf{c}})[X]$$

$$= (-1)^{n/2} \frac{c_n(TX_{\mathbf{c}})}{e(X)}[X] \qquad (4.61)$$

However, bearing in mind the relation 3.124 between the Euler class and the top Chern class, we have $c_n(TX_{\mathbf{c}}) = (-1)^{n/2} e(TX \oplus TX) = (-1)^{n/2} e^2(X)$ so that we obtain immediately

$$index\ d = (-1)^{n/2}(-1)^{n/2} \frac{e^2(X)}{e(X)}[X] = e(X)[X] = \int_X e(X) = \chi(X)$$

$$(4.62)$$

where we have represented the evaluation of the Euler class on $[X]$ as an integral over the Euler form for which we use the same symbol $e(X)$. We have thus accomplished our task of showing that the index theorem applied to the de Rham complex is equivalent to the Gauss–Bonnet theorem.

Example *The Dolbeault complex*

Let us now suppose that X is a complex manifold of complex dimension $n/2$ so that n is even as usual. We saw in chapter 3 that the Dolbeault complex is a kind of complex-analytic version of the de Rham complex with the operator $\bar{\partial}$ substituted for the exterior derivative d. The Dolbeault complex and their attendant cohomology groups are given by

$$\cdots \xrightarrow{\bar{\partial}_{p,q-1}} \mathcal{E}^{p,q} \xrightarrow{\bar{\partial}_{p,q}} \mathcal{E}^{p,q+1} \longrightarrow \cdots$$

$$H_{\bar{\partial}}^{p,q}(M) = \frac{ker\left(\bar{\partial} : \mathcal{E}^{p,q} \longrightarrow \mathcal{E}^{p,q+1}\right)}{Im\left(\bar{\partial} : \mathcal{E}^{p,q-1} \longrightarrow \mathcal{E}^{p,q}\right)} \qquad (4.63)$$

It is also easy to verify that the symbol complex is exact. The above is actually a separate complex for each value of p. It is common practice to set $p = 0$. Let us do this; then we can write the index of the resulting complex in terms of the index of the appropriate elliptic operator. It is natural to denote this operator by $\bar{\partial}$ so that we have

$$index\, \bar{\partial} = \sum_q (-1)^q \dim H_{\bar{\partial}}^{0,q}(X) = \sum_q (-1)^q h_{0,q} \qquad (4.64)$$

where $h_{p,q} = \dim H_{\bar{\partial}}^{p,q}(X)$. So if we denote the Dolbeault complex by $\mathcal{E}$ we have found that its Euler characteristic is given by

$$index\, \bar{\partial} = \sum_q (-1)^q h_{0,q} = \chi(\mathcal{E}) \qquad (4.65)$$

The integer $\chi(\mathcal{E})$ is also called the *arithmetic genus* of the complex manifold X. Turning to the index formula 4.48 allows us to express $\chi(\mathcal{E})$ as the integer obtained by evaluating a cohomology class. This is a calculation very like the one just carried out for the de Rham complex.

The symbol complex for the Dolbeault case is

$$\cdots \xrightarrow{\sigma_1(\bar{\partial}_{0,q-1})} \pi^*(\wedge^{0,q}T^*X) \xrightarrow{\sigma_1(\bar{\partial}_{0,q})} \pi^*(\wedge^{0,q+1}T^*X) \xrightarrow{\sigma_1(\bar{\partial}_{0,q+1})} \cdots \qquad (4.66)$$

where $\wedge^{0,q}T^*X$ is the bundle whose sections are forms ω, say, of type $(0,q)$: a typical ω in local coordinates is written as $\omega = \omega_{i_1 \cdots i_q} d\bar{z}^{i_1} \wedge \cdots \wedge d\bar{z}^{i_q}$; this means that $\wedge^{0,q}T^*X$ is the qth anti-symmetric power of the bundle $\overline{T^*X}$, i.e. the *anti-holomorphic* cotangent bundle. Since the conjugate and the dual bundle of a complex vector bundle coincide it is clear that $\overline{T^*X}$ is isomorphic to the holomorphic tangent bundle TX, i.e. the bundle whose sections are of the form $a_i(z_1, \ldots z_n)\partial/\partial z_i$. For the index calculation we need to compute

$$ch\,(\sigma_1(\bar{\partial})) = \sum_{q=0}^{n/2} (-1)^p ch\,(\wedge^q F) = \frac{c_{n/2}(F^*)}{td\,(F^*)}, \quad \text{with } F = TX \qquad (4.67)$$

Still maintaining $F = TX$ we get

$$index\, \bar{\partial} = (-1)^{n/2} \frac{c_{n/2}(F^*)}{e(X)\, td\,(F^*)}\, td\,(TX_{\mathbf{c}})[X] \qquad (4.68)$$

But

$$c_{n/2}(F^*) = (-1)^{n/2}c_{n/2}(F) = (-1)^{n/2}e(X),$$

$$\text{and} \quad td\,(TX_\mathbf{c}) = td\,(TX \oplus \overline{TX}\,) \qquad\qquad (4.69)$$

$$= td\,(TX)\,td\,(\overline{TX}\,) = td\,(F)\,td\,(F^*)$$

so that the index reduces to

$$index\,\overline{\partial} = (-1)^{n/2}\frac{(-1)^{n/2}e(X)}{e(X)\,td\,(F^*)}td\,(F)\,td\,(F^*)[X] = td\,(F)[X] \qquad (4.70)$$

We follow the usual convention and abbreviate $td\,(TX)$ by $td\,(X)$. Thus we obtain

$$index\,\overline{\partial} = td\,(X)[X] \qquad\qquad (4.71)$$

Finally, if we combine this with the calculation $index\,\overline{\partial} = \chi(\mathcal{E})$ computed above, we have

$$\int_X td\,(X) = \chi(\mathcal{E}) \qquad\qquad (4.72)$$

and this is the seminal *Riemann–Roch theorem* proved now for an *arbitrary* complex manifold. Note that we have again represented the evaluation of the relevant characteristic class as an integral in the standard fashion.

If we wish we can ignore the complex-analyticity of X and apply the de Rham complex to calculate its Euler characteristic $\chi(X)$. Since $d = \partial + \overline{\partial}$ then the usual Hodge theory arguments show directly that the classical Betti numbers b_q and the integers $h_{p,q}$ (which we shall now call Hodge numbers) are related by

$$\chi(X) = \sum(-1)^q b_q = \sum_{p,q}(-1)^{p+q}h_{p,q} \qquad\qquad (4.73)$$

thus showing that the Hodge numbers $h_{p,q}$ are a refinement of the Betti numbers b_q.

Example *The signature complex*

The signature complex is defined when the manifold X is orientable and has even dimension, but before writing down the complex itself we have something to say about the signature $Sign\,(X)$ of X.

Since $n = \dim X$ is even then set $n = 2l$. This means that, in the sequence of cohomology groups $H^0(X;\mathbf{Z}), H^1(X;\mathbf{Z}),\ldots H^{2l}(X;\mathbf{Z})$, there is a *middle* cohomology group $H^l(X;\mathbf{Z})$. This fact allows us to use the cohomology cup product to provide a pairing between elements of the middle cohomology. That is we take $u, v \in H^l(X;\mathbf{Z})$ and form $u \cup v$ which belongs to $H^{2l}(X; \mathbf{Z})$. If we evaluate $u \cup v$ on the generator of $H_{2l}(X; \mathbf{Z})$ then this defines the quadratic form $q(u, v)$ over the integers given by

$$q(u, v) = (u \cup v)[X], \qquad u, v \in H^l(M; \mathbf{Z}) \tag{4.74}$$

(For the case dim $X = 4$ we have already met $q(u, v)$ in chapter 1.) The form q is non-degenerate and so has only non-zero eigenvalues. Let the number of its positive and negative eigenvalues be b^+ and b^- respectively. The difference $b^+ - b^-$ is called the signature of X and is denoted by $Sign(X)$. That is

$$Sign(X) = b^+ - b^- \tag{4.75}$$

Now, if l is *odd* then $q(u, v)$ is anti-symmetric since the standard symmetry properties of the cup product (which are identical to those of differential forms because of the existence of de Rham theory) give

$$q(u, v) = u \cup v = (-1)^{l^2} v \cup u = (-1)^{l^2} q(v, u) \tag{4.76}$$

But a real anti-symmetric form has no positive or negative eigenvalues, thus $Sign(X)$ is automatically zero for l odd. From now on we assume that l is even, which means that n is divisible by 4.

The signature $Sign(X)$ will change sign if the orientation of X is reversed, thus $Sign(X)$ is a topological invariant of an *oriented* X. $Sign(X)$ can be expressed as the index of an appropriately chosen elliptic operator. To describe this operator we return to the complex valued forms $\Omega^p(X)$ of the de Rham example. To define the signature complex we wish to introduce an operator τ which maps $\Omega^p(X)$ into $\Omega^{n-p}(X)$; τ is actually just a certain multiple of the Hodge star $*$. We define τ as the map

$$\begin{aligned} \tau : \Omega^p(X) &\longrightarrow \Omega^{n-p}(X) \\ \omega &\longmapsto i^{(p(p-1)+n/2)} * \omega, \qquad \omega \in \Omega^p(X) \end{aligned} \tag{4.77}$$

The Hodge $*$, for complex valued differential forms, is defined via the *Hermitian* inner product by

$$< \omega, \nu > = \int_X \omega \wedge \overline{*\nu}, \qquad \omega, \nu \in \Omega^p(X)$$

This implies that $*^2 = (-1)^p$ which in its turn implies that $\tau^2 = 1$. To check this set $\lambda(p) = (p(p-1) + n/2)$ and observe that, if $\omega \in \Omega^p(X)$, then

$$\begin{aligned} \tau^2 \omega &= i^{\lambda(p)} \tau(*\omega) = i^{\lambda(p)+\lambda(n-p)} *^2 \omega \\ &= i^{2p} *^2 \omega = i^{2p}(-1)^p \omega = \omega \end{aligned} \tag{4.78}$$

Since $\tau^2 = 1$ τ has eigenvalues ∓ 1 and we wish to decompose the exterior algebra into the corresponding pair of eigenspaces. Thus we write

$$\Omega^*(X) = \bigoplus_p \Omega^p(X) = \Omega^+ \oplus \Omega^- \tag{4.79}$$

Next we need an elliptic operator whose index gives the signature $Sign\,(X)$. The appropriate choice turns out to be a restriction of $d + d^*$ to a well chosen subspace of Ω^*. To find this subspace notice that $\tau d_+ = -d_+\tau$. This anti-commuting property means that $d + d^*$ maps Ω^+ into Ω^- and vice versa. Thus we define the operator d_+ as the restriction of $d + d^*$ to Ω^+. That is, we have

$$d_+ : \Omega^+(X) \longrightarrow \Omega^-(X) \tag{4.80}$$

The adjoint of d_+ is easily seen to d_-, by which we mean the restriction of $d + d^*$ to Ω^-. Recall that $\Omega^p(X)$ are the sections of the bundle $\wedge^p T^* X \otimes_{\mathbf{R}} \mathbf{C}$; in the same way the spaces $\Omega^{\mp}$ are the sections of two bundles which we denote by $\wedge^{\mp} T^* X \otimes_{\mathbf{R}} \mathbf{C}$. With this notation the signature complex is thus the two term complex

$$0 \longrightarrow \Gamma(X, \wedge^+ T^* X \otimes_{\mathbf{R}} \mathbf{C}) \xrightarrow{\ d_+\ } \Gamma(X, \wedge^- T^* X \otimes_{\mathbf{R}} \mathbf{C}) \longrightarrow 0 \tag{4.81}$$

and its associated symbol complex is

$$0 \longrightarrow \pi^* \wedge^+ T^* X \otimes_{\mathbf{R}} \mathbf{C} \xrightarrow{\ \sigma_1(d_+)\ } \pi^* \wedge^- T^* X \otimes_{\mathbf{R}} \mathbf{C} \longrightarrow 0 \tag{4.82}$$

Finally we have to show that the index of this complex is $Sign\,(X)$.

Since d_+ is elliptic we can use Hodge theory to identify the cohomology of the complex with the harmonic sections. Let

$$H^{\mp}(X; \mathbf{C}) = \frac{ker\,d_{\mp}}{Im\,d_{\mp}} \quad \text{and} \quad \dim H^{\mp}(X; \mathbf{C}) = h_{\mp}$$
$$H^*(X; \mathbf{C}) = H^+(X; \mathbf{C}) \oplus H^-(X; \mathbf{C}) \tag{4.83}$$

Then the index of the complex is given by simply

$$index\,d_+ = h_+ - h_- \tag{4.84}$$

We need to refine somewhat the decomposition of $H^*(X; \mathbf{C})$ into the spaces $H^{\mp}(X; \mathbf{C})$. To accomplish this consider the spaces I^p where

$$I^p = H^p(X; \mathbf{C}) \oplus H^{n-p}(X; \mathbf{C}), \quad 0 \le p < n/2 \tag{4.85}$$

Poincaré duality asserts that the two summands $H^p(X;\mathbf{C})$ and $H^{n-p}(X;\mathbf{C})$ have the same dimension; also I^p is invariant under τ and the action of τ on I^p is simply to exchange $H^p(X;\mathbf{C})$ with $H^{n-p}(X;\mathbf{C})$. The upshot of this is that I^p decomposes into two eigenspaces $I^p_{\pm}$ according to

$$I^p = I^p_+ \oplus I^p_-, \quad \text{and} \quad \dim I^p_+ = \dim I^p_- \tag{4.86}$$

Notice that I^p is only defined for $0 \leq p < n/2$; if $p = n/2$ the natural object is just $H^{n/2}(X;\mathbf{C})$ itself. Let us decompose $H^{n/2}(X;\mathbf{C})$ into the subspaces on which τ is positive or negative. To this end let

$$H^{n/2}(X;\mathbf{C}) = H^{n/2}_+(X;\mathbf{C}) \oplus H^{n/2}_-(X;\mathbf{C}) \quad \text{and} \quad \dim H^{n/2}_{\mp}(X;\mathbf{C}) = h^{n/2}_{\mp} \tag{4.87}$$

By representing the cup product as the wedge product it is possible to check that the signature $Sign(X)$ of the quadratic form q is just the difference $h^{n/2}_+ - h^{n/2}_-$. Putting together these various decompositions we can write

$$\begin{aligned} H^*(X;\mathbf{C}) &= H^0(X;\mathbf{C}) \oplus \cdots \oplus H^n(X;\mathbf{C}) \\ &= I^0 \oplus \cdots \oplus I^{n/2-1} \oplus H^{n/2}(X;\mathbf{C}) \\ &= I^0_+ \oplus I^0_- \cdots \oplus I^{n/2-1}_+ \oplus I^{n/2-1}_- \oplus H^{n/2}_+(X;\mathbf{C}) \oplus H^{n/2}_-(X;\mathbf{C}) \\ &= H^+(X;\mathbf{C}) \oplus H^-(X;\mathbf{C}) \end{aligned}$$

$$\Rightarrow h_{\mp} = h^{n/2}_{\mp} + \sum_{p=0}^{n/2-1} \dim I^p_{\mp} \tag{4.88}$$

Then, since $\dim I^p_+ = \dim I^p_-$, we can immediately verify that the index of d_+ is given by

$$index\, d_+ = h_+ - h_- = h^{n/2}_+ - h^{n/2}_- = Sign(X) \tag{4.89}$$

Having shown that the index of d_+ is $Sign(X)$ we wish to calculate $index\, d_+$ using the index theorem. Since the symbol complex for the signature is just the two term complex 4.82 above, this means that the usual alternating sum has reduced to just two terms and we just have to compute

$$ch(\sigma_1(d_+)) = ch(\wedge^+ T^*X \otimes_{\mathbf{R}} \mathbf{C}) - ch(\wedge^- T^*X \otimes_{\mathbf{R}} \mathbf{C}) \tag{4.90}$$

This computation (cf. Atiyah and Singer [3]) requires the combining of the splitting principle with a little group theory for the group $SO(n)$. The result is that

$$ch(\wedge^+ T^*X \otimes_{\mathbf{R}} \mathbf{C}) - ch(\wedge^- T^*X \otimes_{\mathbf{R}} \mathbf{C}) = \prod_{i=1}^{n/2}(e^{-x_i} - e^{x_i}) \tag{4.91}$$

We now insert this into the index formula 4.48 while also applying the splitting principle to express $e(X)$ as $e(X) = x_1 \cdots x_{n/2}$. For the term in the Todd class we use $td\,(TX_{\mathbf{c}}) = td\,(TX)\,td\,(\overline{TX}\,)$. In this way we obtain

$$
\begin{aligned}
index\,d_+ &= (-1)^{n/2} \left\{ \prod_{i=1}^{n/2} \left(\frac{e^{-x_i} - e^{x_i}}{x_i} \frac{x_i}{1 - e^{x_i}} \frac{-x_i}{1 - e^{-x_i}} \right) \right\} [X] \\
&= \prod \frac{x_i(e^{x_i} + 1)}{(e^{x_i} - 1)} [X] \\
&= \prod \frac{x_i \cosh(x_i/2)}{\sinh(x_i/2)} [X] = 2^{n/2} \prod \frac{x_i/2}{\tanh(x_i/2)} [X]
\end{aligned}
\tag{4.92}
$$

The polynomial $\prod (x_i/2)/\tanh(x_i/2)$ is denoted by $\mathcal{L}$. Thus our result can be summarised as

$$
index\,d_+ = 2^{n/2} \mathcal{L}(X)[X] \tag{4.93}
$$

We shall denote the integer $2^{n/2}\mathcal{L}(X)[X]$ by $L(X)$; it is known as the L-genus of X.

There is some variation in the way the L-polynomial is defined: an alternative definition (Hirzebruch [1]) is to replace the polynomial $\mathcal{L}$ by $\prod x_i/\tanh(x_i)$; it is easy to see that if one expands this latter expression to order $n/2$ then its $n/2$-order term coincides with ours. Since it is precisely this term which is evaluated in $L(X)$ above, then either definition would do there. On the other hand, any use of the lower order terms would be sensitive to which definition was used, and, in such situations, one of the definitions has to be discarded. We shall illustrate this point in §3 below.

Thus our application of the index theorem to the operator d_+ has established the following result.

Theorem (Hirzebruch signature theorem) *Let X be a compact oriented manifold of dimension n where n is divisible by 4. Then $Sign\,(X)$, the signature of X, is given by*

$$
L(X) = Sign\,(X) \tag{4.94}
$$

As before if we represent the characteristic classes using curvature forms then we have the usual formula

$$
\int_X 2^{n/2}\mathcal{L}(X) = Sign\,(X) \tag{4.95}
$$

where the symbol $\mathcal{L}(X)$ when placed under an integral sign stands for the same characteristic class expressed in terms of curvature.

Example *The Dirac operator*

We now come to the calculation of the index of the Dirac operator $\displaystyle{\not{D}}$. This is an elliptic differential operator which acts on spinors and to describe it we introduce the spin complex over a spin-manifold X.

Spinors first arise when one encounters the universal cover $Spin(n)$ of the orthogonal group $SO(n)$. If X is an oriented n-dimensional manifold then the structure group of its tangent bundle TX reduces to $SO(n)$; if, in addition, this structure group can be lifted to the univeral cover $Spin(n)$ then X is said to be a *spin-manifold* and the particular choice of lifting is called a *spin structure*. Thus the existence of a spin structure requires of X two things: X should be orientable and the structure group of its tangent bundle lifts to $Spin(n)$. These two requirements are met by the vanishing of the first two Stiefel–Whitney classes $w_1(X)$ and $w_2(x)$ of TX. In fact we have

$$w_1(X) = 0 \Leftrightarrow \ X \text{ is orientable}$$
$$w_2(X) = 0 \Leftrightarrow \ \text{spinors are globally defined on } X \tag{4.96}$$

Each choice of a spin structure determines an element of $H^1(X; \mathbf{Z}_2)$, and vice versa, so that the set of spin structures are parametrised by $H^1(X; \mathbf{Z}_2)$. In the event of $H^1(X; \mathbf{Z}_2)$ vanishing the spin structure is unique—for example, such is the case if X is simply connected. On the other hand, if $X = \mathbf{C}P^n$, then $w_2(\mathbf{C}P^n) = 0 \iff n$ is odd. Thus only 'half' of the complex projective spaces are spin-manifolds.

There is a well known basic representation of $Spin(n)$ called the spin representation, we denote its representation space by S. This representation has dimensions $2^{n/2}$ and $2^{(n-1)/2}$ for n even and odd respectively. For odd n the representation S is irreducible but for n even S is reducible into two irreducible components of equal dimension. Denoting these components by $S^{\mp}$ we have

$$S = S^+ \oplus S^-, \ \text{ where } \dim S = 2^{n/2} \text{ and } \dim S^{\mp} = 2^{(n/2-1)} \tag{4.97}$$

When n is even, which we shall assume from now on, spinors on X are built out of S^+ and S^-: one starts with the principal $SO(n)$-bundle P, say, of TX. Next we lift P to its double cover, which we denote by $\widetilde{P}$. Then the spinor bundles E and $E^{\mp}$ are the associated complex vector bundles defined by

$$E^+ = \widetilde{P} \times_{Spin(n)} S^+, \quad E^- = \widetilde{P} \times_{Spin(n)} S^-, \quad E = E^+ \oplus E^- \tag{4.98}$$

Sections of E and $E^{\mp}$ are called spinors over X.

The construction of the Dirac operator requires the famous Dirac γ-matrices. These are a set of n anti-commuting $2^{n/2} \times 2^{n/2}$ matrices which generate the Clifford algebra of $Spin(n)$. Central to the index problem for the Dirac operator is the chirality operator (for whose existence we need n to be even) which as usual we denote by γ_5; we give its definition below together with the anti-commutation relation of the γ-matrices

$$\gamma_a \gamma_b + \gamma_b \gamma_a \equiv \{\gamma_a, \gamma_b\} = -2\delta_{ab}I$$
$$\gamma_5 = i^{n(n+1)/2}\gamma_1 \cdots \gamma_n, \quad \text{note } \gamma_5^2 = I, \quad \text{and} \quad \{\gamma_5, \gamma_\mu\} = 0 \tag{4.99}$$

The γ-matrices act on the sections of E, and the composition of this Clifford multiplication with covariant differentiation is the Dirac operator $\not{D}$. More precisely, introduce the usual orthonormal frame (vielbein) of tangent vectors $e_a^\mu(x)$ at x, and let ω denote the Riemannian spin connection on X with $\partial + \omega$ the corresponding covariant derivative. The *Dirac operator* $\not{D}$ is then given by the map

$$\not{D} : \Gamma(X, E) \longrightarrow \Gamma(X, E) \text{ with } \not{D} = \gamma^a e_a^\mu(x)(\partial_\mu + \omega_\mu)$$
$$\text{and } e_\mu^a(x) \text{ satisfies } e_\mu^a(x)e_\nu^b(x)\delta_{ab} = g_{\mu\nu}(x) \tag{4.100}$$

where $g_{\mu\nu}(x)$ are the components of the Riemannian metric on X. We employ the standard conventions that Latin and Greek letters label flat space and curved space indices respectively.

The chirality operator γ_5 anti-commutes with $\not{D}$ and satisfies $\gamma_5^2 = 1$ so that, with respect to the fibrewise decomposition $E_x = E_x^+ \oplus E_x^-$, the chirality operator is block diagonal of the form

$$\gamma_5 = \begin{pmatrix} I & 0 \\ 0 & -I \end{pmatrix} \tag{4.101}$$

Thus the restriction to E^+ can be achieved by multiplying expressions involving γ-matrices by the projection operator $(1 + \gamma_5)/2$. In any case the *chiral* Dirac operator $\not{\partial}$ is given by $\not{\partial} = \not{D}(1 + \gamma_5)/2$ and is a differential operator connecting the sections of E^+ and E^-. We define the spin complex as the two term complex which records this fact; writing down this spin complex, together with its associated symbol complex, we have

$$0 \longrightarrow \Gamma(X, E^+) \overset{\not{\partial}}{\longrightarrow} \Gamma(X, E^-) \longrightarrow 0$$
$$0 \longrightarrow \pi^* E^+ \overset{\sigma_1(\not{\partial})}{\longrightarrow} \pi^* E^- \longrightarrow 0 \tag{4.102}$$

The Dirac operator is elliptic because its symbol is given locally by the matrix $\sigma_1^{\partial}(x,p) = i\gamma^a e_a^\mu p_\mu$ and the inverse of this matrix is $i\gamma^a e_a^\mu p_\mu/p^2$, which exists for all $p \neq 0$; note that we need to use the Euclidean signature of the metric here: in a space of Lorentzian signature the Dirac operator is not elliptic. If we interchange E^+ and E^- in the definition of $\partial\!\!\!/$ we obtain a second elliptic operator, which we denote by $\partial\!\!\!/^*$. It is clear that, with respect to the natural inner product on the fibres of $E^\mp$, $\partial\!\!\!/^*$ is the adjoint of $\partial\!\!\!/$. The index of the spin complex is given by

$$index\ \partial\!\!\!/ = \dim ker\ \partial\!\!\!/ - \dim ker\ \partial\!\!\!/^* \tag{4.103}$$

If a spinor ψ satisfies $\partial\!\!\!/\psi = 0$ then we shall call it a *harmonic spinor*. Let us denote the space of harmonic spinors by Ψ. We can decompose Ψ into its subspaces Ψ^+ and Ψ^- of positive and negative chirality respectively, giving

$$\Psi = \Psi^+ \oplus \Psi^- \tag{4.104}$$

Also let us define $n^\mp$ by $n^\mp = \dim \Psi^\mp$. If we make our usual use of Hodge theory then we have

$$ker\ \partial\!\!\!/ = \Psi^+, \quad ker\ \partial\!\!\!/^* = \Psi^- \tag{4.105}$$

Then we can write

$$index\ \partial\!\!\!/ = n^+ - n^- \tag{4.106}$$

Thus *index* $\partial\!\!\!/$ can also be thought of as measuring the difference between the numbers of harmonic spinors of positive and negative chirality respectively.

Finally we wish to find a formula for *index* $\partial\!\!\!/$ in terms of characteristic classes. As usual the object to compute is the appropriate Chern character. In this case some properties of the characters of the maximal torus of $Spin(n)$ are required. The details are in Atiyah and Singer [3] and the result is that

$$ch\left(\sigma_1(\partial\!\!\!/)\right) = ch\left(E^+\right) - ch\left(E^-\right) = \prod_{i=1}^{n/2}(e^{x_i/2} - e^{-x_i/2}) \tag{4.107}$$

and on inserting this into the index formula we obtain

$$index\ \partial\!\!\!/ = (-1)^{n/2}\left\{\prod_{i=1}^{n/2}\left(\frac{e^{x_i/2} - e^{-x_i/2}}{x_i}\frac{x_i}{1 - e^{x_i}}\frac{-x_i}{1 - e^{-x_i}}\right)\right\}[X]$$

$$= (-1)^{n/2}\left\{\prod_{i=1}^{n/2}\left(\frac{x_i/2}{\sinh(x_i/2)}\right)\right\}[X] \tag{4.108}$$

We now define the $\hat{A}$ polynomial by

$$\hat{A} = \prod_{i=1}^{n/2} \left(\frac{x_i/2}{\sinh(x_i/2)} \right) = det \left(\frac{x_i/2}{\sinh(x_i/2)} \right) \tag{4.109}$$

The $\hat{A}$ polynomial has an expansion in terms of Pontrjagin classes given by

$$\hat{A}(X) = 1 - \frac{1}{24}p_1(X) + \frac{1}{5760}(7p_1^2 - p_2)(X) + \cdots \tag{4.110}$$

Note that the presence of the Pontrjagin classes means that $index\ \partial\!\!\!/ = 0$ unless n is divisible by 4; and if n *is* divisible by 4 then the factor $(-1)^{n/2}$ may be omitted. We shall denote the integer $\hat{A}[X]$ by $\hat{A}(X)$; it is known as the $\hat{A}$-genus of X. The index theorem for the Dirac operator can now be stated in the more concise form

$$index\ \partial\!\!\!/ = \begin{cases} \hat{A}(X) & \text{if } \dim X \equiv 0 \bmod 4 \\ 0 & \text{otherwise} \end{cases} \tag{4.111}$$

§ 3. Twisted complexes

The examples of the previous section can be generalised to include elliptic operators with coefficients in an auxiliary bundle. This is a simple idea and we shall illustrate it by taking some examples.

Return to the de Rham complex which we used to compute $index\ d$ where d is the restriction of $d + d^*$ to even forms. The de Rham complex is

$$\cdots \xrightarrow{d_{p-2}} \Omega^{p-1}(X) \xrightarrow{d_{p-1}} \Omega^p(X) \xrightarrow{d_p} \Omega^{p+1}(X) \xrightarrow{d_{p+1}} \cdots \tag{4.112}$$

where $\Omega^p(X) = \Gamma(X, \wedge^p T^* X_{\mathbf{c}})$. Now in physical applications we often need to consider *matrix valued* differential forms. These will be sections, not of $\wedge^p T^* X_{\mathbf{c}}$, but of $F \otimes \wedge^p T^* X_{\mathbf{c}}$ where F is a vector bundle of rank m (m gives the size of the matrices in question). If we introduce the notation $\Omega^p(X; F) = \Gamma(X, F \otimes \wedge^p T^* X_{\mathbf{c}})$ then we obtain the twisted de Rham complex

$$\cdots \xrightarrow{d_{p-2}} \Omega^{p-1}(X; F) \xrightarrow{d_{p-1}} \Omega^p(X; F) \xrightarrow{d_p} \Omega^{p+1}(X; F) \xrightarrow{d_{p+1}} \cdots \tag{4.113}$$

Once we have obtained this twisted de Rham complex the calculation of the index can proceed just as before. The index of this new complex gives the index of $d + d^*$ restricted to even forms with coefficients in F. We denote this new operator by d_F while continuing to denote its untwisted version by d. The formula for $index\, d_F$ is closely related to that for $index\, d$. This relationship is easy to discover. Remember that the formula we used to calculate $index\, d$ was

$$index\, d = (-1)^{n/2}\frac{ch\left(\sum_p(-1)^p[E^p]\right)}{e(X)}\cdot td\,(TX_{\mathbf{c}})[X], \quad \text{with } E^p = \wedge^p T^*X_{\mathbf{c}}$$

(4.114)

The only change needed on passing from d to d_F is that E^p is replaced by $F \otimes E^p$. We can immediately calculate $index\, d_F$ and the result is that

$$
\begin{aligned}
index\, d_F &= (-1)^{n/2}\frac{ch\left(\sum_p(-1)^p[F\otimes E^p]\right)}{e(X)}\cdot td\,(TX_{\mathbf{c}})[X] \\
&= (-1)^{n/2}ch\,(F)\frac{ch\left(\sum_p(-1)^p[E^p]\right)}{e(X)}\cdot td\,(TX_{\mathbf{c}})[X]
\end{aligned}
$$

(4.115)

where we have used the property $ch\,(A \otimes B) = ch\,(A)ch\,(B)$ of the Chern character. We see that the new bundle F enters multiplicatively in a relatively uncomplicated way. This has the elementary consequence that the calculation of $index\, d_F$ can proceed in an almost identical manner to that of $index\, d$, providing us with the result that

$$index\, d_F = ch\,(F)e(X)[X] = \int_X ch\,(F) \wedge e(X) \tag{4.116}$$

However, because the Euler class $e(X)$ is an n-dimensional class then the product $ch\,(F)e(X)$ is rather simple—all the higher terms in the expansion $ch\,(F) = m + c_1(F) + \cdots$ give contributions above dimension n and so must vanish. The final formula for $index\, d_F$ is therefore just

$$index\, d_F = m\int_X e(X) = m(index\, d), \quad \text{where } m = rk\,(F) \tag{4.117}$$

We move on to the other examples where the same technique is used but with more interesting results. If we take the Dolbeault complex then E^p is now $\wedge^{0,p}T^*X$ where X is now a complex manifold. To twist the Dolbeault complex we consider matrix valued complex forms of type $(0,p)$; we met these already in chapter 3 where they were called the sheaf of holomorphic p-forms with coefficients in F and were denoted by $\Omega^p(F)$. The twisting causes E^p to be replaced by $F \otimes E^p$. The index of the twisted Dolbeault operator $\overline{\partial}_F$ is easily verified to be

$$index\, \overline{\partial}_F = ch\,(F)td\,(X)[X] = \int_X ch\,(F) \wedge td(X) = \chi(X,F) \tag{4.118}$$

where the integer $\chi(X, F)$ is the Euler characteristic of the sheaf of holomorphic sections of F. In this case since both $td\,(X)$ and $ch\,(F)$ are polynomials then $ch\,(F)$ makes its presence felt in a non-trivial way. For example, let X have complex dimension 2 and use the expansion

$$ch\,(F)td\,(X) = (rk\,(F) + c_1(F) + \frac{1}{2}(c_1^2(F) - 2c_2(F)) + \cdots)$$
$$(1 + \frac{1}{2}c_1(X) + \frac{1}{12}(c_1^2(X) + c_2(X)) + \cdots)$$
(4.119)

To evaluate this expansion on X means picking out the terms of real dimension 4 and this gives the result

$$index\,\overline{\partial}_F = \frac{rk\,(F)}{12}\int_X \{c_1^2(X) + c_2(X)\}$$
$$+ \frac{1}{2}\int_X \{c_1^2(F) - 2c_2(F) + c_1(F) \wedge c_1(X)\}$$
(4.120)

and the non-trivial effect of the bundle F is self-evident.

The signature complex and the spin complex can be subjected to the same treatment with similar results.

For the signature complex, tensoring with the auxiliary bundle F gives rise to the generalised signature operator d_+^F. The appropriate index calculation can be readily carried out and yields the formula

$$index\,d_+^F = 2^{n/2}\int_X ch\,(F) \wedge \mathcal{L}(X)$$
(4.121)

An important consequence of this formula is that, in contrast to the ordinary signature operator d_+ which has zero index in dimensions not divisible by 4, $index\,d_+^F$ can be non-zero in any even dimension.

Finally the twisted spin complex has a Dirac operator with coefficients in F. This Dirac operator will contain a covariant derivative extended to take account of the bundle F. In local coordinates it will be of the form $\gamma^a e_a^\mu(x)(\partial_\mu + \omega_\mu + A_\mu)(1 + \gamma_5)/2$ where A_μ is a Yang–Mills connection. If we denote this Dirac operator by $\partial\!\!\!/_F$ we find that

$$index\,\partial\!\!\!/_F = (-1)^{n/2}\int_X ch\,(F) \wedge \hat{A}$$
(4.122)

It is also clear that this latest Dirac operator $\partial\!\!\!/_F$, unlike its untwisted counterpart, will have a non-zero index in dimensions not divisible by 4 (thus the factor $(-1)^{n/2}$ mentioned on p. 116 has now reappeared); in addition, even in dimensions which are divisible by 4, it can happen that $index\,\partial\!\!\!/ = 0$ but at the same time $index\,\partial\!\!\!/_F \neq 0$.

§ 4. The index theorem for families of operators

In this section we suppose that we have not *one* elliptic operator but a *family* of elliptic operators parametrised by the elements of a set Y. This means that instead of a single operator D we have the family $\{D_y, y \in Y\}$ where each D_y is a separate elliptic operator acting on a manifold X. We shall defer giving examples of such families until chapter 10.

We can consider an index problem for elliptic families except that we no longer have a discrete object such as an integer for the index, rather we define a mathematical quantity which measures the difference between $ker\,D_y$ and $coker\,D_y$ as y varies over Y. Since $ker\,D_y$ and $coker\,D_y$ are vector spaces parametrised by y then we expect that they can be considered as vector bundles over Y; given this state of affairs, and the fact that K-theory is a theory of differences of vector bundles, it is natural to define this 'difference' as an element of $K(Y)$. This leads us to define an analytic index for the elliptic family $\{D_y, y \in Y\}$ in the following manner. Let $ker\,D$ and $coker\,D$ denote the vector bundles $\bigcup_y ker\,D_y$ and $\bigcup_y coker\,D_y$ respectively, and denote the analytic index by $index\,D$. Then we define

$$index\,D = [ker\,D] - [coker\,D] \tag{4.123}$$

so that $index\,D \in K(Y)$. Now $ker\,D$ and $coker\,D$ will only be vector bundles over Y *provided* both $dim\,ker\,D$ and $dim\,coker\,D$ are constant as y varies; in practice the dimension of $ker\,D_y$ can jump at certain y so that $ker\,D$ is not a vector bundle over Y, nevertheless, the definition of $index\,D$ can be generalised slightly so as to deal with this eventuality, cf. Atiyah and Singer [4].

A further important point about elliptic families is that we have assumed that all the operators D_y act on the *same* manifold X. This assumption is unnecessary and the proper thing to do is to allow D_y to act on a manifold X_y which varies with y, the dimension of X_y, however, remaining fixed. It is natural to allow the X_y to form a fibre bundle Z, say, over Y. Thus each D_y is a pseudo-differential elliptic operator of the form

$$D_y : \Gamma(X_y, E_y) \longrightarrow \Gamma(X_y, F_y) \tag{4.124}$$

where E_y and F_y are vector bundles over the manifold X_y and the manifolds X_y are the fibres of the bundle Z over Y. Also E_y, F_y and X_y vary continuously with y. The generalisation made by the introduction of the bundle

Z does not drastically complicate things since, locally, Z is a product over Y and the analysis is hardly altered. However, to make proper invariant sense of any local calculations, the content of these calculations must be invariant under diffeomorphisms of the manifold X. This requirement can be encoded into the fibration of Z over X if we demand that the structure group of the fibration be $Diff(X)$: the diffeomorphisms of X.

Granted an elliptic family of the kind just described then there is an index theorem, Atiyah and Singer [**4**], which provides a topological formula for *index* D. The main ideas used in the proof run along the same lines as the index theorem for a single operator: one gives two definitions of the index—one analytical, and one topological—and then shows that they coincide. Nevertheless, the introduction of the new spaces Z and Y and the elliptic family $\{D_y, y \in Y\}$ does produce some new features, which we now describe.

There is a leading symbol $\sigma_m(D_y)$ for each operator D_y on X_y, but recall that the symbol class $[\sigma_m(D_y)]$ defines an element of $K(T^*X_y)$; and, since X_y is a fibre of the bundle Z, then it is useful to introduce the (co)tangent bundle along the fibres of Z, by which we mean $\{T^*X_y, y \in Y\}$. We denote this bundle by $T_F Z$. The leading symbols $\sigma_m(D_y)$ determine an element $[\sigma_m(D_y)]$ of $K(T_F Z)$ which we denote by $[\sigma_m(D)]$. The analytical index is then a homomorphism

$$ind_a : K(T_F Z) \longrightarrow K(Y)$$
$$[\sigma_m(D)] \longmapsto index\, D \tag{4.125}$$

The topological index is defined using an embedding just as in the single operator case: the main difference is that, instead of embedding the manifold X into the Euclidean space $\mathbf{R}^n$, one embeds Z into the space $Y \times \mathbf{R}^n$ for some n. When the topological index has been defined as a homomorphism $ind_t : K(T_F Z) \to K(Y)$ it is shown (Atiyah and Singer [**4**]) to coincide with the analytical index in the axiomatic manner described in § 1.

There is also a cohomological formula for the topological index which is obtained, as it was in the single operator case, by applying the Chern character to the K-theory definition. The formula for families does not compute an integer, rather it computes an element of the cohomology ring $H^*(Y; \mathbf{Q})$: let *index* $D \in K(Y)$ be the index of an elliptic family then the cohomological formula for *index* D says that

$$ch(index\, D) = (-1)^n \pi_* \{ch([\sigma_m(D)]) \cdot td(T_F Z_{\mathbf{c}})\} \tag{4.126}$$

where $T_F Z_{\mathbf{c}}$ is the complexification of $T_F Z$ and π_* denotes integration along the fibre of Z. Integration along the fibre should be viewed as a map which decreases cohomology dimension by the dimension of the fibre: if we represent a typical element of the (compactly supported) cohomology of $T_F Z$ by the class $[\omega]$, where a ω is a closed p-form $\omega \in \Omega^p(T_F Z)$, then, bearing in mind that the fibres of Z are copies of X and that $\dim X = n$, we have

$$\pi_* : H^*(T_F Z) \longrightarrow H^*(Y)$$

$$[\omega] \longmapsto [\nu], \quad \text{where} \quad \nu = \int_X \omega, \quad \text{so that} \quad \nu \in \Omega^{p-n}(Y) \tag{4.127}$$

Note, too, that the formula only computes the rational cohomology class $ch\,(index\,D) \in H^*(Y;\mathbf{Q})$ rather than $index\,D \in K(Y)$; thus we lose any torsion that may be present.

§ 5. The index for real families

So far we have considered fairly general elliptic families. We now want to draw attention to a rather special type of family consisting only of real elliptic operators—by a real elliptic operator D we mean an operator whose symbol, expressed in local coordinates, satisfies the reality condition $\bar{\sigma}^D(x, p) = \sigma^D(x, -p)$ where bar denotes complex conjugate; in the case where D is a differential operator this just means that it has real coefficients. A family $\{D_y, y \in Y\}$ of elliptic operators is said to be *real* if the operators D_y are real. If the family is not real we call it *complex*.

First of all, to obtain something new about the index of real elliptic operators it is not sufficient to consider just one operator. This is because if D is a single real elliptic operator its symbol complex will consist of real vector bundles over T^*X. It is then straightforward to check that one can always calculate the index of this complex by first complexifying everything. However, for an elliptic family this is not so. The point is that the map that assigns the index of a real family to that of its complexification is *not injective*. Thus one loses information if one only works with complex families.

The index of a complex elliptic family $\{D_y, y \in Y\}$ belongs to $KU(Y)$ the group of complex vector bundles over Y (until now in this chapter we have written $K(Y)$ for simplicity). To define the index of a real elliptic family $\{D_y, y \in Y\}$ one might expect that $KU(Y)$ would be replaced by $KO(Y)$—this is not so. Instead of $KO(Y)$ the appropriate K-theory to use is $KR(Y)$ (Atiyah [2]). We shall not describe $KR(Y)$ here except to say that it is the K-theory to use when Y is a space with an involution—should the involution be trivial then $KR(Y)$ can be identified with the standard real K-theory $KO(Y)$. The origin of the involution in the case of real elliptic families is that the reality condition $\bar{\sigma}^D(x,p) = \sigma^D(x,-p)$ leads to consideration of the antipodal involution $p \mapsto -p$ on the bundle $T_F Z$; this in turn leads to the construction of a *real* symbol class belonging to $KR(T_F Z)$ instead of $K(T_F Z)$, and so the topological index naturally becomes a map of the form

$$ind_t : K(T_F Z) \to KR(Y)$$

An important class of real elliptic operators is the class of real differential operators of *odd* order. These operators are automatically skew-adjoint and so, taken singly, have index zero. Nevertheless, the kernel of a fixed skew-adjoint operator D does contain interesting topological data; it is just that the relevant object to compute is $\dim \ker D$ mod 2 rather than $index\, D$. From now on it is convenient to follow Atiyah and Singer [6] and use the notation $ind_1 D$ is to denote $\dim \ker D$ mod 2. Moreover, this quantity $ind_1 D$ is a deformation invariant just like the ordinary index. It is this invariance of $ind_1 D$ under (continuous) deformations that suggests that it might be computable topologically from data determined by its symbol. In fact the point of mentioning this deformation invariant of the *single* real operator D in the present context is that it can be shown to be equal to the index of an associated real *family* $\tilde{D}$ where $\tilde{D}$ is a family of real elliptic operators parametrised by the circle S^1. The family $\tilde{D}$ has an index which we denote by $index\, \tilde{D}$ and, from our previous paragraph, we know that

$$index\, \tilde{D} \in KR(S^1)$$

However, it is known that $index\, \tilde{D}$ lies in the reduced part of $KR(S^1)$, i.e. in $\widetilde{K}R(S^1)$; it is also known that $\widetilde{K}R(S^1) = \mathbf{Z}_2$. Let us denote the generator of $\widetilde{K}R(S^1)$ by $[\eta]$, say, then $index\, \tilde{D}$ must be a multiple of $[\eta]$, thus

$$index\, \tilde{D} = m[\eta], \quad m \in \mathbf{Z} \qquad (4.128)$$

A simple calculation given in Atiyah and Singer [5] shows that $m = \dim \ker D$; but, since $[\eta]$ is the generator of $\mathbf{Z}_2$, the above formula for $index\, \tilde{D}$ only computes the dimension of $\ker D$ mod 2. Thus, identifying $[\eta]$ with the generator of $\mathbf{Z}_2$, we can write

$$index\, \tilde{D} = \dim \ker D \text{ mod } 2 = ind_1 D \qquad (4.129)$$

where the LHS is the index of a real family $\tilde{D}$ and the RHS is the mod 2 index of a single skew-adjoint operator D.

In view of the existence of cohomological expressions for computing the index it might be expected that the index of real families over Y might be expressible directly in terms of the mod 2 cohomology of Y. Unfortunately, $index\ \widetilde{D}$ is an element of $KR(Y)$ rather than $H^*(Y)$. Also, in the present case this element is a torsion element. Therefore the standard application of the Chern character to pass between the K-theory of Y and its cohomology will just give zero, since the Chern character only produces elements of the rational cohomology group $H^*(Y; \mathbf{Q})$.

§ 6. Index theory and fixed points

Fixed points play an important rôle in topology. The extent of this importance can be eloquently illustrated by showing that, in suitable circumstances, the index theorem can be interpreted as a fixed point theorem. In fact, if one considers the index of an elliptic operator D invariant under a compact group G, then a more general index theorem holds and this latter theorem can be interpreted as a generalised fixed point theorem. Such generalised fixed point theorems are often referred to as generalised Lefschetz fixed point theorems; the main results were obtained in Atiyah and Bott [1,2]. The index-theoretic formulation is in Atiyah and Segal [1]. In this section we give a brief account of some of the relevant index theory.

We first recall the classical Lefschetz fixed point theorem: Let $g : X \rightarrow X$ be a continuous map. If $H^p(X; \mathbf{Z})$ is a cohomology group then g induces pullback maps g_p^* between the cohomology groups of X. If we give a basis to the $H^p(X; \mathbf{Z})$ then the pullbacks g_p^* can be represented as matrices (with integer entries) and so one can compute their traces $tr(g_p^*)$. The alternating sum of these traces is necessarily an integer and is known as the Lefschetz number $L(g)$ of g. In summary, we have

$$g : X \longrightarrow X, \text{ and } g_p^* : H^p(X; \mathbf{Z}) \longrightarrow H^p(X; \mathbf{Z})$$

$$\text{with} \qquad L(g) = \sum_{p=0}^{n} (-1)^p tr(g_p^*) \qquad (4.130)$$

Then when $L(g) \neq 0$ the map g has at least one fixed point. For example, if g is just the identity map then g_p^* is the identity matrix acting on $H^*(X; \mathbf{Z})$, so $tr(g_p^*) = \dim H^*(X; \mathbf{Z}) = b_p$. Thus we have

$$L(g) = \sum_{p=0}^{n} (-1)^p b_p = \chi(X) \qquad (4.131)$$

showing that the Lefschetz number of the identity map is just the Euler characteristic of X—actually since g_p^* is only sensitive to the homotopy type of g then we know that $L(g) = \chi(X)$ for any map *homotopic* to the identity. In addition, if $\chi(X) \neq 0$, such maps have at least one fixed point; and if we assume that the fixed points are isolated then $\chi(X)$ is equal to the number of fixed points.

Now we know that the Euler characteristic $\chi(X)$ is the index of an elliptic complex, namely the de Rham complex. This suggests that there may be a more general fixed point result obtained by passing from the de Rham complex to a general elliptic complex. To this end let E be the usual elliptic complex

$$0 \longrightarrow \Gamma(X, E^0) \xrightarrow{d_0} \Gamma(X, E^1) \xrightarrow{d_1} \cdots \xrightarrow{d_n} \Gamma(X, E^{n+1}) \longrightarrow 0 \qquad (4.132)$$
$$d_i \circ d_{i-1} = 0$$

Assume that the map $g : X \to X$ induces maps T_i, say, between each space $\Gamma(X, E^i)$ and itself. We shall say that the complex E is *invariant* under a map g if the T_i *commute* with the elliptic operators d_i. That is we require

$$0 \longrightarrow \Gamma(X, E^0) \xrightarrow{d_0} \Gamma(X, E^1) \xrightarrow{d_1} \cdots \xrightarrow{d_n} \Gamma(X, E^{n+1}) \longrightarrow 0 \qquad (4.133)$$
$$\Gamma(X, E^i) \xrightarrow{T_i} \Gamma(X, E^i), \quad \text{and } d_i T_i = T_{i+1} d_i$$

Then, denoting the cohomology of the elliptic complex in the usual way by $H^p(E)$ and the corresponding pullbacks by $g_p^*(E)$, we define $L(g, E)$, the generalised Lefschetz number of the g-invariant elliptic complex E, by

$$L(g, E) = \sum_{p=0}^{n} (-1)^p tr(g_p^*(E)) \qquad (4.134)$$

We want to enlarge the number of maps g under consideration. A good way to do this is to let a (compact) group G act on X. This action is a continuous map $\phi : G \times X \to X$, and so for each fixed $g \in G$ we get a continuous map from X to X. Next we require the complex E to be G-invariant; this means that E is invariant under all $g \in G$. Thus we can define a Lefschetz number $L(g, E)$ for each $g \in G$. Now a trace function $tr(g_p^*(E))$ considered as a function of the group element g is a little reminiscent of the character of a group representation. This turns out to be a fruitful observation: the Lefschetz number [4] $L(g, E)$, for $g \in G$, can be regarded as a character valued index.

---

[4] The maps g are all invertible and tied together by their common membership of G. In general fixed point theory we would not impose such restrictions on the maps allowed. However, when a group G is naturally present index theory provides a powerful method for calculating Lefschetz numbers.

The way to formulate an index theorem for such G-invariant elliptic complexes is to combine the relevant group theory and bundle theory into a K-theoretic framework. This is accomplished by defining an equivariant K-theory (Segal [1]) which takes into account the group action of G on X. The K-theoretic object that we wish to define is denoted by $K_G(X)$ and is defined as follows. Let us call X a G-space if it is provided with a continuous action of G on X. A vector bundle F over X will be called a G-vector bundle if it also possesses a continuous action by G subject to two natural conditions. These conditions are that the action on F commutes with the bundle projection $\pi : F \to X$, that is $\pi(g \cdot f) = g \cdot \pi(f)$ (the dot denotes the group action both on F or X); and that the group action on X induces a *linear* map between the fibres of F, that is the map

$$F_x \longrightarrow F_{g \cdot x} \qquad\qquad (4.135)$$

is linear. Given all the G-vector bundles over a G-space X we can imitate the ordinary K-theory construction using equivalence classes of G-vector bundles. The resulting object is what we call $K_G(X)$ and is a ring just like its counterpart $K(X)$.

There are two rather simple cases for which it is easy to identify $K_G(X)$ in terms of something familiar. These occur when we take either G or X to be trivial. If we take G to be trivial, i.e. G consists of just the identity e, then $K_G(X)$ reverts to $K(X)$

$$K_{\{e\}}(X) = K(X)$$

If we take X to be the trivial space consisting of a single point then, referring to the condition 4.135 above, we see that an element of $K_G(X)$ is just a finite dimensional representation of G; thus we can identify $K_G(X)$ with the character ring $R(G)$ of G

$$K_G(\{x_0\}) = R(G)$$

Passing on to index theory we introduce the analytic index in this G-invariant framework. We take a general elliptic operator D with symbol complex $\sigma(E)$ over T^*X. Assume that we have smooth actions of a compact Lie group G on X and the vector bundles over X. Then if the symbol complex $\sigma(E)$ of D is G-invariant it gives rise to an element of $K_G(T^*X)$ which we write as $[\sigma_m(D)]$ in the usual way. Now consider the two null spaces *ker D* and *coker D*; because of the action of G on vector bundles these spaces carry (finite dimensional) representations of G. If we denote the corresponding characters of these representations by $[ker\,D]$ and $[coker\,D]$

respectively then we can compute their difference which is also an element of $R(G)$. That is we have

$$([ker\, D] - [coker\, D]) \in R(G) \qquad (4.136)$$

Then we replace the integer $\dim ker\, D - \dim coker\, D$, used in the definition of the non G-invariant index, by the difference $[ker\, D] - [coker\, D]$. This gives the definition of the G-invariant analytical index ind_a as the map

$$ind_a : K_G(T^*X) \longrightarrow R(G)$$
$$[\sigma_m(D)] \longmapsto [ker\, D] - [coker\, D] \qquad (4.137)$$

The topological index ind_t for the G-invariant case can also be defined using a construction very close to that of §1. Some noteworthy differences are that the embedding of X into $\mathbf{R}^n$ must be replaced by an embedding into a representation space V of G—the map $i : X \to \mathbf{R}^n$ of §1 now becomes $i : X \to V$, while the map $j : P \to \mathbf{R}^n$ of 4.26 becomes a map of the form $j : P \to V$ where P is a point. Provided we use these new versions of the maps i and j the definition of the topological index ind_t is otherwise unchanged. Thus we still write $ind_t = (j_!)^{-1} \circ i_!$. Displaying both of the maps $i_!$ and $j_!$ together we have an analogous diagram to that of 4.27

$$K_G(T^*X) \xrightarrow{i_!} K_G(T^*V) \xleftarrow{j_!} K_G(T^*P), \;\; \text{and } ind_t = (j_!)^{-1} \circ i_! \qquad (4.138)$$

But P is a point so that $T^*P \simeq P$, and we know that $K_G(P) = R(G)$ rather than $\mathbf{Z}$, so we see that $R(G)$ naturally enters the picture, thus rendering the topological index character valued. The topological index is now the map

$$ind_t : K_G(T^*X) \longrightarrow R(G)$$
$$[\sigma_m(D)] \longmapsto ((j_!)^{-1} \circ i_!)([\sigma_m(D)]) \qquad (4.139)$$

Finally the Atiyah–Singer index theorem (Atiyah and Singer [1]) still asserts that the topological index and the analytical index coincide.

We can now return to the fixed point situation. It is shown in Atiyah and Segal [1] that the generalised Lefschetz number $L(g, E)$ for a G-invariant elliptic complex E is given by evaluating the (character valued) index of its symbol complex at the element $g \in G$. That is to say, if $[\sigma(E)] = u$ is the element of $K_G(T^*X)$ determined by the symbol complex of E then $ind_t(u)$ is an element of $R(G)$, and evaluating this character on the element g gives

$$L(g, E) = \chi(g) \;\; \text{with } \chi = ind_t(u) \text{ and } u = [\sigma(E)] \qquad (4.140)$$

Finally there is a cohomological formula for the index character, albeit a relatively complicated one. We shall not be making any explicit use of this formula so, rather than quote it, we refer the interested reader to Atiyah and Singer [3]. We limit ourselves to a few observations.

Recall that the cohomological formula for the ordinary integer index consists of characteristic classes evaluated on TX. Now from the present group-theoretic point of view the ordinary index is obtained by choosing the group G to be trivial and, in that case, there is only one map $g : X \to X$ given by $g = e$, the identity of G. Thus, if we define X^g to be the fixed point set of a map g, then the fixed point set of g is the whole space X or, $X^g = X$. Therefore we can think of the cohomological formula as an evaluation on the space TX^g. This is also what happens in the case where G is non-trivial: the cohomological formula for $L(g, E) = (ind_t(u))(g)$ is an evaluation on TX^g where X^g is the fixed point set of the map $g : X \to X$. For a given g the fixed point set X^g may be a finite set of points or a set of positive dimension. In the cases where X^g is a finite set of points the formula for $L(g, E)$ will be much simpler.

§ 7. Index theory for manifolds with boundary

In this section we allow the manifold X to have a non-empty boundary. We can still study elliptic operators acting on X but we will now have to supply appropriate elliptic boundary conditions.

The search for appropriate boundary conditions is central to the study. The main problem here is that, if one chooses a specific elliptic operator and desires to impose certain boundary conditions, such boundary conditions may not be allowed because of the presence of topological obstructions to their existence, cf. Atiyah and Bott [3] and Palais *et al.* [1]. In particular one may have to abandon *local* boundary conditions in favour of *global* ones. A general index theorem for manifolds with boundary will involve the relative K-theory $K(X, \partial X)$ rather than $K(X)$ and has to provide for non-local boundary conditions.

Boundary conditions for many of the classical partial differential elliptic problems are indeed local. For example, if we choose our elliptic operator to be the Laplacian Δ acting on an open set Ω in $\mathbf{R}^n$ then the Dirichlet problem requires us to solve

$$\Delta u = 0, \text{ subject to } u = g \text{ on } \partial\Omega \tag{4.141}$$

the Dirichlet boundary condition being the requirement that $u = g$ on $\partial\Omega$. Let n be the normal to the boundary $\partial\Omega$, then we can change the problem to

$$\Delta u = 0, \text{ subject to } \frac{\partial u}{\partial n} = g \text{ on } \partial\Omega \tag{4.142}$$

and we have the Neumann problem where the boundary condition is the first order equation $\partial u/\partial n = g$ on $\partial\Omega$. Clearly both the Dirichlet and Neumann problems have local boundary conditions.

Now suppose that we replace the Laplacian Δ by the signature operator $d_+ : \Gamma(X, \wedge^+ T^* X \otimes_{\mathbf{R}} \mathbf{C}) \to \Gamma(X, \wedge^- T^* X \otimes_{\mathbf{R}} \mathbf{C})$ acting on a manifold X with boundary. We would like to specify boundary conditions so that the index of d_+ gives the signature of X. This is possible but only if we use non-local boundary conditions. Let us look briefly at what is involved.

Near the boundary we use local coordinates, one of which is the inward normal n to the boundary, and then we shall *assume* that d_+ can be written as the sum of a normal and a tangential part

$$d_+ = A\frac{\partial}{\partial n} + B \cdot d_+^T \tag{4.143}$$

where the symbols A and B are matrices, and d_+^T is the tangential part of d_+. Note that d_+^T acts entirely on the boundary ∂X. Ellipticity guarantees that A is invertible so that we may rewrite the above as

$$d_+ = A\left(\frac{\partial}{\partial n} + A^{-1}B \cdot d_+^T\right)$$
$$= A\left(\frac{\partial}{\partial n} + L\right) \quad \text{where } L = A^{-1}B \cdot d_+^T \tag{4.144}$$

We assume that L is independent of n; this assumption, together with the observation that L is self-adjoint, means that L is an elliptic operator on ∂X. Thus we can use its eigenfunctions as a basis for the space $L^2(\partial X, \wedge^+ T^* X \otimes_{\mathbf{R}} \mathbf{C})$ of square integrable sections on ∂X. Let H_+ and H_- denote the spaces of non-negative and negative eigenfunctions respectively. Now the appropriate elliptic boundary conditions for the index of d_+ to give the signature of X are that the solutions u satisfy

$$u|_{\partial X} \in H_- \tag{4.145}$$

Let Π denote the orthogonal projection onto the space H_+, then this boundary condition should be viewed as the equation

$$\Pi v = 0, \quad v \in H_+ \oplus H_- \tag{4.146}$$

We can see that Π is a zero[th]-order pseudo-differential equation in the following way. First we make an invertible operator from L by defining $M = L+N$ where N is the projection onto the null space of L. This new operator M

is therefore invertible. We can also raise positive pseudo-differential elliptic operators such as M to arbitrary powers (cf. p. 46 of chapter 2 and Seeley [1]) so that, in particular, the operator $\sqrt{M^2}$ is defined—we shall take the positive value of the square root and so use the more suggestive notation $|M|$ to denote $\sqrt{M^2}$. Then it is easy to check that Π is given by the zeroth-order pseudo-differential combination

$$\Pi = \frac{M^{-1}}{2}(M + |M|) \qquad (4.147)$$

In addition, the boundary condition 4.146 is non-local because the action of projection onto the space of positive eigenvalues is non-local. However, as far as the index is concerned we need only consider the homotopy class of the symbol $\sigma_0^L(x, p)$ of L: indeed if we could find a homotopy of $\sigma_0^L(x, p)$ which deformed it into a function of x only, then it would be homotopic to the symbol of a multiplication operator which is local. Unfortunately, a homotopy of this type does not exist, in this case, because of the non-vanishing of the topological obstructions described in Atiyah and Bott [3].

Now we shall investigate how the index of an elliptic operator with elliptic boundary conditions might depend on ∂X. We start by considering the Gauss–Bonnet theorem since this is a case where something is already known about the rôle of boundary corrections. If we take X to be a smooth surface with boundary then the Gauss–Bonnet theorem determines the Euler characteristic and we have

$$\int_X \frac{R}{2\pi} + \int_{\partial X} \frac{R_g}{2\pi} = \chi(X) \qquad (4.148)$$

where R is the curvature of X and R_g is the geodesic curvature of ∂X in X. The presence of the boundary integral in the above expression is not mandatory—if X is isometric to a product near its boundary, i.e. X is isometric to $\partial X \times \mathbf{R}^+$, then the boundary integral vanishes; this means that, for such surfaces, the Euler characteristic is given by the *same* integral as in the case without boundary, that is

$$\int_X \frac{R}{2\pi} = \chi(X) \text{ for } X \text{ isometric to } \partial X \times \mathbf{R}^+ \text{ near } \partial X \qquad (4.149)$$

However, the Euler characteristic is a less subtle invariant than the signature $Sign(X)$; and if we examine the formulae for $Sign(X)$ when X has a boundary a new feature is revealed: Let us first take a four dimensional orientable manifold $\widetilde{X}$ *without boundary*. Then we know from § 2 that the formula for $Sign(\widetilde{X})$ is

$$Sign(\widetilde{X}) = \int_{\widetilde{X}} 2^{n/2} \mathcal{L}(\widetilde{X}) = \frac{1}{3} \int_{\widetilde{X}} p_1(\widetilde{X}) \qquad (4.150)$$

where $p_1(\widetilde{X})$ is the four dimensional term picked out from the expansion of $\mathcal{L}(\widetilde{X})$ in terms of Pontrjagin classes. Now if we take a four dimensional manifold X, *with a boundary*, and with X isometric to a product near its boundary, we find that 4.150 does *not* hold any longer. In fact explicit calculation shows that we can define a real number $\eta(\partial X)$ by

$$Sign\,(X) - \frac{1}{3}\int_X p_1(X) = \frac{-1}{2}\eta(\partial X) \neq 0 \qquad (4.151)$$

The factor $-1/2$ in front of $\eta(\partial X)$ is included for later notational convenience. The interesting point is that the difference $\eta(\partial X)$ only depends on the boundary ∂X and not on X itself; that is, if we take two different manifolds X and Y with the same boundary then the difference $\eta(\partial X)$ is equal to $\eta(\partial Y)$. In sum

If $X \neq Y$ but $\partial X = \partial Y$

then $Sign\,(X) - \dfrac{1}{3}\displaystyle\int_X p_1(X) = Sign\,(Y) - \dfrac{1}{3}\displaystyle\int_Y p_1(Y)$ $\qquad (4.152)$

This can be proved by constructing a new closed manifold Z by glueing X to Y along their common boundary and then applying the formula 4.150 to Z. The precise nature of the difference $\eta(\partial X)$ is difficult to anticipate—it is not a local geometric object, instead it is constructed entirely from the spectrum of the operator L and so is a global object (cf. Atiyah, Patodi and Singer [1–4]). To define $\eta(\partial X)$ let $\{\lambda_n\}$ be the spectrum of L and consider the function $\eta(s)$ where

$$\eta(s) = \sum_{\lambda_n \neq 0} \frac{sign(\lambda_n)}{|\lambda_n|^s} \qquad (4.153)$$

The series defining $\eta(s)$ converges if Re s is large enough; it can also be analytically continued to define a meromorphic function of s with simple poles. Further, though one might expect $\eta(s)$ to have a simple pole at $s = 0$, the residue of this pole vanishes in the present case so that $\eta(0)$ is *finite*. The connection of $\eta(\partial X)$ with $\eta(s)$ is just that

$$\eta(\partial X) = \eta(0) \qquad (4.154)$$

Thus the formula for $Sign\,(X)$ when $\partial X \neq \phi$ can be written as

$$Sign\,(X) = \frac{1}{3}\int_X p_1(X) - \frac{1}{2}\eta(0) \qquad (4.155)$$

Moreover, $Sign(X)$ is the index of the signature operator d_+ on X subject to the global boundary conditions introduced above.

The properties of the spectral function $\eta(s)$ are worthy of further discussion. We discuss first the continuation of $\eta(s)$ to arbitrary values of s and then give an example of the calculation of $\eta(0)$.

The continuation of $\eta(s)$ can be done using the zeta function of a positive elliptic operator in the following way. Let P be an elliptic pseudo-differential operator of positive order on a closed compact manifold. Suppose also that P is self-adjoint and positive with spectrum $\{\mu_n\}$, then it is known (Seeley [1]) that the function

$$\zeta_P(s) = \sum_{\mu_n} \frac{1}{\mu_n^s} \tag{4.156}$$

is a meromorphic function of s with simple poles and that $\zeta_P(0)$ is always finite—for $\mathrm{Re}\, s$ large enough the series defining $\zeta_P(0)$ is convergent and can be thought of as $tr\,(P^{-s})$. We can define $\eta(s)$ in terms of two such zeta functions (Atiyah, Patodi and Singer [4]). We define two elliptic operators $L_\mp$ by

$$L_+ = \frac{3}{2}|L| + \frac{1}{2}L, \qquad L_- = \frac{3}{2}|L| - \frac{1}{2}L \tag{4.157}$$

If we denote the positive and negative eigenvalues of L by $\{\lambda_n^+\}$ and $\{\lambda_n^-\}$ respectively, then the eigenvalues of L_+ and L_- are clearly the pairs $\{2\lambda_n^+, -\lambda_n^-\}$ and $\{\lambda_n^+, -2\lambda_n^-\}$ respectively. Note that L_+ and L_- have been constructed so as to be positive, thus their zeta functions $\zeta_{L_+}(s)$ and $\zeta_{L_-}(s)$ exist as meromorphic functions of s. It turns out that their difference determines $\eta(s)$, for

$$
\begin{aligned}
\zeta_{L_+}(s) - \zeta_{L_-}(s) &= \left\{ \sum_{\lambda_n^+} \frac{1}{(2\lambda_n^+)^s} + \sum_{\lambda_n^-} \frac{1}{(-\lambda_n^-)^s} \right\} \\
&\quad - \left\{ \sum_{\lambda_n^+} \frac{1}{(\lambda_n^+)^s} + \sum_{\lambda_n^-} \frac{1}{(-2\lambda_n^-)^s} \right\} \\
&= (2^{-s} - 1) \left\{ \sum_{\lambda_n^+} \frac{1}{(\lambda_n^+)^s} - \sum_{\lambda_n^-} \frac{1}{(-\lambda_n^-)^s} \right\} \\
&= (2^{-s} - 1)\eta(s) \\
\Rightarrow \eta(s) &= \frac{1}{(2^{-s} - 1)} \left\{ \zeta_{L_+}(s) - \zeta_{L_-}(s) \right\}
\end{aligned}
\tag{4.158}
$$

Thus the analytic continuation of the zeta[5] functions induces the analytic continuation of $\eta(s)$. Also $\eta(s)$ can be thought of as a trace: for Re s large enough it is the trace of the operator $L|L|^{-s-1}$.

If one examines the definition of $\eta(0)$ it appears to be some kind of regularised measure of the asymmetry between the positive and the negative eigenvalues of L. An explicit calculation of $\eta(s)$ helps to bring this to the fore. We shall calculate $\eta(0)$ for the operator $id/d\theta + t$ acting on the circle S^1 with local coordinate θ and $0 < t < 1$. Calculating the eigenvalues we find that

$$\lambda_n = n + t, \ n = \ldots - 2, -1, 0, 1, 2, \ldots$$

$$\Rightarrow \eta(s) = \sum_{n=-\infty}^{\infty} \frac{sign(n+t)}{(n+t)^s} \qquad (4.159)$$

Clearly $\eta(s)$ can be reduced to an expression involving the generalised Riemann zeta function

$$\zeta(t, s) = \sum_{0}^{\infty} \frac{1}{(n+t)^s} \qquad (4.160)$$

Then we can compute $\eta(0)$ by expanding this expression about $s = 0$ and using the fact (Whittaker and Watson [1]) that $\zeta(t, 0) = -t + 1/2$. The result is that

$$\eta(0) = 1 - 2t \qquad (4.161)$$

This shows that $\eta(0)$ is indeed a measure of the asymmetry of the eigenvalues $n + t$ about 0—note that precisely when $t = 1/2$ there is no asymmetry in the eigenvalues and $\eta(0)$ vanishes as it should.

This discussion of the index problem for the signature operator d_+ generalises to other elliptic boundary value problems. We now take a general linear first order elliptic operator $D : \Gamma(X, E) \to \Gamma(X, F)$ where E and F are vector bundles over a general manifold X with boundary. We assume that, in a neighbourhood $\partial X \times \mathbf{R}^+$ of the boundary, D has a form analogous to 4.144. We also impose the same type of global boundary conditions. This means that we have

$$D : \Gamma(X, E) \longrightarrow \Gamma(X, F)$$

$$D = A\left(\frac{\partial}{\partial n} + L\right) \ \text{near} \ \partial X \times \mathbf{R}^+ \qquad (4.162)$$

$$\text{and} \qquad \Pi v = 0, \ v \in L^2(\partial X, E)$$

---

[5] It is also possible to relate $\zeta_P(s)$ to the kernel of the heat equation $\partial u/\partial t = Pu$ and use this fact to compute the index of D where $P = DD^*$. In this way one obtains a heat equation derivation of the index theorem, cf. Atiyah, Bott and Patodi [1].

where L is an operator solely on ∂X and Π is the usual spectral projection onto the positive eigenvalues of L. So far the zero eigenvalue of the operator L has not figured in the discussion. This is not typical: Suppose that L has a non-empty kernel and define h and $\eta(s)$ by

$$h = \dim \ker L, \qquad \eta(s) = \sum_{\lambda_n \neq 0} \frac{sign(\lambda_n)}{|\lambda_n|^s} \tag{4.163}$$

Then we have the following result.

Theorem (Index theory for boundary value problems) *The index of the operator D for the elliptic boundary value problem 4.162 is given by*

$$index\, D = \int_X P - \frac{h + \eta(0)}{2} \tag{4.164}$$

where P is the differential form representing the combination of characteristic classes which occurs when $index\, D$ is calculated on a *closed* manifold; since P is expressed in terms of the curvature R we can write $P = P(R)$.

We have seen that choosing $D = d^+$, the signature operator, provides an example where these boundary corrections are non-zero. On the other hand, we also saw in 4.149 that, at least for surfaces, the Euler characteristic of X can be calculated by the same formula as the case without boundary. This result extends to higher dimensional X; thus we can say that the index of the de Rham complex does not depend on non-local boundary conditions. To see another boundary value problem where non-local conditions are needed we move on to the next example.

Example *The $\bar{\partial}$ operator*

We take a closed Riemann surface Σ_g of genus g and give it a boundary by deleting from it p discs $D_1, \ldots, D_p$. Thus X is a complex manifold whose boundary ∂X consists of p disjoint circles. The two manifolds Σ_g and X are clearly related by

$$X = \Sigma_g - \bigcup_{i=1}^{p} D_i \tag{4.165}$$

To satisfy the condition that X should be isometric to a product near its boundary we assign coordinates $z_1, \ldots, z_p$ to the discs $D_1, \ldots, D_p$ so that $z_i = 0$ corresponds to the centre of the disc D_i; then we choose a metric on X which satisfies

$$ds^2 = \left| \frac{dz_i}{z_i} \right|^2, \quad \text{near the boundary of } D_i, \ i = 1, \ldots, p \tag{4.166}$$

Note that this metric would be singular on Σ_g but is perfectly regular on X. As elliptic operator we take the $\bar{\partial}$ operator. If we use polar coordinates (r_i, θ_i) instead of z_i it is easy to verify that, near each piece ∂D_i of the boundary, $\bar{\partial}$ has the desired form

$$\bar{\partial} = \frac{e^{i\theta_i}}{2} \left\{ \frac{\partial}{\partial r_i} + \frac{i}{r_i} \frac{\partial}{\partial \theta_i} \right\} \tag{4.167}$$

This shows that the tangential operator on each piece of the boundary is $(-i/r_i)\partial/\partial\theta_i$. To apply theorem 4.164 we have to calculate $\eta(0)$ and h for each disc; but the contribution from each disc is identical, thus we obtain the formula

$$index\,\bar{\partial} = \int_X td\,(X) - p\frac{(h + \eta(0))}{2} \tag{4.168}$$

where we have also set the function P of theorem 4.164 equal to $td\,(X)$ since (as we showed in 4.72) this is the characteristic class in the formula for the index of $\bar{\partial}$ on closed manifolds. However, if we use the expansion 3.106 of $td\,(X)$ in terms of Chern classes and the relation 3.124 of the top Chern class to the Euler class we find that

$$td\,(X) = 1 + \frac{c_1(X)}{2}, \quad \text{and} \quad c_1(X) = e(X) \tag{4.169}$$

So the index of $\bar{\partial}$ can be expressed as

$$index\,\bar{\partial} = \frac{1}{2}\int_X e(X) - p\frac{(h + \eta(0))}{2} \tag{4.170}$$

Next we observe that the operator $(-i/r_i)\partial/\partial\theta_i$ has $h = 1$ since its only zero eigenfunction is the constant function; further, we can see that its η-function has $\eta(0) = 0$ since the non-zero eigenvalues are just the non-zero integers which are symmetrically distributed about zero. Now we can evaluate the integral. We write

$$\int_X e(X) = \int_{\Sigma_g} e(\Sigma_g) - \sum_{i=1}^{p} \int_{D_i} e(D_i) \tag{4.171}$$

But we can apply the Gauss–Bonnet theorem to both parts of this expression: For the Riemann surface Σ_g we have immediately the statement

$$\int_{\Sigma_g} e(\Sigma_g) = \chi(\Sigma_g) = 2 - 2g \tag{4.172}$$

while for the disc D_i we use the version 4.148 of Gauss–Bonnet for a surface with boundary and thus obtain

$$\int_{D_i} e(D_i) + \int_{\partial D_i} \frac{R_g}{2\pi} = \chi(D_i) = 1 \qquad (4.173)$$

However, with the metric that we introduced above, the boundary integral is zero. Thus we can straightaway deduce that

$$
\begin{aligned}
index\, \overline{\partial} &= \frac{1}{2}\{\chi(\Sigma_g) - p\chi(D_i)\} - \frac{p}{2}h \\
&= \frac{1}{2}(2 - 2g - p) - \frac{p}{2} \\
&= 1 - g - p
\end{aligned}
\qquad (4.174)
$$

The requirement that X be isometric to $\partial X \times \mathbf{R}^+$ near the boundary can be relaxed. If we omit this requirement then the formula for $index\, D$ acquires an extra term which is an integral over the boundary. We shall briefly indicate the origin of this extra term.

The manifold X is assumed to be a Riemannian manifold with metric g determined locally by an expression of the form $ds^2 = \sum g_{ij}dx^i dx^j$. If we restrict to the boundary, the metric g induces a metric h, say, on ∂X. Near ∂X we introduce coordinates y^i, $i = 2, \ldots, n$ for ∂X, giving coordinates (n, y^i) for X where n is the normal to the boundary. The isometric requirement discussed above means that near the boundary g is a product metric with h. Then using these coordinates we can write

$$ds^2 = g_{11}dn^2 + \sum_{i,j \geq 2} h_{ij}dx^i dx^j \qquad (4.175)$$

With this product metric on X we can calculate the Riemannian connection Γ of g on X and verify that it has no normal component on the boundary. Now we return to the Gauss-Bonnet example 4.148 which, in general, has a boundary integral involving the geodesic curvature R_g; this term, when calculated, is found to be proportional to the normal component of Γ and therefore vanishes if g is a product metric. More generally, if we consider the index of the operator D then there should be an integral over the boundary ∂X. However, this integral vanishes if the normal component of the connection Γ vanishes on the boundary. If we decompose Γ into tangential and normal components on ∂X according to $\Gamma = \Gamma^T + \Gamma^\perp$ then the normal component $\Gamma^\perp$ is called the second fundamental form.

We can see that if we use two different Riemannian connections on X then the difference that this produces to the calculation of $index\ D$ is in general a boundary integral. Let g and g' be two metrics on X with associated connections Γ and Γ' respectively. For each connection we have a characteristic polynomial $P(R)$ or $P(R')$, and we know (Nash and Sen [1]) that the difference between two such polynomials is exact so that $P(R) - P(R') = dQ$ for some differential form Q. If we choose one of these connections to be that due to a metric which reduces to a product near the boundary, then Q will be proportional to the second fundamental form of the other one. Integrating this statement we obtain

$$\int_X P(R) = \int_X P(R') + \int_X dQ = \int_X P(R') + \int_{\partial X} Q \qquad (4.176)$$

Thus a more general index formula for D would be of the form

$$index\ D = \int_X P(R) + \int_{\partial X} Q - \frac{h + \eta(0)}{2} \qquad (4.177)$$

for an appropriate Q expressed in terms of the second fundamental form $\Gamma^\perp$.

The validity of the index theorem above implies that the η-function associated to the operator L has the property that $\eta(0)$ is finite. One could consider *any* elliptic self-adjoint operator B, say, on a (closed compact) manifold Y, and construct its η-function $\eta_B(s)$. If we do this it is not immediately clear whether $\eta_B(0)$ is always finite, as is the case for $\zeta_P(0)$; in fact there are certain features special to the index theorem. These are that L is an operator on a manifold Y which is the boundary of another manifold X, and also that D is related to L by the decomposition $D = A(\partial/\partial n + L)$. Nevertheless, it can be proved that $\eta_B(0)$ is finite if Y is odd dimensional, and also for even dimensional Y if B is a differential operator of odd order.

Part of the delicacy involved in proving finiteness can be understood by realising the following: if one thinks of $\eta_B(s)$ as $tr\,(B|B|^{-s-1})$ then the *operator* $B|B|^{-s-1}$ can be defined for arbitrary s by analytic continuation. When this is done, one can produce an example (Atiyah, Patodi and Singer [4]) where the operator has a pole at $s = 0$ with a non-vanishing *matrix valued* residue; however, the *trace* of this matrix valued residue is zero, giving a finite $\eta_B(0)$.

CHAPTER V

Some Algebraic Geometry

§ 1. Algebraic varieties

Algebraic geometry can be regarded as the study of algebraic varieties. An algebraic variety is a space whose points are the set of zeros of a finite or infinite collection of homogeneous polynomials. These zeros are taken as belonging to some fixed field K; we shall always take K to be the field of complex numbers $\mathbf{C}$—that is, we limit ourselves to the study of complex algebraic varieties.

There is a close connection between algebraic varieties and complex projective spaces. This is because algebraic varieties are defined as the zero locus of homogeneous polynomials and the complex projective spaces $\mathbf{CP}^n$ are endowed with homogeneous coordinates. As far as $\mathbf{CP}^n$ is concerned we write its homogeneous coordinates as $(z_1, z_2, \ldots, z_n, z_{n+1})$ and recall that these are thought of as representing a line in $\mathbf{C}^{n+1}$. The important fact is that a point p in $\mathbf{CP}^n$ is determined by a set of ratios: for example, if $z_{n+1} \neq 0$, then we can consider points such as $(z_1/z_{n+1}, z_2/z_{n+1}, \ldots, z_n/z_{n+1}, 1)$; and, in general, ratios of this type provide $\mathbf{CP}^n$ with a set of local coordinates. These coordinates are complex and they make it possible to check that it is a compact complex manifold of complex dimension n. Now let us take a typical homogeneous polynomial $P(z_1, z_2, \ldots, z_n, z_{n+1})$ of degree d, say, then we know that

$$P(\lambda z_1, \ldots, \lambda z_{n+1}) = \lambda^d P(z_1, \ldots, z_{n+1}), \qquad \lambda \in \mathbf{C}, \ \lambda \neq 0 \qquad (5.1)$$

If we suppose again that $z_{n+1} \neq 0$ we obtain

$$P(z_1, \ldots, z_n, z_{n+1}) = 0 \Rightarrow P(z_1/z_{n+1}, \ldots, z_n/z_{n+1}, 1) = 0 \qquad (5.2)$$

Thus the zero locus of the polynomial $P(z_1, z_2, \ldots, z_{n+1})$ is some subset of $\mathbf{CP}^n$. Applying this idea to a collection of such polynomials shows that an algebraic variety V can be thought of as being given by an embedding into some projective space $\mathbf{CP}^n$ and we write $V \subset \mathbf{CP}^n$.

In general an algebraic variety $V \subset \mathbf{CP}^n$ will not be a sub-manifold of $\mathbf{CP}^n$, this is because V may have singular points. Singular points can be detected in the following way. The idea is to use the Jacobian of the embedding of V into $\mathbf{CP}^n$. Recall that if $f : M \rightarrow N$ is a smooth map between two manifolds M and N which maps the local coordinates $(x_1, \ldots, x_m)$ of M onto the coordinates $(y_1, \ldots, y_n)$ of N, then the Jacobian $J(f)$ is the matrix given in local coordinates by $[\partial y_i / \partial x_j]_{n \times m}$. Now let $V \subset \mathbf{CP}^n$ be specified by the zero locus of the k polynomials $P_1, \ldots, P_k$. Then if the Jacobian matrix of the underlying embedding $e : V \rightarrow \mathbf{CP}^n$ always has its *maximal rank*, namely k, the variety V has no singular points. On the other hand, if there are any points in V for which the rank of $J(e)$ is less than k then these are called *singular points*. In the cases where the variety V does turn out to be a complex sub-manifold of $\mathbf{CP}^n$ then V is called an algebraic sub-manifold of $\mathbf{CP}^n$.

Another interesting property of complex sub-manifolds M of $\mathbf{CP}^n$ is that if one takes a general analytic embedding $e : M \rightarrow \mathbf{CP}^n$ but does not insist that e is specified by the zero locus of a collection of homogeneous polynomials, then M is in fact an algebraic sub-manifold—that is, e is, after all, specified by a collection of homogeneous polynomials and M is thus a non-singular algebraic variety. This result is known as Chow's theorem and is proved in Griffiths and Harris [1].

A general criterion for determining when a compact complex manifold is an algebraic variety is provided by what is known as the Kodaira embedding theorem, which we now quote.

Theorem (Kodaira) *A compact complex manifold M is an algebraic variety if and only if it has a closed, positive 2-form ω of type $(1,1)$ such that the cohomology class $[\omega]$ determined by ω belongs to $H^2(M; \mathbf{Q})$.*

The point to note about the cohomology class $[\omega]$ is that it automatically belongs to $H^2(M; \mathbf{R})$ but not necessarily to $H^2(M; \mathbf{Q})$; also of relevance in the proof of the theorem (cf. Griffiths and Harris [1] or Wells [1]) is the fact that if ω does belong to $H^2(M; \mathbf{Q})$ then some *multiple* of $[\omega]$ will actually belong to $H^2(M; \mathbf{Z})$.

An important class of complex manifolds which are algebraic varieties is formed by the compact Riemann surfaces of genus p which we denote [1] by Σ_p. Riemann surfaces are complex manifolds of (complex) dimension one and are therefore often referred to as algebraic curves. Most of the remainder

---

[1] In this chapter we shall use p to denote genus rather than g; this is because we wish to use g to denote a metric on the surface Σ_p.

of this chapter will be concerned with algebraic geometry in dimension one, that is with Riemann surfaces.

§ 2. Riemann surfaces and divisors

A compact closed Riemann surface Σ_p is topologically a compact closed orientable surface and is thus a sphere with p handles; but, because it is also an algebraic variety, it can be embedded in $\mathbf{CP}^n$ for an appropriate n which depends on its genus p. If $p = 0$ then we know that Σ_p is the Riemann sphere, which is identical to $\mathbf{CP}^1$; thus, in this case we somewhat trivially have $n = 1$. If $p = 1$ then any Σ_p can be represented by a cubic curve in $\mathbf{CP}^2$ of the form

$$x(x - 1)(x - \lambda) - y^2 = 0; \quad \lambda \text{ a constant}, \lambda \in \mathbf{C} \qquad (5.3)$$

with x and y local (complex) coordinates in $\mathbf{CP}^2$. For $p = 2, 3, \cdots$ any Σ_p can be embedded in $\mathbf{CP}^3$.

When the genus p is greater than zero there is not just one Riemann surface but a family of possible surfaces each with a distinct complex structure. For the case when $p = 1$ this family is one dimensional with local coordinate the λ appearing in 5.3 above. For genus $p > 1$ this family has complex dimension $3p - 3$ and is known as the moduli space $\mathfrak{M}_p$; closely related to $\mathfrak{M}_p$ is its covering space, the Teichmüller space $\mathfrak{T}_p$. We shall study both these spaces in § 4 and § 5 below.

Complex manifolds M, say, are often studied by constructing holomorphic bundles over M. A distinguished holomorphic bundle over such an M is what is called the *canonical bundle* K_M. This is defined as follows. The canonical bundle K_M of a complex manifold M of dimension n is the maximum non-trivial exterior power of the holomorphic cotangent bundle, that is, K_M is the bundle $\wedge^n T^* M$. Note that because $T^* M$ is a holomorphic vector bundle of rank n then $K_M = \wedge^n T^* M$ has rank 1, thus K_M is always a line bundle. The canonical bundle can be made more concrete by realising that a holomorphic section of K_M is just a holomorphic n-form of type $(n, 0)$,i.e. an element of the sheaf Ω^n introduced in chapter 3; note too that a holomorphic section of a bundle E is an element of the sheaf $\mathcal{O}(E)$, thus we observe that when $E = K_M$ we have $\mathcal{O}(K_M) = \Omega^n$.

Applying the preceding paragraph to the case where M is a Riemann surface Σ_p we see that, since Σ_p is of dimension 1, $T^* \Sigma_p$ is already a line bundle and hence K_{Σ_p} coincides with $T^* \Sigma_p$. Now in general one has to consider many holomorphic line bundles on Σ_p and holomorphic line bundles on algebraic varieties can be studied by the closely related notion of divisors. We shall now examine some of their basic properties.

Consider an arbitrary holomorphic line bundle L over Σ_p and a meromorphic section s of L. Such a section s will have some zeros and some poles at a set of points $\{P_i\}$, say. Each zero or pole at P_i has an order which is an integer n_i—we take the convention that the order of a pole is a negative integer. The divisor associated to the section s is the formal finite linear combination $div\,(s)$ where

$$div\,(s) = \sum_i n_i P_i \qquad (5.4)$$

The degree of $div\,(s)$ is the integer $deg\,(div\,(s))$ defined by

$$deg\,(div\,(s)) = \sum_i n_i \qquad (5.5)$$

Now if we replace the *section* s by a meromorphic *function* f on Σ_p then we can also construct the divisor $div\,(f)$. However, $div\,(f)$ necessarily has zero degree because the number of poles and zeros of a meromorphic function on a Riemann surface, counted with their multiplicity, are always equal. We digress briefly to remind the reader of the argument that establishes this fact.

Consider the standard integral $(1/2\pi i)\int_C df/f$ which measures $n_0 - n_p$, the number of zeros of f minus the number of poles inside C. For $p = 0$ we choose the contour C on the Riemann sphere Σ_0 so as to enclose all the zeros and poles of f—this is always possible because Σ_0 is compact and so the zeros and poles form a finite set F, say. Having done this observe that the combination df/f is invariant under coordinate transformations and is therefore a closed differential form defined on the set $\Sigma_0 - F$. This means that the integral depends only on the homology class of C, and, since C is homologous to a point on $\Sigma_0 - F$, the integral vanishes. For $p > 0$ we use the standard representation of Σ_p as a polygon having $4p$ sides which we write out in order as $a_1, b_1, a_1^{-1} b_1^{-1}, \ldots, a_p, b_p, a_p^{-1}, b_p^{-1}$ where a_i is identified with a_i^{-1} and b_i with b_i^{-1}; also the minus one in a_i^{-1} means that it has the opposite orientation to a_i. For $p = 1$ the reader can immediately verify that this gives a rectangle with opposite sides identified—that is, a torus. We can assume without loss of generality that the sides of the polygon do not pass through any zeros or poles; in addition, we can always ensure that C contains all the zeros and poles of f by pushing C close enough to the sides of the polygon. Then we consider the limit where the contour expands so as to coincide with the sides of the polygon. The integral now vanishes since the contribution from each side is cancelled by another of opposite orientation.

Now suppose that we take a different section s' of the same line bundle L, then s' will be proportional to s via a meromorphic function f. Thus we write

$$s' = fs, \qquad f \text{ a meromorphic function} \tag{5.6}$$

It is clear from the definition of a divisor that

$$div\,(s') = div\,(fs) = div\,(f) + div\,(s) \tag{5.7}$$

However taking the degree of both sides we find that

$$\begin{aligned} deg\,(div\,(s')) &= deg\,(div\,(f)) + deg\,(div\,(s)) \\ &= deg\,(div\,(s)) \end{aligned} \tag{5.8}$$

since f is meromorphic. Thus $div\,(s')$ and $div\,(s)$ have the same degree, and, because of this, we define two divisors D_1 and D_2 to be *equivalent* if $D_1 - D_2 = div\,(f)$ for some meromorphic function f. This means that the line bundle L gives rise to just one divisor modulo this equivalence.

Conversely, given any divisor D over Σ_p it gives rise to a holomorphic line bundle over Σ_p, which we denote by $[D]$. We have

$$D = \sum_i n_i P_i \tag{5.9}$$

Now let $\{U_\alpha\}$ be a covering of Σ_p, then the restriction of D to any U_α is a meromorphic function f_α on U_α. Further, in the overlap $U_\alpha \cap U_\beta$ the two divisors $div\,(f_\alpha)$ and $div\,(f_\beta)$ coincide; as a consequence the *ratio*

$$\frac{f_\alpha}{f_\beta}$$

is non-vanishing and holomorphic on $U_\alpha \cap U_\beta$. This means that we can define the holomorphic line bundle $[D]$ over Σ_p by defining its transition functions $g_{\alpha\beta}$ to be this ratio. That is, we define

$$g_{\alpha\beta} = \frac{f_\alpha}{f_\beta} \tag{5.10}$$

The basic invariants of divisors and line bundles are their degree and first Chern class respectively. It is useful to note that if we take the divisor D and its associated line bundle $[D]$ then these two invariants are the same, or, in short

$$c_1([D]) = deg\,(D) \tag{5.11}$$

This can be proved by starting with D and constructing its transition function according to 5.10; then one applies the usual curvature formula to calculate the first Chern class. Another useful fact about divisors which is easily checked is that, if $deg(D) > 0$, then its associated bundle $[D]$ has a global holomorphic section and vice versa. Of course such a section will vanish somewhere, otherwise the bundle $[D]$ would be trivial.

This whole discussion shows that one may work either with (equivalence classes of) divisors or with holomorphic line bundles depending on which is more natural for the problem at hand.

A sheaf theoretic viewpoint can be deployed to formulate a divisor as an element of an appropriate sheaf: let $\mathcal{M}^*$ be the sheaf of non-trivial meromorphic functions on Σ_p where the group operation is multiplication of sections. Then if we take the other multiplicative sheaf that we have encountered, namely $\mathcal{O}^*$, we can form their quotient $\mathcal{D}$ given by

$$\mathcal{D} = \frac{\mathcal{M}^*}{\mathcal{O}^*} \tag{5.12}$$

A section of this sheaf $\mathcal{D}$ is a divisor D. Hence if we let $Div(M)$ be the set of all divisors on M then we have

$$Div(M) = H^0(M; \mathcal{M}^*/\mathcal{O}^*) \tag{5.13}$$

If we replace the Riemann surface Σ_p by a complex manifold M of dimension n greater than one then we can still define divisors. Instead of taking formal linear combinations of points we take formal linear combinations of analytic subvarieties V_i which have dimension $n-1$. Thus we write a divisor D as

$$D = \sum_i n_i V_i, \quad n_i \in \mathbf{Z} \tag{5.14}$$

In general the term codimension of a variety V is used to denote $\dim M - \dim V$. The subvarieties V_i have codimension one and so in the Riemann surface case they are just points.

One can also represent D locally by a meromorphic function just as we did in the Riemann surface case—a meromorphic function in n complex variables is defined as the ratio of two holomorphic functions, its singular set will generally have codimension one. We see that when $\dim M$ is large the use of divisors requires us to consider singularities which are not just points or lines but have codimension one. This has its unfortunate aspect— we would expect fewer technical problems to be caused by singularities of a higher codimension.

Moving on to the homology of the Riemann surface we recall that because Σ_p is topologically a sphere with p handles then its first homology group is given by $H_1(\Sigma_p; \mathbf{Z}) = \mathbf{Z} \oplus \cdots \oplus \mathbf{Z}$ ($2p$ times). A standard basis or set of generators for $H_1(\Sigma_p; \mathbf{Z})$ consists of p pairs of 1 cycles $(a_1, b_1), \ldots, (a_p, b_p)$ where a typical pair (a_i, b_i) encircles the ith handle of Σ_p so that a_i intersects b_i orthogonally. This means that the intersection number of any two of these cycles is particularly simple: denoting the number of intersections of two oriented cycles a and b by (a, b) we have

$$
\begin{aligned}
(a_i, a_j) &= (b_i, b_j) = 0 \\
(a_i, b_j) &= -(b_i, a_j) = \delta_{ij}
\end{aligned}
\tag{5.15}
$$

Next we pass to the dual and introduce a basis for $H^1(\Sigma_p; \mathbf{Z})$ consisting of p holomorphic 1-forms [2] $\omega_1, \ldots, \omega_p$ and their complex conjugates $\bar{\omega}_1, \ldots, \bar{\omega}_p$. These homology and cohomology bases are used to construct a $p \times 2p$ matrix Ω known as the period matrix and defined by

$$
\Omega = \begin{pmatrix} \int_{a_1} \omega_1 & \cdots & \int_{a_p} \omega_1 & \int_{b_1} \omega_1 & \cdots & \int_{b_p} \omega_1 \\ \vdots & & \vdots & \vdots & & \vdots \\ \int_{a_1} \omega_p & \cdots & \int_{a_p} \omega_p & \int_{b_1} \omega_p & \cdots & \int_{b_p} \omega_p \end{pmatrix}_{p \times 2p}
\tag{5.16}
$$

The $2p$ columns of Ω are referred to as periods. Each column can be regarded as a p-dimensional vector Π_i with $\Pi_i \in \mathbf{C}^p$ and, when this is done, it can be verified that the Π_i are linearly independent over $\mathbf{R}$. Now we specialise from real linear combinations of the Π_i to *integral* linear combinations. In this way we construct the lattice Λ of all integral linear combinations of the Π_i. This lattice is a subspace of the vector space $\mathbf{C}^p$ and so we can form a complex torus by passing to the quotient. This torus is called the Jacobian variety $J(\Sigma_p)$ of Σ_p. We have

$$
\Lambda = \{ m_1 \Pi_1 + \cdots + m_{2p} \Pi_{2p} : m_i \in \mathbf{Z} \} \quad \text{and} \quad J(\Sigma_p) = \frac{\mathbf{C}^p}{\Lambda}
\tag{5.17}
$$

Actually there is still some freedom left to simplify the bases chosen somewhat: the 'a-periods' correspond to a $p \times p$ sub-matrix consisting of the first p rows and columns of Ω and this sub-matrix is invertible; this allows us to normalise the a-periods so as to satisfy

$$
\int_{a_i} \omega_j = \delta_{ij}
\tag{5.18}
$$

---

[2] Holomorphic 1-forms are also known as Abelian differentials or differentials of the first kind.

The period matrix can now be written in the form

$$
\Omega = \begin{pmatrix} 1 & \cdots & 0 & \int_{b_1}\omega_1 & \cdots & \int_{b_p}\omega_1 \\ \vdots & \ddots & \vdots & \vdots & & \vdots \\ 0 & \cdots & 1 & \int_{b_1}\omega_p & \cdots & \int_{b_p}\omega_p \end{pmatrix}_{p\times 2p}
$$
$$
\equiv \begin{pmatrix} I_{p\times p} & \widetilde{\Omega}_{p\times p} \end{pmatrix}_{p\times 2p}
\tag{5.19}
$$

where $\widetilde{\Omega}$ is the matrix of b-periods and can also be referred to as the period matrix. $\widetilde{\Omega}$ has the property that it is symmetric and possesses a positive imaginary part. Two Riemann surfaces with the same homology basis are isomorphic if their period matrices are equal.

§ 3. Serre duality, line bundles and Kähler manifolds

In this section we do not restrict ourselves to Riemann surfaces and the symbol M denotes a closed compact complex manifold of any complex dimension n.

The first topic that we deal with is a generalisation of Poincaré duality which is appropriate for sheaf cohomology. This is known as Kodaira–Serre duality or Serre duality for short. The statement of Serre duality is as follows: Let $\Omega^p(E)$ denote the sheaf of holomorphic E-valued p-forms over M for some vector bundle E—we know from chapter 3 that this is the sheaf whose sections are given by the holomorphic elements of $\Gamma(M, \wedge^p T^* M \otimes E)$. Then, if M has complex dimension n, Serre duality is the isomorphism

$$
H^p(M, \Omega^q(E)) \simeq \left\{ H^{n-p}(M, \Omega^{n-q}(E^*)) \right\}^*
\tag{5.20}
$$

where, on the RHS, $*$ denotes the dual of any object to which it is applied. Notice that, when p is zero or n, referral to the space of sections which defines $\Omega^p(E)$ shows that

$$
\Omega^0(E) = \mathcal{O}(E) \qquad \text{and} \qquad \Omega^n(E) = \mathcal{O}(K_M \otimes E)
\tag{5.21}
$$

This gives us an alternative way of stating Serre duality when $q = 0$ namely

$$
H^p(M, \mathcal{O}(E)) \simeq \left\{ H^{n-p}(M, \mathcal{O}(K_M \otimes E^*)) \right\}^*
\tag{5.22}
$$

The detailed construction of the Serre dual uses wedge product and integration over M in the following fashion. Suppose that

$$
\theta \in H^p(M, \Omega^q(E)) \quad \text{and} \quad \nu \in H^{n-p}(M, \Omega^{n-q}(E^*))
$$

Then using the Dolbeault isomorphism

$$H^{(p,q)}_{\overline{\partial}_E}(M) = H^q(M, \Omega^p(E))$$

we can realise θ and ν as appropriate bundle valued forms on M. Having done this the action of the Serre dual is that the form ν acts on the form θ to produce the complex number

$$\int_M < \theta, \nu >$$

where the inner product $< \, , \, >$ denotes evaluation with respect to the Hermitian inner product used to define the ordinary dual E^* of a bundle E.

The language of sheaf cohomology used for the discussion is also suited to examining line bundles. This is particularly so if we wish to expose the difference between holomorphic line bundles and ordinary differentiable line bundles. We shall have a brief look at this topic. A complex line bundle L over a smooth real or complex manifold M is completely determined by its transition functions $g_{\alpha\beta}$. This transition function satisfies the cocycle condition $g_{\alpha\beta}g_{\beta\gamma} = g_{\alpha\gamma} = g_{\gamma\alpha}^{-1}$ and so provides an element $[g_{\alpha\beta}]$, say, of the cohomology group $H^1(M; \mathcal{E}^*)$ where $\mathcal{E}^*$ is the sheaf of smooth non-vanishing sections of M. We can insert $\mathcal{E}^*$ into a sheaf exact sequence in order to relate the transition function of L to its Chern class $c_1(L)$. This exact sequence is

$$0 \longrightarrow \mathbf{Z} \overset{i}{\longrightarrow} \mathcal{E} \overset{\exp}{\longrightarrow} \mathcal{E}^* \longrightarrow 0 \tag{5.23}$$

According to the sheaf cohomology treated in chapter 3 this exact sequence induces a cohomology sequence of the form

$$\cdots \longrightarrow H^1(M; \mathcal{E}) \longrightarrow H^1(M; \mathcal{E}^*) \overset{c_1}{\longrightarrow} H^2(M; \mathbf{Z}) \longrightarrow H^2(M; \mathcal{E}) \longrightarrow \cdots \tag{5.24}$$

However, we saw in chapter 3 that $\mathcal{E}$ is a fine sheaf and therefore has $H^i(M; \mathcal{E}) = 0$ for $i > 0$. Hence we have the isomorphism

$$H^1(M; \mathcal{E}^*) \simeq H^2(M; \mathbf{Z}) \tag{5.25}$$

Furthermore this isomorphism is realised by assigning each cocycle class $[g_{\alpha\beta}]$ in $H^1(M; \mathcal{E}^*)$ to its Chern class $c_1(L)$ in $H^2(M; \mathbf{Z})$. This means that all smooth complex line bundles L are completely determined, up to a differentiable isomorphism, by their Chern class $c_1(L)$. Thus if $c_1(L) = 0$ then L is trivial. Note that there is no such simple property possessed by vector bundles of higher rank: it does not follow that a vector bundle E of

rank $k > 1$, all of whose Chern classes vanish, is a product. Still more to
the point, in this chapter, a line bundle L with $c_1(L) = 0$ is differentiably
(and topologically) trivial but not necessarily *holomorphically* trivial. We
can delve deeper into this question using sheaves if we replace the various
sheaves of differentiable objects by their holomorphic counterparts.

Carrying out this replacement produces the two exact sequences

$$0 \longrightarrow \mathbf{Z} \overset{i}{\longrightarrow} \mathcal{O} \overset{\exp}{\longrightarrow} \mathcal{O}^* \longrightarrow 0$$

$$\to H^1(M; \mathbf{Z}) \to H^1(M; \mathcal{O}) \to H^1(M; \mathcal{O}^*) \overset{c_1}{\longrightarrow} H^2(M; \mathbf{Z}) \to H^2(M; \mathcal{O}) \to$$
(5.26)

However, $\mathcal{O}$ is not a fine sheaf and so, in general, $H^i(M; \mathcal{O}) \neq 0$; thus, for a
given M, $H^1(M; \mathcal{O}^*)$ need not be isomorphic to $H^2(M; \mathbf{Z})$. Now let us take
the transition functions $g_{\alpha\beta}$ to be holomorphic rather than just smooth;
that is, L is a *holomorphic* line bundle. Then L is not holomorphically determined by its Chern class $c_1(L)$. Moreover, holomorphic line bundles with
vanishing Chern class, though topologically and differentiably trivial, can
have rich holomorphic properties: for example, thinking of a holomorphic
line bundle L as a divisor, we saw in section §2 that a global holomorphic
section of L is a meromorphic function, and this is not holomorphically
trivial.

The set of all holomorphic line bundles forms a group which is a subset
of the group of all smooth line bundles. This group is called the Picard
group and is denoted by $Pic\,(M)$. From the above arguments it is clear
that

$$Pic\,(M) = H^1(M; \mathcal{O}^*) \tag{5.27}$$

The group of smooth line bundles is of course $H^1(M; \mathcal{E}^*)$. The Picard
group itself has a subgroup $Pic_0\,(M)$ which consists of those holomorphic
line bundles with zero Chern class. Using the exact sequence 5.26 above
enables us to identify $Pic_0\,(M)$. We have

$$Pic_0\,(M) = \frac{H^1(M; \mathcal{O})}{H^1(M; \mathbf{Z})} \tag{5.28}$$

We now turn our attention to Kähler manifolds. Manifolds of Kähler
type occupy an important position in the theory of complex manifolds. We
shall first give their definition and then consider some examples.

Definition (Kähler manifold) *Let M be a complex manifold with a Hermitian metric g expressed locally in the form $ds^2 = g_{ij}dz^i \otimes d\bar{z}^j$; using the metric we form an associated 2-form ω of type $(1,1)$ defined by*

$\omega = (i/2)g_{ij}dz^i \wedge d\bar{z}^j$. Then M is called Kähler if there exists at least one metric of this kind for which the 2-form ω is closed. That is, if

$$ds^2 = g_{ij}dz^i \otimes d\bar{z}^j, \qquad \omega = \frac{i}{2}g_{ij}dz^i \wedge d\bar{z}^j$$

$$and \qquad d\omega = 0 \tag{5.29}$$

A metric for which ω is closed is called a Kähler metric; if J is the usual matrix associated with the complex structure on M then ω can also be defined using the expression $\omega(X, Y) = g(JX, Y)$ where $\omega(X, Y)$ denotes ω evaluated on any two vector fields X and Y on M. Since ω is a closed non-degenerate two form it is symplectic; thus Kähler manifolds are examples of symplectic manifolds.

A simple example of a Kähler manifold is a complex torus $\mathbf{C}^n/\mathbf{Z}^n$ furnished with the Euclidean metric $g_{ij} = \delta_{ij}$—of course in this case ω trivially satisfies the closure requirement. Another easy class of Kähler manifolds to identify are the Riemann surfaces Σ_p. These are all automatically Kähler because $d\omega$ will always be zero, being a 3-form.

A less obvious example of a Kähler manifold is the projective space $\mathbf{CP}^n$ when it is given the Fubini–Study metric. To define the Fubini–Study metric g_{ij} let $(z_0, z_1, \ldots, z_n)$ be homogeneous coordinates on $\mathbf{CP}^n$ and let us use local coordinates z_i/z_0 for $i = 1, \ldots, n$; on setting $z_0 = 1$ these coordinates become $(z_1, \ldots, z_n)$. The Fubini–Study metric g_{ij} is the Hermitian metric on $\mathbf{CP}^n$ defined by

$$ds^2 = g_{ij}dz_i \otimes d\bar{z}_j = \frac{i}{2}\partial_i\bar{\partial}_j \ln(1 + z_1^2 + \cdots z_n^2)\, dz_i \otimes d\bar{z}_j$$

$$\Rightarrow g_{ij} = \frac{i}{2}\frac{\delta_{ij}(1 + |z|^2) - z_i\bar{z}_j}{(1 + |z|^2)^2} \tag{5.30}$$

$$\omega = \frac{i}{2}\partial\bar{\partial} \ln(1 + |z|^2) \qquad \text{where} \qquad |z|^2 = \sum_{k=1}^{n} z_k\bar{z}_k$$

It is immediate that the form ω is closed. It is interesting to note that the Fubini–Study metric may be written as $g_{ij} = \delta_{ij} + h_{ij}$ and, when this is done, h_{ij} is of order 2 in the coordinates z_i. This is a general property of Kähler metrics and is even enough to characterise them. More precisely, if a complex manifold M has a Hermitian metric g_{ij} which, in some coordinate system, can be written as $g_{ij} = \delta_{ij} + h_{ij}$, with h_{ij} of order 2 in these coordinates, then this metric is Kähler, rendering M a Kähler manifold.

If we embed a manifold M in a Kähler manifold N then the Kähler metric on N induces a metric on M. It is not difficult to check that the induced metric on N is a Kähler metric—all one has to do is to check that the induced 2-form ω on N is closed. Thus N is also Kähler. It therefore follows that any complex manifold embeddable in projective space $\mathbf{CP}^n$ is automatically Kähler.

Kählerity implies certain facts about the cohomology of M which are expressed in terms of Hodge theory. In Hodge theory one uses operators of Laplacian type to investigate cohomology. For a complex Hermitian manifold the three differential operators d, ∂ and $\bar{\partial}$ give rise to three separate Laplacians, namely

$$\Delta_d = (dd^* + d^*d), \quad \Delta_\partial = (\partial\partial^* + \partial^*\partial), \quad \Delta_{\bar{\partial}} = (\bar{\partial}\bar{\partial}^* + \bar{\partial}^*\bar{\partial}) \qquad (5.31)$$

There is no particular relation between these Laplacians but on a Kähler manifold they are identical up to an innocuous factor of 2. One has (Wells [1])

$$\Delta_d = 2\Delta_\partial = 2\Delta_{\bar{\partial}} \qquad (5.32)$$

In chapter 4 we encountered the Hodge numbers $h_{p,q}$ defined by

$$h_{p,q} = \dim H_{\bar{\partial}}^{p,q}(M) \qquad (5.33)$$

Let us decompose the cohomology of M according to

$$H^i(M;\mathbf{C}) = \bigoplus_{p+q=i} H^{(p,q)}(M;\mathbf{C}) \qquad (5.34)$$

Now if we apply Hodge theory in the usual way to represent the above cohomology using harmonic forms, and combine this with the Laplacian properties just mentioned, we obtain some strong results. Among these are the following

(i) $b_i = \sum_{p+q=i} h_{p,q}$
(ii) b_2 is always positive
(iii) b_{2i+1} is always even
(iv) $h_{p,q} = h_{q,p}$
(v) $h_{p,p}$ is always positive.

§ 4. The Teichmüller space $\mathfrak{T}_p$

We have already referred to the existence of families of Riemann surfaces with the same genus; that is to say, families of Riemann surfaces with different complex structures. In this section we initiate a study of these families.

To do this consider an arbitrary Riemannian metric on a Riemann surface Σ, i.e. specify [3]

$$ds^2 = Adx^2 + 2Bdxdy + Cdy^2$$

It is always possible to rewrite ds^2 in terms of z and $\bar{z}$ as

$$\begin{aligned} ds^2 &= \lambda(dz + \mu d\bar{z})(d\bar{z} + \bar{\mu}dz) \\ &= \lambda \left| dz + \mu d\bar{z} \right|^2 \end{aligned} \tag{5.35}$$

for some appropriate functions λ and μ where $\lambda > 0$. Thus a metric also determines a complex structure on the Riemann surface.

Now our object is to single out distinct complex structures on Σ; and here we point out that there is a one to one correspondence between complex structures and conformal equivalence classes of metrics. This is because a conformal change of metric consists merely of multiplying the function λ of 5.35 by some positive function f, say, and this leaves the complex structure unchanged. Consequently, this means that if we have a *second metric* specified by

$$ds_1^2 = \lambda_1 \left| dz + \mu_1 d\bar{z} \right|^2$$

then when $\mu = \mu_1$ these metrics are in the same conformal equivalence class. Thus the parameters μ label the complex structures on Σ and comprise the space which we wish to investigate.

Now let us see in a more geometric fashion what this space looks like. Let $Met(\Sigma)$ denote the space of metrics on Σ. Now if two metrics g_1 and g_2 are conformally equivalent it means that $g_1 = fg_2$ for some positive function f. If we define $C_+^\infty(\Sigma)$ to be the space of all smooth positive functions on Σ then $C_+^\infty(\Sigma)$ is a group that acts on $Met(\Sigma)$ and we wish to form the quotient of $Met(\Sigma)$ by this action. We shall denote this quotient by $Conf(\Sigma)$ and we summarise the definition of this action, which we call A, and the definition of $Conf(\Sigma)$ below.

$$\begin{aligned} A : C_+^\infty(\Sigma) \times Met(\Sigma) &\longrightarrow Met(\Sigma) \\ (f(x), g(x)) &\longmapsto f(x)g(x) \\ Conf(\Sigma) = \frac{Met(\Sigma)}{C_+^\infty(\Sigma)} \end{aligned} \tag{5.36}$$

In searching for metrics on Σ, we do not wish to count two metrics as different if one is mapped into the other under a diffeomorphism. To ensure

---

[3] From now on we often simplify the notation and denote a Riemann surface by Σ rather than Σ_p.

that this does not happen we shall replace $Conf(\Sigma)$ by the space that results when we quotient out by the action of such diffeomorphisms. It is natural to restrict ourselves to orientation *preserving* diffeomorphisms. This is because we are studying distinct complex structures and an orientation *reversing* diffeomorphism will change the complex structure to its complex conjugate structure.

To this end we introduce the notation $Diff^+(\Sigma)$ to denote the space of orientation preserving diffeomorphisms of Σ. This space $Diff^+(\Sigma)$ has a normal subgroup consisting of those diffeomorphisms continuously connected to the identity and we denote this subgroup by $Diff_0(\Sigma)$. For the moment we temporarily restrict ourselves to only those diffeomorphisms in $Diff_0(\Sigma)$; we shall lift this restriction in §5. The quotient that we obtain is well behaved if the genus p of the Riemann surface Σ is 2 or more; and in that case we have indeed obtained the space of complex structures on Σ: the Teichmüller space $\mathcal{T}_p$. Thus we have

$$\frac{Conf(\Sigma)}{Diff_0(\Sigma)} = \mathcal{T}_p, \quad p \geq 2 \tag{5.37}$$

Unfortunately, for $p < 2$ this quotient is singular because, for these values of p, $Diff_0(\Sigma)$ does not act freely [4] on $Conf(\Sigma)$. The existence of fixed points for the action of $Diff_0(\Sigma)$ on $Conf(\Sigma)$ is sometimes referred to as the existence of *conformal Killing vectors* on Σ. Geometrically speaking, these fixed points correspond to metrics which are mapped onto a conformally equivalent counterpart under the action of a (non-trivial) coordinate transformation in $Diff_0(\Sigma)$.

We can resolve this difficulty by giving an alternative definition of $\mathcal{T}_p$ valid for all p. The key to our being able to do this is that, for each $g \in Met(\Sigma)$, there exists a constant curvature metric g_c conformal to g. Also, if $p \neq 1$, this metric g_c is unique, and when $p = 1$ we supply an additional condition which brings about uniqueness. These constant curvature

---

[4] As usual a free action of a group G on a manifold M is one for which there are no fixed points. More precisely, if we represent the action by the smooth map $A : G \times M \to M$, and use the customary notation to write $A(g, x) = g \cdot x$, then A must satisfy $e \cdot x = x$ where e is the identity in G and $(gh) \cdot x = g \cdot (h \cdot x)$. The action A is *free* if $g \cdot x = x \Rightarrow g = e$. The quotient M/G is the space of orbits and has dimension $\dim M - \dim G$. If the action A is not free consider the following *example* of what can go wrong: A has at least one point x_0 with a non-trivial stability group $G_0 \subset G$ and we may have $\dim G_0 > 0$. In this case x_0 belongs to an orbit of dimension $\dim M - \dim G - \dim G_0$ and this renders the orbit space M/G singular since it is no longer of the same dimension everywhere.

metrics g_c are obtained by using the property of Riemann surfaces known as Riemann uniformisation.

Riemann uniformisation can be introduced as follows. Any compact Riemann surface Σ always has a simply connected universal cover $\tilde{\Sigma}$. Also the standard theory of covering spaces tells us that Σ is realised as the quotient of $\tilde{\Sigma}$ by a discrete group which must be isomorphic to its fundamental group $\pi_1(\Sigma)$. In short

$$\Sigma = \tilde{\Sigma}/\pi_1(\Sigma) \tag{5.38}$$

However, uniformisation says that there are only three simply connected (compact or non-compact) Riemann surfaces $\tilde{\Sigma}$ and that each of them has a unique complex structure, this complex structure being determined by a metric of *constant curvature*. We display below these three simply connected surfaces $\tilde{\Sigma}$ together with the Riemann surfaces Σ that they cover. We have

$$\tilde{\Sigma} = \begin{cases} S^2 & \text{if } p = 0, \text{ i.e. } \Sigma \simeq S^2 \\ \mathbf{C} & \text{if } p = 1, \text{ i.e. } \Sigma \simeq T^2 \\ H^2 & \text{if } p \geq 2, \text{ i.e. } \Sigma \text{ is homeomorphic to a sphere with } p \text{ handles} \end{cases} \tag{5.39}$$

where H^2 is the hyperbolic plane (or Poincaré upper-half-plane) given by $Im\, z > 0$. The constant curvature metrics possessed by the $\tilde{\Sigma}$ are

for S^2 the natural metric that S^2 acquires on being embedded in $\mathbf{R}^3$

for $\mathbf{C}$ the ordinary flat Euclidean metric on the plane

for H^2 the usual hyperbolic or Poincaré upper-half-plane metric,

i.e. the curvature -1 metric determined by $ds^2 = dzd\bar{z}/(z - \bar{z})^2$

$$\tag{5.40}$$

Thus, no matter what the genus of Σ, its universal cover must be one of the $\tilde{\Sigma}$ just described; also, the covering projection $\pi : \tilde{\Sigma} \to \Sigma$ induces a constant curvature metric on Σ.

Now we take a section s of $Met(\Sigma)$, considering the latter as a bundle over $Conf(\Sigma)$. The definition of s is that each conformal equivalence class $c \in Conf(\Sigma)$ has

$$s(c) = g_c \in Met(\Sigma) \tag{5.41}$$

where each g_c is required to have constant curvature equal to: $+1$ for $p = 0$, 0 for $p = 1$, and -1 for $p \geq 2$—note that the existence of such an s follows straight from the statement of Riemann uniformisation. Also g_c is not yet unique—it can still be changed by an element of $Diff_0(\Sigma)$. Thus, using $R(g)$ to denote the Gaussian curvature of a metric g, we define the space

$Met_{const}(\Sigma)$ by,

$$Met_{const}(\Sigma) = \begin{cases} \{g \in Met(\Sigma) : R(g) = 1\}, & \text{if } p = 0 \\ \{g \in Met(\Sigma) : R(g) = 0, vol(\Sigma) = 1\}, & \text{if } p = 1 \\ \{g \in Met(\Sigma) : R(g) = -1\}, & \text{if } p \geq 2 \end{cases} \quad (5.42)$$

where the specification $vol(\Sigma) = 1$ for the $p = 0$ case is because in that case Σ is the torus T^2. But T^2 has zero Euler characteristic χ and therefore the Gauss–Bonnet theorem imposes no normalisation restriction on its volume. Finally, our alternative definition of $\mathcal{T}_p$, valid for all genera p, is simply

$$\mathcal{T}_p = \frac{Met_{const}(\Sigma)}{Diff_0(\Sigma)}, \quad p = 0, 1, \ldots \quad (5.43)$$

§ 5. The moduli space $\mathcal{M}_p$

In this section we no longer restrict the space of diffeomorphisms to the connected component of the identity but consider all orientation preserving diffeomorphisms. In short we use the group $Diff^+(\Sigma)$ instead of its normal subgroup $Diff_0(\Sigma)$. If we make this sole change in the definition 5.43 above then the new space that we obtain is the moduli space $\mathcal{M}_p$. Displaying the Teichmüller space and the moduli space together we have

$$\mathcal{T}_p = \frac{Met_{const}(\Sigma)}{Diff_0(\Sigma)}, \qquad \mathcal{M}_p = \frac{Met_{const}(\Sigma)}{Diff^+(\Sigma)} \quad (5.44)$$

In addition it is useful to define the mapping class group Γ_Σ by taking the quotient of $Diff^+(\Sigma)$ by its normal subgroup $Diff_0(\Sigma)$. Then we can exhibit the following relationship between the Teichmüller space and the moduli space.

$$\mathcal{M}_p = \frac{\mathcal{T}_p}{\Gamma_\Sigma}, \qquad \Gamma_\Sigma = \frac{Diff^+(\Sigma)}{Diff_0(\Sigma)} \quad (5.45)$$

The mapping class group Γ_Σ is a *discrete* group; more abstractly, Γ_Σ is the group of connected components of $Diff^+(\Sigma)$.

The moduli space is the complete space of complex structures on Σ; however, it is not a manifold since, as we shall see in this section, Γ_Σ does not act freely on $\mathcal{T}_p$ but has fixed points—equivalently one could say that $Diff^+(\Sigma)$ does not act freely on $Met_{const}(\Sigma)$. A singular quotient space such as $\mathcal{M}_p$, which fails to be a smooth space of orbits, is referred to as an *orbifold*.

We can obtain some insight into the structure of $\mathcal{M}_p$ by using the notion of Riemann uniformisation as we did for $\mathcal{T}_p$. We know from the previous section that any Σ is given by a quotient of the form

$$\Sigma = \frac{\widetilde{\Sigma}}{\pi_1(\Sigma)} \tag{5.46}$$

To construct the RHS of 5.46 we need an action of $\pi_1(\Sigma)$ on $\widetilde{\Sigma}$. This action must satisfy two requirements. To ensure Σ is a smooth manifold this action must be free (no fixed points); and to provide Σ with a complex structure this action must also preserve the complex structure on $\widetilde{\Sigma}$. We satisfy these two requirements by demanding that the action be a holomorphic isomorphism of $\widetilde{\Sigma}$. Thus, all possible Σ are constructed by finding all possible group actions of this kind. The space of all these group actions will be the moduli space.

Let us start with genus zero. Then $\widetilde{\Sigma}$ and Σ coincide, which is consistent with the fact that $\pi_1(S^2) = 0$. On the other hand, elementary complex analysis tells us that there are no holomorphic automorphisms of S^2 without fixed points; thus the moduli space at genus zero consists of a single point.

Moving on to genus one, the covering space $\widetilde{\Sigma}$ is the complex plane $\mathbf{C}$ and Σ is the torus T^2 whose fundamental group is $\mathbf{Z} \oplus \mathbf{Z}$. Now an obvious way that we can make $\mathbf{Z} \oplus \mathbf{Z}$ act holomorphically and invertibly on $\mathbf{C}$ is to let $\mathbf{Z} \oplus \mathbf{Z}$ act as a pair of translations. We can always adjust one of the translations to be unity; the remaining translation λ, say, is a local modular parameter and labels complex structures on the torus. It is easy to check that we can take $Im\,\lambda > 0$. A general translation is now of the form $z \mapsto z + m + n\lambda$, with m and n integers, and $\lambda \in \mathbf{C}^+$ where $\mathbf{C}^+$ is the upper-half-plane of $\mathbf{C}$. This space $\mathbf{C}^+$ is actually the Teichmüller space for genus one, that is

$$\mathcal{T}_1 = \mathbf{C}^+ \tag{5.47}$$

However, we have not yet proved that distinct complex structures on a torus correspond to distinct λ—in fact there is still some freedom to change λ without changing the complex structure. This freedom exists because there are discrete automorphisms of the torus and we shall now find these automorphisms.

Let T_1^2 and T_2^2 be two tori with complex structures labelled by λ_1 and λ_2 respectively. Suppose the complex structure on T_1^2 is mapped into that on T_2^2 by a conformal map

$$f : T_1^2 \longrightarrow T_2^2 \tag{5.48}$$

Lift f to the covering space of each torus; it then becomes a conformal map $\tilde{f} : \mathbf{C} \to \mathbf{C}$ which will be required to commute with the covering projections p_1 and p_2, say, of T_1^2 and T_2^2. It is immediate that $\tilde{f}(z) = pz + q$ since any conformal map of $\mathbf{C}$ is of this form. This gives us the commutative diagram

$$
\begin{array}{ccc}
\mathbf{C} & \overset{\tilde{f}}{\longrightarrow} & \mathbf{C} \\
{\scriptstyle p_1}\downarrow & & \downarrow{\scriptstyle p_2} \\
T_1^2 & \overset{f}{\longrightarrow} & T_2^2
\end{array}
\qquad
\begin{array}{l}
\text{with } f \circ p_1 = p_2 \circ \tilde{f} \\
\text{and } \tilde{f}(z) = pz + q
\end{array}
\qquad (5.49)
$$

Now let us denote the groups of translations on the tori T_1^2 and T_2^2 by G_1 and G_2 respectively. These two groups are, of course, both isomorphic to the fundamental group $\mathbf{Z} \oplus \mathbf{Z}$ of an abstract torus; the difference between G_1 and G_2 in the present context lies in their actions on $\mathbf{C}$. A general element of G_1 acts on $\mathbf{C}$ as a translation $z \mapsto z + m + n\lambda_1$ and the corresponding element for G_1 is of the form $z \mapsto z + m + n\lambda_2$. The way we progress to something concrete is to observe that the condition 5.49 above on the map $\tilde{f}$ becomes more explicit when stated in terms of group actions. We have the following: given a $g_2 \in G_2$, and denoting each group action by a dot, we can say that

$$
\begin{aligned}
\tilde{f}(g_1 \cdot z) &= g_2 \cdot \tilde{f}(z) \\
\Rightarrow g_1 \cdot z &= \tilde{f}^{-1}(g_2 \cdot \tilde{f}(z))
\end{aligned}
\qquad \text{for some } g_1 \in G_1 \qquad (5.50)
$$

Now let us choose g_2 to be the two generators $z \mapsto z + \lambda_2$ and $z \mapsto z + 1$ of G_2 in turn. This provides us with two equations, namely

$$
\begin{aligned}
z + b + a\lambda_1 &= \frac{\{(pz + q) + \lambda_2 - q\}}{p} = z + \frac{\lambda_2}{p} \\
z + d + c\lambda_1 &= \frac{\{(pz + q) + 1 - q\}}{p} = z + \frac{1}{p}
\end{aligned}
\qquad \text{with } a, b, c, d \in \mathbf{Z} \quad (5.51)
$$

We can straightaway deduce that

$$
\lambda_2 = \frac{a\lambda_1 + b}{c\lambda_1 + d}, \qquad a, b, c, d \in \mathbf{Z} \qquad (5.52)
$$

But the invertibility of $\tilde{f}$ imposes the restriction

$$
det \begin{pmatrix} a & b \\ c & d \end{pmatrix} \neq 0 \qquad (5.53)
$$

Now if we note that this determinant is an integer, and the symmetry between T_1^2 and T_2^2 demands the same of its inverse, then it follows that $(ad - bc) = \mp 1$. Furthermore, if we evaluate the imaginary part of both sides of 5.52 and use the positivity of $Im\,\lambda_1$ and $Im\,\lambda_2$, then we find that $(ad - bc) = +1$.

Thus we have found a criterion for testing whether two toroidal complex structures, labelled by $\lambda_1, \lambda_2 \in \mathbf{C}^+$, are the same. The criterion is that λ_1 and λ_2 are related by

$$\lambda_2 = \frac{a\lambda_1 + b}{c\lambda_1 + d} \qquad \text{with } a, b, c, d \in \mathbf{Z}, \quad \text{and } ad - bc = 1 \qquad (5.54)$$

The maps defined in 5.54 generate the modular group $SL(2, \mathbf{Z})/\{\mp I\}$—the modular group is a discrete subgroup of the group of Möbius transformations which are the set of transformations

$$z \longmapsto \frac{az + b}{cz + d}, \qquad a, b, c, d \in \mathbf{C}, \; ad - bc = 1 \qquad (5.55)$$

All the signs of a, b, c, d can be changed in the above definition without changing the transformation. Therefore the Möbius group is isomorphic to $SL(2, \mathbf{C})/\{\mp I\}$; we can now see that the modular group is the discrete subgroup of the Möbius group obtained by restricting a, b, c, d to be integers.

Having obtained the full description of the complex structures on the torus we can display the moduli space at genus one. We have

$$\mathcal{M}_1 = \frac{\mathcal{J}_1}{\Gamma_{T^2}} = \frac{\mathbf{C}^+}{SL(2, \mathbf{Z})/\{\mp I\}} \qquad (5.56)$$

The mapping class group of the torus has now been identified: we have $\Gamma_{T^2} \simeq SL(2, \mathbf{Z})/\{\mp I\}$. We also know that the non-triviality of the mapping class group means that the torus has non-trivial discrete automorphisms, i.e. those automorphisms which, considered as elements of $Diff^+(\Sigma)$, belong to a *disconnected* component. We can write down these automorphisms: Let (x, y) be coordinates of a point on the torus considered as a square with opposite ends identified; then the discrete automorphisms of T^2 are just the maps

$$(x, y) \longmapsto (ax + by, cx + dy), \qquad a, b, c, d \in \mathbf{Z}, \quad ad - bc = 1 \qquad (5.57)$$

When we identify λ's which differ by a modular transformation we convert $\mathcal{J}_1$ into $\mathcal{M}_1$. The actual identifications are that λ is identified with $-1/\lambda$ and also with $\lambda + 1$ (actually the maps $\lambda \mapsto -1/\lambda$ and $\lambda \mapsto \lambda + 1$ generate the modular group, they are commonly denoted by S and T respectively; also, the inversion carried out by S has the effect of interchanging the two radii of the torus.). Furthermore, in carrying out this quotient of $\mathbf{C}^+$ by the modular group we can easily check that the transformations $\lambda \mapsto -1/\lambda$ and $\lambda \mapsto (1 - 1/\lambda)$ each have a fixed point. These fixed points are responsible for giving $\mathfrak{M}_1$ the singularities to which we alluded above. The resulting moduli space $\mathfrak{M}_1$ has cusps or point singularities.

Notice that these discrete automorphisms also possess fixed points when they act on the space $Met_{const}(\Sigma)$. For example, represent T^2 by a unit square provided with the Euclidean metric δ_{ij}. Subject T^2 to the discrete automorphism which corresponds to $\lambda \mapsto -1/\lambda$. Referring to 5.54 and 5.57 we see that we must select $a = d = 0$ and $b = -c = 1$, and, having done this, we find that $(x, y) \mapsto (-y, -x)$. However, this latter automorphism of T^2 is an isometry since it preserves the Euclidean metric; therefore δ_{ij} is a fixed point of this automorphism. Thus we see again that the quotient

$$Met_{const}(T^2)/Diff^+(T^2) = \mathfrak{M}_1$$

is an orbifold.

For a general genus $p \geq 2$ a similar situation pertains. The group of automorphisms of the hyperbolic plane H^2 is $SL(2, \mathbf{R})/\{\mp I\}$ acting on $z \in H^2$ by

$$z \longmapsto \frac{az + b}{cz + d}, \qquad a, b, c, d \in \mathbf{R}, \qquad ad - bc = 1 \qquad (5.58)$$

To produce a Riemann surface of genus $p \geq 2$ we select a discrete fixed point free subgroup G, say, of $SL(2, \mathbf{R})/\{\mp I\}$ and form the quotient H^2/G. These G are called Fuchsian groups. Now if we consider the moduli space we know that the existence of Riemann surfaces with metrics possessing a discrete isometry group gives rise to the singularities in the moduli space $\mathfrak{M}_p$. As the genus p increases, things improve somewhat and it is known (Griffiths and Harris [1]) that, for $p \geq 3$, the singularities of $\mathfrak{M}_p$ have codimension at least two. In addition, the total number of automorphisms possible is no longer infinite and can be bounded from above. More precisely, if $p \geq 2$ then this bound is the number $84(p - 1)$.

As well as $\mathfrak{M}_p$ being an orbifold it also possesses non-trivial topology. For example, $\mathfrak{M}_1 = \mathfrak{T}_1/\Gamma_{T^2}$ from which it follows that $\pi_1(\mathfrak{M}_1) = SL(2, \mathbf{Z})/\{\mp I\}$. When $p \geq 2$ the Teichmüller space $\mathfrak{T}_p$ is a contractible non-compact complex manifold homeomorphic to $\mathbf{C}^{3p-3}$; it is also the universal covering space for the moduli space $\mathfrak{M}_p$.

In general the mapping class group of a Riemann surface Σ can be generated by a series of 2π rotations known as Dehn twists. To define a Dehn twist we take a simple closed curve C on Σ and thicken it to a neighbourhood of C; then we excise this neighbourhood, give one of its boundaries a 2π rotation, and glue it back in again. The result is a surface diffeomorphic to Σ and the diffeomorphism is a Dehn twist. The set of all these Dehn twists generate the mapping class group of Σ although they are clearly not a minimal set of generators. Important curves are those which are not contractible, such as those which encircle a handle of Σ, or, in the case of punctured Riemann surfaces to which this result extends, curves which enclose a puncture.

We shall return to both the moduli space and the Teichmüller space in chapter 9 when we discuss string theory. In that discussion we shall supply $\mathcal{T}_p$ with a natural metric known as the Weil–Petersson metric. This metric is rather special since it turns out that, with respect to this metric, $\mathcal{T}_p$ is actually a Kähler manifold.

§ 6. The dimension of the moduli space

Having defined $\mathcal{M}_p$ in a geometrical way we now set ourselves the task of calculating its dimension. This will entail a return to complex analysis and cohomology techniques. We have already seen that the metric on Σ can be written as

$$ds^2 = \lambda \left| dz + \mu d\bar{z} \right|^2 \tag{5.59}$$

and that the function μ labels the possible complex structures on Σ. We begin our task by perturbing the metric g on Σ and in so doing produce a perturbation $\delta\mu$ in the parameter μ. This perturbation $\delta\mu$ will belong to the tangent space $T_{[g]}\mathcal{M}_p$ to $\mathcal{M}_p$ at $[g]$ ($[g]$ denotes the conformal equivalence class of the metric g). Hence if we can calculate the dimension of this space of $\delta\mu$'s we have calculated the dimension of $T_{[g]}\mathcal{M}_p$, which is the same as $\dim \mathcal{M}_p$. To make progress we shall need to know the transformation properties of λ and μ under a coordinate change. The invariance of ds^2 under such a change allows us to deduce their transformation laws from that of the metric. If f is a holomorphic coordinate change then we find that

$$\lambda(f(z))f_z\bar{f}_z = \lambda(z)$$
$$\mu(f(z))\frac{\bar{f}_z}{f_z} = \mu(z) \tag{5.60}$$

The obvious geometrical invariant quantity determined by μ is given by the expression

$$\mu(z,\bar{z})\frac{\partial}{\partial z} \otimes d\bar{z} \tag{5.61}$$

which is known as a Beltrami differential (Incidentally for a general metric, parametrised by λ and μ, a diffeomorphism $f(z, \bar{z})$ is conformal if $f_{\bar{z}}/f_z = \mu$. This is called the Beltrami equation.).

We begin the calculation of $\dim \mathfrak{M}_p$ by perturbing the metric g_{ab} by an amount δg_{ab}. To carry out this calculation we express g_{ab} in isothermal coordinates. These are coordinates which have the property that, in some patch on Σ, the corresponding line element takes the form

$$ds^2 = \lambda |dz|^2$$

The infinitesimally perturbed line element, $d\tilde{s}^2$ say, is

$$d\tilde{s}^2 = (\lambda + \delta\lambda)|dz + \delta\mu d\bar{z}|^2$$

Thus the perturbation from which one obtains δg_{ab} is simply

$$\delta(ds^2) = d\tilde{s}^2 - ds^2$$

If we neglect higher order terms then we obtain

$$\delta(ds^2) = \delta\lambda dz d\bar{z} + \lambda \delta\mu d\bar{z}d\bar{z} + \lambda \delta\bar{\mu} dz dz$$

a formula from which one can read off δg_{ab}. It is now clear that we simply require the component $\delta g_{\bar{z}\bar{z}}$ and in fact

$$\delta\mu = \frac{\delta g_{\bar{z}\bar{z}}}{g_{\bar{z}z}} \tag{5.62}$$

Also, if we write $\delta g_{ab} = \epsilon h_{ab}$ where ϵ is small and h_{ab} is another symmetric tensor, then 5.62 becomes

$$\delta\mu = \epsilon \frac{h_{\bar{z}\bar{z}}}{g_{\bar{z}z}} \tag{5.63}$$

and so the perturbation $\delta\mu$ is determined by the ratio $h_{\bar{z}\bar{z}}/g_{\bar{z}z}$. But this ratio has the same tensorial structure as μ itself and so determines a tensor like that in 5.61. In other words, we can represent it by the tensor

$$\nu(z, \bar{z})\frac{\partial}{\partial z} \otimes d\bar{z} \tag{5.64}$$

for some function ν. It is now opportune to demonstrate that this tensor belongs to $ker\, \bar{\partial}_{K_{\Sigma}^*}^{(0,1)}$. This is because the space $ker\, \bar{\partial}_{K_{\Sigma}^*}^{(0,1)}$ consists of $\bar{\partial}$-closed forms of type (0,1) on Σ with coefficients in K_{Σ}^*. Now Σ has complex dimension 1, so $K_\Sigma = T^*\Sigma$. Thus it follows that $K_\Sigma^* = T\Sigma$; and, for the same reason, all 1-forms of type (0,1) on Σ are $\bar{\partial}$-closed. Also a $(0,1)$ form with coefficients in $K_\Sigma^* = T\Sigma$ is simply a vector valued form representable in local coordinates as

$$\nu d\bar{z} \equiv \nu(z, \bar{z}) \frac{\partial}{\partial z} \otimes d\bar{z} \tag{5.65}$$

Thus any element of $ker\,\bar{\partial}_{K_\Sigma^*}^{(0,1)}$ gives rise to a perturbation in μ.

But from our work earlier in this section we know that we must quotient the space of $\delta\mu$'s by the space $Diff_0(\Sigma)$ of diffeomorphisms. Let us now see how this is done. To make progress we need a more explicit description of the group $Diff_0(\Sigma)$. This is found by considering the Lie algebra of $Diff_0(\Sigma)$. The Lie algebra of $Diff_0(\Sigma)$ is just the space of smooth vector fields on Σ; we write this space as $vect(\Sigma)$ (not to be confused with $Vect(\Sigma)$ which is the set of vector bundles over Σ). To check this fact let $V \in vect(\Sigma)$, then we can exponentiate to $Diff_0(\Sigma)$ to obtain the one parameter family $\alpha_t = \exp[tV]$ of diffeomorphisms. The meaning of the map α_t is that it represents the flow on Σ of the vector field V: consider the integral curve of V that passes through a particular point x, and let $x(t)$ be the point labelled by t on this curve. Then α_t is defined by

$$\begin{aligned} \alpha_t : \Sigma &\longrightarrow \Sigma \\ x &\longmapsto x(t) \end{aligned} \tag{5.66}$$

This map α_t induces an action A_t, say, on $Met(\Sigma)$ via a push-forward, i.e. via $(\alpha_t)_*$. For notational convenience we summarise this below

$$\begin{aligned} A_t : Diff_0(\Sigma) \times Met(\Sigma) &\longrightarrow Met(\Sigma) \\ (\alpha_t, g) &\mapsto \alpha_t \cdot g \end{aligned} \tag{5.67}$$

with $(\alpha_t \cdot g)(x) = ((\alpha_t)_* g)(x) = g(\alpha_t(x)) = g(x(t))$. Now denote the metric $\alpha_t \cdot g$ by $g(t)$, which can be thought of as an integral curve on $Met(\Sigma)$. Then we have

$$g(t) = g(0) + tL_V g(0) + \cdots \tag{5.68}$$

where L_V denotes the Lie derivative with respect to V. Thus

$$g'(0) = L_V g(0) \tag{5.69}$$

and $g'(0)$ is the set of components of a tangent vector to $Met(\Sigma)$ (compare the finite dimensional case where we replace $g(t)$ by $x(t)$ an integral curve on the manifold Σ generated by a tangent vector $Y = Y^i(\partial/\partial x^i)$. Then $L_Y x^j = Y x^j = Y^i(\partial x^j/\partial x^i) = Y^j)$. In any case we have found that [5]

$$L_V g_{ab} \in T_g Met(\Sigma) \tag{5.70}$$

Applying this to the calculation at hand means that we must quotient out those $\delta\mu$ whch are obtained from any δg_{ab} of the form

$$\delta g_{ab} = L_V g_{ab}$$

This means that $\delta\mu$ takes the form

$$\delta\mu = \frac{L_V g_{\bar{z}\bar{z}}}{g_{z\bar{z}}}$$

The formula for the Lie derivative of a tensor gives

$$g'_{ab}(0) = L_V g_{ab} = \frac{\partial g_{ab}}{\partial x^i}V^i + g_{ib}\frac{\partial V^i}{\partial x^a} + g_{ai}\frac{\partial V^i}{\partial x^b} \tag{5.71}$$

Thus we find for $\delta\mu$ the equation

$$\delta\mu = \frac{\{g_{\bar{z}\bar{z},z}V^z + g_{\bar{z}\bar{z},z}V^z + g_{zz}V^z_{,\bar{z}} + g_{\bar{z}z}V^{\bar{z}}_{,\bar{z}} + g_{\bar{z}z}V^z_{,\bar{z}} + g_{\bar{z}z}V^{\bar{z}}_{,z}\}}{g_{\bar{z}z}}$$

But a glance at ds^2 shows that the only non-zero components of g_{ab} are $g_{\bar{z}z}$ and so we obtain the simple result that

$$\delta\mu = 2V^z_{,\bar{z}}$$
$$= 2\partial_{\bar{z}}V^z \tag{5.72}$$

Thus we must quotient by those $\delta\mu$ of the form $\partial_{\bar{z}}V^z$ for any vector field $V = V^z\partial/\partial z + V^{\bar{z}}\partial/\partial\bar{z}$ (the factor 2 in 5.72 is not of course significant). But $V^z\partial/\partial z$ is just a section of the bundle $T\Sigma = K^*_\Sigma$. Thus a $\delta\mu$ given by 5.72 is just the image under $\bar{\partial}$ of a section of K^*_Σ; that is, it is an element of

$$Im\,\bar{\partial}^{(0,0)}_{K^*_\Sigma} \tag{5.73}$$

---

[5] In general if a metric g satisfies $L_K g = Cg$, for some C, then K is an example of a conformal Killing vector for the metric g.

It is therefore expressible locally as $V_{\bar{z}}^{z}\partial/\partial z \otimes d\bar{z}$. The cohomological interpretation of the μ perturbations can now be given since we have identified the space of infinitesimal $\delta\mu$'s to be the quotient

$$\frac{ker\ \bar{\partial}_{K_{\Sigma}^{*}}^{(0,1)}}{Im\ \bar{\partial}_{K_{\Sigma}^{*}}^{(0,0)}} \tag{5.74}$$

However, this is immediately recognisable as a Dolbeault cohomology group. In terms of Dolbeault cohomology we have

$$H_{\bar{\partial}_{K_{\Sigma}^{*}}}^{(0,1)}(\Sigma) = \frac{ker\ \bar{\partial}_{K_{\Sigma}^{*}}^{(0,1)}}{Im\ \bar{\partial}_{K_{\Sigma}^{*}}^{(0,0)}} \tag{5.75}$$

In view of the fact that this space is also the tangent space to the moduli space we write

$$H_{\bar{\partial}_{K_{\Sigma}^{*}}}^{(0,1)}(\Sigma) = T_{[g]}\mathcal{M}_{p} \tag{5.76}$$

where, as before, $[g]$ denotes the point in moduli space determined by the conformal equivalence class $[g]$ of the metric g.

We have just seen that the Beltrami differentials form the tangent space to $\mathcal{M}_{p}$. The cohomology language that we employed therein is of great help in identifying the corresponding cotangent space $T_{[g]}^{*}\mathcal{M}_{p}$. All we have to do is to use Serre duality.

First of all we have the sheaf cohomological fact (cf. chapter 3) that

$$H_{\bar{\partial}_{K_{\Sigma}^{*}}}^{(0,1)}(\Sigma) = H^{1}(\Sigma, \mathcal{O}(K_{\Sigma}^{*})) \equiv H^{1}(\Sigma, \Omega^{0}(K_{\Sigma}^{*}))$$

and, since $\dim_{\mathbf{C}} \Sigma = 1$, its Serre dual as given by 5.20 and 5.22 is

$$H^{0}(\Sigma, \Omega^{1}(K_{\Sigma})) = H^{0}(\Sigma, \mathcal{O}(K_{\Sigma}^{2}))$$
$$= H_{\bar{\partial}_{K_{\Sigma}^{2}}}^{(0,0)}(\Sigma) \tag{5.77}$$

So the tangent and cotangent spaces to the moduli space are given by

$$T_{[g]}\mathcal{M}_{p} = H_{\bar{\partial}_{K_{\Sigma}^{*}}}^{(0,1)}(\Sigma) \qquad \text{and} \qquad T_{[g]}^{*}\mathcal{M}_{p} = H_{\bar{\partial}_{K_{\Sigma}^{2}}}^{(0,0)}(\Sigma) \tag{5.78}$$

This space $H_{\bar{\partial}_{K_{\Sigma}^{2}}}^{(0,0)}(\Sigma)$, which is dual to the space of Beltrami differentials, is known as the space of (holomorphic) quadratic differentials. Quadratic differentials are easily represented using local coordinates. Let ϕ be a holomorphic quadratic differential, then in local coordinates we must have[6]

$$\phi = \phi(z)dz \otimes dz$$
$$\text{and} \quad \bar{\partial}\phi = 0$$

Before continuing with the calculation of $\dim \mathcal{M}_p$ we take time off to consider an example of Serre duality in action. We have just used the appropriate definitions to deduce that the spaces of quadratic differentials and Beltrami differentials are dual to one another. This can be examined in a bit more detail.

Example *Serre duality and quadratic differentials*

Let us take a Beltrami differential μ and a quadratic differential ϕ so that

$$\mu = \mu(z, \bar{z})\frac{\partial}{\partial z} \otimes d\bar{z}$$
$$\phi = \phi(z)dz \otimes dz$$

Serre duality gives the action of ϕ on μ as

$$\int_\Sigma <\mu, \phi> = \int_\Sigma \mu(z, \bar{z})\phi(z) < \frac{\partial}{\partial z} \otimes d\bar{z}, dz \otimes dz >$$
$$= \int_\Sigma \mu(z, \bar{z})\phi(z)dzd\bar{z}$$

using the usual rules for evaluating inner products on tangent spaces. Further, had μ been of the *cohomologically trivial* form $V_{,\bar{z}}^z(z, \bar{z})\partial/\partial z \otimes d\bar{z}$ then the above calculation would have given the answer zero. This is indeed so, for we would then have

$$\int_\Sigma <\mu, \phi> = \int_\Sigma \frac{\partial V^z}{\partial \bar{z}}(z, \bar{z})\phi(z)d\bar{z}dz = -\int_\Sigma V^z(z, \bar{z})\bar{\partial}\phi(z)d\bar{z}dz = 0$$

by holomorphicity of ϕ. Thus we see that μ and ϕ do have encoded into them some of the properties that the validity of Serre duality requires.

---

[6] For ϕ to be invariant it is clear that under a holomorphic transformation $z \mapsto h(z)$ we require $\phi(z) = \phi(h(z))(h'(z))^2$—more generally if $\phi = \phi(z) \otimes^q dz$, then ϕ is a holomorphic q-differential, i.e. $\phi(z) = \phi(h(z))(h'(z))^q$. Further, if a discrete automorphism group Γ such as a Fuchsian group acts on Σ and leaves ϕ invariant, then ϕ is a Γ-invariant holomorphic q-differential and is called an automorphic form of weight q; when Γ is the modular group $SL(2, \mathbf{Z})/\{\mp I\}$, ϕ is known as a modular form of weight q.

Returning now to the main discussion we observe that it is obvious that

$$H^{(0,0)}_{\bar{\partial}_{K^2_\Sigma}}(\Sigma) = ker\, \bar{\partial}_{K^2_\Sigma} \tag{5.79}$$

This observation is a key one for us since it immediately allow us to expresses the dimension of $\mathcal{M}_p$ as the dimension of the kernel of an elliptic operator. The precise statement is clearly

$$\dim \mathcal{M}_p = \dim T^*_{[g]}\mathcal{M}_p = \dim ker\, \bar{\partial}_{K^2_\Sigma} \tag{5.80}$$

It is at once tempting to try and use the Atiyah–Singer index theorem to derive an expression for $\dim ker\, \bar{\partial}_{K^2_\Sigma}$. This does indeed work and we proceed at once to the calculation. Referral to 4.118 of chapter 4 shows that we require the Riemann–Roch theorem for an elliptic operator of the form $\bar{\partial}_F$. In our particular case the index formula says that

$$
\begin{aligned}
index\, \bar{\partial}_{K^2_\Sigma} &= \dim ker\, \bar{\partial}_{K^2_\Sigma} - \dim ker\, \bar{\partial}^*_{K^2_\Sigma}\\
&= \int_\Sigma ch(K^2_\Sigma) \wedge td(T\Sigma)
\end{aligned}
\tag{5.81}
$$

But the bundles K^*_Σ and $T\Sigma$ coincide, and K^2_Σ is a line bundle, so that

$$ch(K^2_\Sigma) = 1 + c_1(K^2_\Sigma), \qquad td(T\Sigma) = td(K^*_\Sigma) = 1 + \frac{1}{2}c_1(K^*_\Sigma)$$

$$\Rightarrow ch(K^2_\Sigma) \wedge td(T\Sigma) = (1 + c_1(K^2_\Sigma)) \wedge (1 + \frac{1}{2}c_1(K^*_\Sigma))$$

$$= (1 + c_1(K^2_\Sigma) + \frac{1}{2}c_1(K^*_\Sigma) + \cdots)$$

$$= (1 + 2c_1(K_\Sigma) - \frac{1}{2}c_1(K_\Sigma) + \cdots)$$

$$= (1 + \frac{3}{2}c_1(K_\Sigma) + \cdots)$$

$$\Rightarrow index\, \bar{\partial}_{K^2_\Sigma} = \frac{3}{2}\int_\Sigma c_1(K_\Sigma) \tag{5.82}$$

However, the Gauss–Bonnet theorem determines this integral according to

$$\int_\Sigma c_1(K^*_\Sigma) = 2 - 2p = -\int_\Sigma c_1(K_\Sigma) \tag{5.83}$$

so we have as an immediate consequence

$$index\, \bar{\partial}_{K^2_\Sigma} = -\frac{3}{2}(2 - 2p) = 3p - 3 \tag{5.84}$$

To complete our task all that remains is to compute $\dim \ker \bar{\partial}^*_{K^2_\Sigma}$. We shall show that, provided $p \geq 2$, $\ker \bar{\partial}^*_{K^2_\Sigma}$ vanishes; if $p < 2$ then $\ker \bar{\partial}^*_{K^2_\Sigma}$ does *not* vanish and we shall use a slightly modified argument to deal with these cases. The main tool is the Kodaira vanishing theorem. This is a theorem concerned with positive line bundles (positive meaning $c_1(E) > 0$) which states that

Theorem (Kodaira) *If E is a positive line bundle then*

$$H^p(M, \Omega^q(E)) = 0, \qquad p + q > n \tag{5.85}$$

The proof of this theorem uses what is called a Weitzenböck positivity argument. We shall briefly summarise the logic of the proof since variations of it are found in other cohomology vanishing theorems. The cohomology group $H^p(M, \Omega^q(E))$ is expressed as the kernel of an appropriate Laplacian in the standard Hodge fashion. This Laplacian can be expressed as the sum of the ordinary Euclidean Laplacian plus a curvature term. Then the Chern class $c_1(E)$ is written in terms of the curvature. Finally, the positivity of $c_1(E)$ implies the positivity of the Laplacian's curvature term; and thus the Laplacian itself is positive and so has a vanishing kernel.

Before we can use this theorem we must interpret $\ker \bar{\partial}^*_{K^2_\Sigma}$ as a sheaf cohomology group. This is a straightforward application of Hodge theory from which it is immediate that

$$
\begin{aligned}
\ker \bar{\partial}^*_{K^2_\Sigma} &\simeq H^{(0,1)}_{\bar{\partial}_{K^2_\Sigma}}(\Sigma) \\
&\simeq H^1(\Sigma, \mathcal{O}(K^2_\Sigma))
\end{aligned}
\tag{5.86}
$$

We would like to apply this to the case above where we have $E = K^2_\Sigma$ and $n = 1$; but the condition $p + q > n$ is *not* fulfilled. However, Serre duality comes to our rescue since, when applied to $H^1(\Sigma, \mathcal{O}(K^2_\Sigma))$, it yields

$$
\begin{aligned}
\ker \bar{\partial}^*_{K^2_\Sigma} \simeq H^1(\Sigma, \mathcal{O}(K^2_\Sigma)) &\simeq \left\{ H^0(\Sigma, \mathcal{O}(K^{-2}_\Sigma \otimes K_\Sigma)) \right\}^* \\
&\simeq \left\{ H^0(\Sigma, \mathcal{O}(K^*_\Sigma)) \right\}^* \\
&\simeq \left\{ \ker \bar{\partial}_{K^*_\Sigma} \right\}^*
\end{aligned}
\tag{5.87}
$$

This has the effect of replacing the bundle $E = K^2_\Sigma$ by the bundle K^*_Σ and the latter will be negative because the former is positive. There is now a vanishing theorem for *negative* line bundles, which we obtain by applying Serre duality to the vanishing theorem above. The result of this is clearly the statement: if E is a positive line bundle then

$$H^{n-p}(M, \Omega^{n-q}(E^*)) = 0, \qquad p + q > n \tag{5.88}$$

But this is equivalent to

Theorem (Kodaira) *If E is a negative* line bundle then

$$H^i(M, \Omega^j(E)) = 0, \qquad i + j < n \tag{5.89}$$

Now if we consult 5.87 we have $E = K_\Sigma^*$, $i = j = 0$, and $n = 1$, so

$$H^0(\Sigma, \mathcal{O}(K_\Sigma^*)) = 0 \tag{5.90}$$

provided $c_1(K_\Sigma^*) < 0$. But

$$c_1(K_\Sigma^*) = 2 - 2p \tag{5.91}$$

so we deduce that

$$H^0(\Sigma, \mathcal{O}(K_\Sigma^*)) = 0, \qquad p \geq 2 \tag{5.92}$$

thus

$$ker \, \bar{\partial}_{K_\Sigma^2}^* = \emptyset, \qquad p \geq 2$$

Therefore we have the result

$$\dim_{\mathbf{C}} \mathcal{M}_p = 3p - 3, \qquad p \geq 2 \tag{5.93}$$

For the remaining two cases we have to make separate arguments. We deal first with $p = 0$. In that case Riemann–Roch says that

$$index \, \bar{\partial}_{K_\Sigma^2} = -3 \tag{5.94}$$

so clearly this time $\dim ker \, \bar{\partial}_{K_\Sigma^2}^* > 0$. Now we apply the Kodaira vanishing theorem to $ker \, \bar{\partial}_{K_\Sigma^2}$ itself. We have

$$ker \, \bar{\partial}_{K_\Sigma^2} = H^0(\Sigma, \mathcal{O}(K_\Sigma^2)) \tag{5.95}$$

but now the criteria of 5.89 are satisfied since, for genus zero, K_Σ^2 is a negative line bundle. Thus

$$ker \, \bar{\partial}_{K_\Sigma^2} = \emptyset \tag{5.96}$$

and so [7]

$$\dim_{\mathbf{C}} \mathcal{M}_p = 0, \qquad p = 0 \tag{5.97}$$

---

[7] In this case we have a 3-complex parameter family of conformal Killing vectors for the Riemann sphere S^2—these are given by the action of the Möbius group $SL(2, \mathbf{C})/\{\mp I\}$ (which is of the correct dimension) on the complex plane.

Finally, for genus $p = 1$

$$index \, \bar{\partial}_{K_\Sigma^2} = 0$$
$$\Rightarrow \quad \dim ker \, \bar{\partial}_{K_\Sigma^2}^* = \dim ker \, \bar{\partial}_{K_\Sigma^2} \tag{5.98}$$

Thus the only question remaining is what are the sizes of the kernels in 5.98. Note that vanishing theorems are useless here since now K_Σ^2 is trivial and so is neither positive nor negative. However, in § 5 we have already constructed $\mathcal{J}_p$ for the case $p = 1$ and found it to be 1-dimensional. Hence this is also the dimension of $\mathcal{M}_p$ when $p = 1$.

Nevertheless, the index theorem is still of some value when $p = 1$ since $\dim \mathcal{M}_p = 1$ implies that $\dim ker \, \bar{\partial}_{K_\Sigma^2}^* = 1$. But the space $ker \, \bar{\partial}_{K_\Sigma^2}^*$ is the dual of the space of conformal Killing vectors. To see this consider a perturbation of the metric which is only a conformal change of metric, that is, impose the restriction $\delta\mu = 0$. Now we have seen that, locally, $\delta\mu = \partial_{\bar{z}} V^z(z, \bar{z})$ with V^z one of the components of a vector field V. The restriction $\delta\mu = 0$ is therefore just $\partial_{\bar{z}} V^z(z, \bar{z}) = 0$. In other words, V^z must be holomorphic and can be thought of as a holomorphic section of the bundle $T\Sigma = K_\Sigma^*$. This space of sections is the cohomology group $H^0(\Sigma, \mathcal{O}(K_\Sigma^*))$ and we showed in 5.92 that this group is isomorphic to the dual of $ker \, \bar{\partial}_{K_\Sigma^2}^*$. Hence our result follows and the index theorem has shown us that the space of conformal Killing vectors is 1-dimensional for genus one.

We have completed our calculation of $\dim_{\mathbf{C}} \mathcal{M}_p$ for arbitrary genus p. In summary, we have found that

$$\dim_{\mathbf{C}} \mathcal{M}_p = \begin{cases} 0, & \text{if } p = 0 \text{ (in this case } \mathcal{M}_p \text{ has just one point)} \\ 1, & \text{if } p = 1 \\ 3p - 3, & \text{if } p \geq 2 \end{cases} \tag{5.99}$$

§ 7. Weierstrass gaps and Weierstrass points

A particularly interesting property of meromorphic functions on a Riemann surface is the nature of their singularities. On the Riemann sphere a meromorphic function f with a *single* pole at the point z_0 may have leading behaviour, near z_0, of the form

$$f(z) \rightarrow \frac{c}{(z - z_0)^n} + \cdots \tag{5.100}$$

with *no restriction* on n [8]. For a surface of higher genus p there are restrictions on the values of n allowed. What happens depends on the point z_0 where the pole is located. For most points z_0 on Σ a pole at z_0 has $n > p$; but there are *certain special points* of Σ at which poles may have $n \leq p$. These points are always finite in number and are called *Weierstrass points*. However, even if a pole does occur at a Weierstrass point, not all $n \leq p$ are allowed, an example being $n = 1$. In general an integer n_i which is not allowed is called a *Weierstrass gap* for the point z_0. The following theorem describes the general situation.

Theorem (Weierstrass gap theorem) *Let Σ_p be a Riemann surface of genus p, and let f be any meromorphic function whose only singularity is at z_0 where it has a pole of order n. If $p = 0$ then f may have a pole of any order n at any point z_0. If $p \geq 1$ then a pole of f at z_0 always has order $n > p$ except at a finite number of z_0's known as Weierstrass points. For a general point z_0 there are p gap values $n_1, \ldots, n_p$ satisfying $1 = n_1 < n_2 < \cdots < n_p < 2p$ such that none of the integers n_i may be the order of the pole at z_0 of the meromorphic function f.*

The first gap value n_1 is always 1 and this is easy to understand: the sum of the residues of a meromorphic function on a Riemann surface Σ_p of positive genus is always zero. This is because we can use a variant of the argument of p. 140 where we represented Σ_p as a $4p$-sided polygon. Using the same contour C, applied to f rather than df/f, shows that the sum of the residues of f is zero. Hence if the function has only one pole its residue must vanish. Of course for zero genus this is not true: the function $f(z) = 1/z$ is an example of a meromorphic function on the Riemann sphere with just a simple pole. In other words, there are no gap values for genus zero.

Points which are not Weierstrass points are referred to as normal points, the gap values for normal points are the simple sequence $1, 2, \ldots, p$. Notice, too, that the theorem asserts that for genus $p = 0$ and $p = 1$ there can be *no* Weierstrass points.

We can outline a proof of the Weierstrass gap theorem by applying the Riemann–Roch theorem to divisors at z_0.

Let a divisor D be given by

$$D = nz_0, \qquad \text{with } n \geq 1 \qquad (5.101)$$

---

[8] In this section we shall call the positive integer n the order of the pole. This is the opposite convention to that used in § 2 when we discussed divisors; in that discussion the order of a pole was a negative integer.

Note that $deg(D) = n$ which is positive. Now consider the line bundle $[D]$ associated to D. Then, because the degree of D is positive, $[D]$ has a global holomorphic section. This being so let us write down the space of *all* holomorphic sections of $[D]$; sheaf cohomology identifies this straightaway as

$$H^0(\Sigma_p; \mathcal{O}([D])) = H^0(\Sigma_p; \mathcal{O}([nz_0])), \quad \text{since } D = nz_0 \qquad (5.102)$$

Let s be that holomorphic section such that, in the patch U_α containing z_0, it has a zero of order n (s is really the localisation of the divisor D to U_α). Now multiply s by a meromorphic function f whose sole singularity (if it has one) is a pole at z_0. Then, as long as the order of this pole is at most n, the product fs is another holomorphic section; also all holomorphic sections are obtained in this manner. Thus we can identify $H^0(\Sigma_p; \mathcal{O}([nz_0]))$ with the space of all meromorphic functions whose sole singularity is a pole at z_0 of order at most n. Let us denote this latter space by $\mathcal{L}(nz_0)$.

Note that a function $f \in \mathcal{L}(nz_0)$ does not have to be singular at z_0; but if f is holomorphic at z_0 then it is holomorphic everywhere on Σ_p and is thus a constant. Note, too, that if *all* of the $f \in \mathcal{L}(nz_0)$ were holomorphic everywhere, and thus constant, then $\dim \mathcal{L}(nz_0)$ would have attained its *minimum* value, namely 1.

Now if $f \in \mathcal{L}(nz_0)$ then we can expand f in a Laurent series about z_0 so that near z_0 we have

$$f(z) \to a_0 + \frac{a_1}{(z - z_0)} + \cdots + \frac{a_n}{(z - z_0)^n} \qquad (5.103)$$

Since this expression depends on $n+1$ parameters we might conjecture that

$$\dim \mathcal{L}(nz_0) = \dim H^0(\Sigma_p; \mathcal{O}([nz_0])) = n + 1 \qquad (5.104)$$

However, the Weierstrass gap theorem *contradicts* this conjecture. For example, if $n \leq p$, and z_0 is *not* a Weierstrass point, then the theorem tells us that

$$\dim \mathcal{L}(nz_0) = 1 \qquad (5.105)$$

More generally, if $n \geq (2p - 1)$, there are p *gaps* in the parameters a_i, i.e. p of them are zero; in terms of dimensions, this is the assertion that

$$\dim \mathcal{L}(nz_0) = n + 1 - p \qquad (5.106)$$

which should be contrasted with our conjecture 5.104 above.

Now since $\mathcal{L}(nz_0) = H^0(\Sigma_p; \mathcal{O}([nz_0]))$, and $H^0(\Sigma_p; \mathcal{O}([nz_0]))$ is the sort of space to which the Riemann–Roch theorem applies, it is sensible to use the Riemann–Roch theorem here.

The Riemann–Roch theorem for the operator $\bar{\partial}_{[nz_0]}$ says that

$$index\ \bar{\partial}_{[nz_0]} = \dim ker\ \bar{\partial}_{[nz_0]} - \dim ker\ \bar{\partial}^*_{[nz_0]} = \int_{\Sigma_p} ch([nz_0]) \wedge td(T\Sigma_p)$$

$$(5.107)$$

However, using the same algebra as we did in our calculation of the dimension of the moduli space, we find

$$ch([nz_0]) \wedge td(T\Sigma_p) = (1 + c_1([nz_0]))(1 + \frac{1}{2}c_1(T\Sigma_p))$$

$$\Rightarrow index\ \bar{\partial}_{[nz_0]} = \int_{\Sigma_p} c_1([nz_0]) + \frac{1}{2}\int_{\Sigma_p} c_1(T\Sigma_p) \qquad (5.108)$$

$$= n + \frac{1}{2}(2 - 2p) = n + 1 - p$$

Now we can identify $ker\ \bar{\partial}^*_{[nz_0]}$ with $H^1(\Sigma_p; \mathcal{O}([nz_0]))$. Then, using Serre duality, and the Kodaira vanishing theorem for negative line bundles, we find

$$\text{Serre} \Rightarrow H^1(\Sigma_p; \mathcal{O}([nz_0])) \simeq \left\{ H^0(\Sigma_p; \mathcal{O}(K_{\Sigma_p} \otimes [nz_0]^*)) \right\}^*$$

$$\text{Kodaira} \Rightarrow H^0(\Sigma_p; \mathcal{O}(K_{\Sigma_p} \otimes [nz_0]^*)) = 0, \quad \text{if } n \geq (2p - 1)$$

$$(5.109)$$

The inequality $n \geq (2p - 1)$ comes from satisfying the criterion for the vanishing theorem, which says that the bundle $K_{\Sigma_p} \otimes [nz_0]^*$ must have negative Chern class. But $c_1(K_{\Sigma_p} \otimes [nz_0]^*) = c_1(K_{\Sigma_p}) + c_1([nz_0]^*) = 2p - 2 - n$ and our inequality now follows. Thus we have found that, for $n \geq (2p-1)$,

$$\dim H^0(\Sigma_p; \mathcal{O}([nz_0])) = \dim \mathcal{L}(nz_0) = n + 1 - p \qquad (5.110)$$

in agreement with 5.106 above.

The statement that the number of Weierstrass points is finite can be phrased as

$$\text{if } n \geq p, \quad \dim \mathcal{L}(nz_0) > 1 \text{ for a finite number of } z_0 \qquad (5.111)$$

This can be proved (Griffiths and Harris [1]) by counting singular points on certain algebraic curves associated to Σ_p.

Instead of considering functions we can also consider meromorphic q-differentials; a very similar proof establishes that these also have Weierstrass points. The main difference in the proof is that the sheaf $\mathcal{O}([nz_0])$ is replaced by $\mathcal{O}(K^q_{\Sigma_p} \otimes [nz_0])$.

The number of Weierstrass points on Σ_p is finite and any automorphism of Σ_p permutes its Weierstrass points. This fact can be used to show that, if $p \geq 2$, a Riemann surface has only a finite number of automorphisms, cf. Griffiths and Harris [1]. Indeed, as we mentioned at the end of §5, the number of automorphisms for $p \geq 2$ is at most $84(p - 1)$.

CHAPTER VI

Infinite Dimensional Groups

§ 1. Some infinite dimensional groups

Finite dimensional groups are widely used in mathematics and physics. In dimension zero we have finite and infinite discrete groups; if the dimension is positive then we can turn to Lie groups and obtain thereby a multitude of important compact and non-compact groups.

Physicists have made extensive use of finite discrete groups in crystallography. Infinite discrete groups also occur—for example, there is the modular group $SL(2, \mathbf{Z})/\{\mp I\}$ which is fundamental to string theory. The continuous groups used by physicists are usually non-Abelian simple and semi-simple Lie groups and the Abelian group $U(1)$. On the compact side, the five exceptional groups E_6, E_7, E_8, F_4 and G_2, and the various classical groups $U(n)$, $O(n)$ and $Sp(n)$ are frequently symmetries of the action of some quantum field theory. It is also necessary to use non-compact, locally compact, groups. For instance the group $SL(2, \mathbf{C})$ is needed in order to construct Lorentz-invariant quantum field theories.

The representations of these finite dimensional groups have been systematically studied and, especially in the compact case, an extensive theory has been worked out. However it is also the case that various infinite dimensional groups occur naturally in physics. This being so they ought to be studied and the construction of their representation theory is an important part of any such study. In this chapter we give some examples of infinite dimensional groups, largely restricting ourselves to those groups which occur in subsequent chapters. We shall also say a limited amount about representations. These infinite dimensional groups also provide us with various concrete examples of infinite dimensional manifolds. We shall see that most of these examples are, at least locally, of the form $Map\,(X, Y)$ for appropriate finite dimensional X and Y.

The infinite dimensional groups that we shall study can be given a smooth structure and, when this is done, they become infinite dimensional smooth manifolds.

An infinite dimensional manifold M is defined in a similar way to the finite dimensional case: M is a topological space which is sewn together from open sets of some infinite dimensional topological vector space E; then M is said to be *modeled* on E. In finite dimensions E is just $\mathbf{R}^n$ or $\mathbf{C}^n$ but in the infinite dimensional case E is a topological vector space such as a Hilbert space, a Sobolev space, a general Banach space, or some other locally convex space. One then refers to M as a Hilbert manifold, or a Sobolev manifold and so on. In finite dimensions if a manifold is a Lie group then its Lie algebra, which is also a vector space, is the model space $\mathbf{R}^n$. In infinite dimensions if M is a Lie group its Lie algebra is also the model space E. The interested reader can find a detailed account of infinite dimensional manifolds in Hamilton [1], Milnor [2] and Abraham and Marsden [1].

Example *The group* $Map\,(X, G)$

A very easy infinite dimensional group to describe is the group of smooth maps from a compact space X to a finite dimensional Lie group G. In accordance with the notation used in chapter 1 for spaces of maps we denote this group by $Map\,(X, G)$. Actually in chapter 1 on p. 8 we introduced the closely related group $Map_0(X, G)$ which just consists of the base point preserving maps in $Map\,(X, G)$. In order to make $Map\,(X, G)$ into a group we have to define the group operation, but to do this we can repeat the 'pointwise evaluation' definition used for $Map_0(X, G)$. This means that if $\alpha, \beta \in Map\,(X, G)$ we denote their product by $\alpha \bullet \beta$ which we define by

$$\begin{aligned} \alpha \bullet \beta : X &\longrightarrow G \\ x &\longmapsto \alpha(x) * \beta(x) \end{aligned} \tag{6.1}$$

where $*$ stands for the product in the finite dimensional group G.

The product $(\alpha, \beta) \mapsto \alpha \bullet \beta$ is a *smooth* map as is the operation of inversion $\alpha \mapsto \alpha^{-1}$. This means that $Map\,(X, G)$ is an infinite dimensional *Lie* group. The smoothness of $Map\,(X, G)$ means that we can construct its tangent space at the identity. This tangent space $T_e Map\,(X, G)$ is its Lie algebra. It is not difficult to check that this tangent space is defined in terms of the tangent space $T_e G$ to the finite dimensional group G. This means that, denoting the Lie algebra of $Map\,(X, G)$ by $\mathfrak{L}Map\,(X, G)$, and that of G by $\mathfrak{g}$, we have

$$\mathfrak{L}Map\,(X, G) = Map\,(X, \mathfrak{g}) \tag{6.2}$$

Thus the Lie algebra of $Map(X, G)$ is also given by a space of maps.

The group $Map(X, G)$ can be rather complicated if $\dim X > 1$; for example, $Map(X, G)$ generally has more than one connected component. To see this we compare $Map(X, G)$ to the group of based maps $Map_0(X, G)$. Now, for each $g \in G$, let $\alpha_g \in Map(X, G)$ denote the constant map defined by $\alpha_g(x) = g$, $\forall x \in X$. Now suppose that an arbitrary $\alpha \in Map(X, G)$ maps the base point x_0 of X to $g \in G$, i.e. $\alpha(x_0) = g$. Then α can be written in a unique way as a product of a constant map and a based map: one simply writes $\alpha = \alpha_g \bullet (\alpha_g^{-1} \bullet \alpha)$. This shows that, viewed as topological spaces rather than groups, we have

$$Map(X, G) \simeq G \times Map_0(X, G) \tag{6.3}$$

The connected components of $Map(X, G)$ are given by $\pi_0(Map(X, G))$ and so we obtain

$$\pi_0(Map(X, G)) = \pi_0(G) \oplus \pi_0(Map_0(X, G))$$

But $\quad \pi_0(Map_0(X, G)) = [S^0, Map_0(X, G)]_0 \tag{6.4}$

$$= [S^0 \wedge X, G]_0 \quad \text{using 1.18}$$

Now if we take the specific example where $X = S^m$ and $G = U(n)$ we know that $\pi_0(U(n)) = 0$ and $[S^0 \wedge X, G,]_0 = [S^m, U(n)]_0 = \pi_m(U(n))$. Thus we find that, if $X = S^m$ and $G = U(n)$, then

$$\pi_0(Map(X, G)) = \pi_m(U(n)) \neq 0 \quad \text{in general} \tag{6.5}$$

More specifically still, we can set $S^m = S^3$; then, using the standard fact for $U(n)$ that $\pi_3(U(n)) = \mathbf{Z}$, $n > 1$, we deduce that $\pi_0(Map(X, U(n))) = \mathbf{Z}$ and so $Map(X, U(n))$ has infinitely many components.

Slightly more generally we can replace S^m by any manifold X but keep $G = U(n)$. Then we find that

$$\pi_0(Map(X, U(n))) = [S^0 \wedge X, G,]_0$$

$$= [S^0 \wedge X, \Omega BU(n)]_0 \quad \text{cf. p. 73}$$

$$= [SX, BU(n)]_0 \tag{6.6}$$

$$= \widetilde{K}U^{-1}(X) \text{ if } n > \frac{\dim X}{2} \text{ cf. pp. 74 and 76}$$

The non-triviality of $\widetilde{K}U^{-1}(X)$ in general ensures that $Map(X, U(n))$ has many components. [1]

---

[1] Actually a K-theoretic calculation shows that $\widetilde{K}U^{-1}(X) = \mathbf{Z} \oplus \cdots \oplus \mathbf{Z} \oplus T$ where T is a torsion group; also if there are m copies of $\mathbf{Z}$ in the summation then $m = b_1 + b_3 + \cdots$ where the b_i are the Betti numbers of X.

For our next example we select a group of the form $Map(X, G)$ where X is just one dimensional. The group in question is known as the loop group.

Example *The loop group*

The loop group LG is the group $Map(X, G)$ when $X = S^1$. That is

$$LG = Map(S^1, G) \tag{6.7}$$

It is clear that, for simply connected G, LG has only one component. The loop group is of considerable interest to both mathematicians and physicists. For a very good mathematical account of LG and its representations the reader should consult Pressley and Segal [1].

The Lie algebra of LG is the space

$$Map(S^1, \mathfrak{g}) \tag{6.8}$$

It has been extensively studied via its representation theory. All the interesting representations are actually representations of a central extension of $Map(S^1, \mathfrak{g})$. We denote this central extension of LG by $\overline{LG}$. The resulting algebra has generators T_m^a which satisfy the commutation relations

$$[T_m^a, T_n^b] = i f^{abc} T_{m+n}^c + km\delta^{ab}\delta_{m,-n}, \quad m, n \in \mathbf{Z} \tag{6.9}$$

where k is the central term and f^{abc} are the structure constants of the Lie algebra $\mathfrak{g}$ of G. Notice that the generators $\{T_0^a\}$ form a finite dimensional sub-algebra isomorphic to $\mathfrak{g}$; these are the generators of the constant loops in the product 6.3 above. These commutation relations are those of an (untwisted) affine Kac–Moody algebra. A useful basis for the algebra is given by

$$T_n^a = \exp[in\theta] T^a \tag{6.10}$$

where θ is a coordinate on the S^1 contained in $Map(S^1, \mathfrak{g})$. The mathematical theory of Kac–Moody algebras, as well as being of interest in its own right, has connections with other branches of mathematics such as integrable systems, the theory of modular forms and the representations of finite simple groups such as the Monster group. A few useful references are Kac [1], Segal and Wilson [1], Mason [1] and Conway and Norton [1]. The physical applications are also numerous and have close ties with the mathematical ones. Areas of application include statistical mechanics, conformal quantum field theories, current algebra and string theories. Some of these are described in Lepowsky, Mandelstam and Singer [1], Goddard and Olive [1] and Mickelsson [1]. Another important application of loop groups is in the construction of knot invariants via conformal quantum field theory—cf. chapter 12.

Example *The diffeomorphism group*

The diffeomorphisms of a manifold X form a group which we denote by $Diff(X)$. The product $\alpha \bullet \beta$ of two diffeomorphisms $\alpha, \beta \in Diff(X)$ is just their composition as maps. That is we have

$$\alpha \bullet \beta : X \longrightarrow X$$
$$x \longmapsto \alpha(\beta(x)) \tag{6.11}$$

$Diff(X)$ is a Lie group and its Lie algebra is identified as follows. Let V be a (smooth) vector field on X with integral curves $v(t)$; then flowing X a distance t along these integral curves produces a family of diffeomorphisms α_t labelled by t. This identifies the Lie algebra of $Diff(X)$ as the smooth vector fields on X which we write as $vect(X)$. That is

$$\mathfrak{L}(Diff(X)) = vect(X) \tag{6.12}$$

In local coordinates we can write

$$V = \sum_i v_i \frac{\partial}{\partial x^i}, \qquad \alpha_t = \exp[tV] \tag{6.13}$$

The Lie bracket of $vect(X)$ corresponds to the ordinary Lie bracket for vector fields.

Example *The Virasoro group*

A special case of $Diff(X)$ is the group obtained when $X = S^1$. This group $Diff(S^1)$ is fundamental to string theory (Green, Schwarz and Witten [1,2]). More precisely the group of importance in string theory is a central extension of $Diff(S^1)$ known as the Virasoro group which we write as $\overline{Diff(S^1)}$. In this case the Lie algebra is the famous Virasoro algebra (Virasoro [1] and Gel'fand and Fuks [1]). Its commutation relations are usually quoted in the centrally extended form

$$[L_m, L_n] = (m - n)L_{m+n} + \frac{c}{12}m(m^2 - 1)\delta_{m,-n}, \qquad m, n \in \mathbf{Z} \tag{6.14}$$

where c is the central element also known as the central charge.

The Virasoro algebra is closely connected with conformal invariance: In two dimensions, unlike the case in higher dimensions, the conformal group is infinite dimensional because any holomorphic or anti-holomorphic transformation preserves angles and hence is conformal. Holomorphic transformations can be determined by their boundary values on the unit circle and thus by maps from S^1 to S^1. The Lie algebra of the conformal group is thereby easily calculated by a Fourier expansion on S^1 and is found to consist of two commuting ($c = 0$) Virasoro algebras, cf. Green, Schwarz and Witten [1]. In this way one can find a formula for the generators, which is

$$L_n = -z^{n+1}\frac{d}{dz} \equiv \exp[in\theta]i\frac{d}{d\theta} \quad \text{on } S^1 \qquad (6.15)$$

Example *The group of gauge transformations*

The term gauge transformation is well known to physicists and a typical gauge transformation is often written as $g(x)$ where x is a point on X and $g(x)$ is an element of G. This suggest that, at least locally, gauge transformations are elements of $Map(X, G)$. However, gauge transformations are not globally elements of $Map(X, G)$ and some additional structure is required in order to construct the group of gauge transformations—a group that we shall denote by $\mathcal{G}$.

The additional structure referred to above is simply a fibre bundle. Suppose that we have a G-connection A over X, then we know that this is constructed over a principal G-bundle P on X (Nash and Sen [1]). Any bundle automorphism of P produces a bundle with identical physical content, and so such automorphisms are symmetries of physical theories. These automorphisms are essentially the gauge transformations.

A short account of $\mathcal{G}$ is as follows. Let $Aut(P)$ denote the group of *smooth* bundle automorphisms of P: a bundle automorphism $\alpha \in Aut(P)$ is a map which commutes with the defining group action on P, i.e. it leaves the fibration intact. This means that

$$\alpha(g \cdot p) = g \cdot \alpha(p) \qquad (6.16)$$

where the group action on P is denoted by a dot. For a principal bundle P the group G acts on the right and therefore consists of group multiplication by g^{-1}; that is, we have

$$g \cdot p = pg^{-1} \qquad (6.17)$$

Using this in our equation for α, with g interchanged with g^{-1} for convenience, gives

$$\alpha(pg) = \alpha(p)g \qquad (6.18)$$

The group product in $Aut\,(P)$ is given by composition: $(\alpha \bullet \beta)(p) = \alpha(\beta(p))$. Now, since the elements of $Aut\,(P)$ are fibre preserving, we can use the bundle projection $\pi : P \longrightarrow X$ to project them onto the base X. This gives us a group of smooth coordinate transformations of X into itself. This group is none other than $Diff\,(X)$. On the other hand, this projection is actually a group homomorphism h, say, and so its kernel $ker\,h$ is also a group; $ker\,h$ is the group of gauge transformations. It will be convenient in chapter 8 to distinguish between the group of *all* gauge transformations $ker\,h$, and a *subgroup* which is defined by imposing a base point condition. This condition is that each of its elements α satisfies $\alpha(p_0) = I$ (the identity) for some fixed $p_0 \in P$. We shall use $\widetilde{\mathcal{G}}$ to denote $ker\,h$, and $\mathcal{G}$ to denote its subgroup. Let $\pi(p_0) = x_0$ so that the fibre of P to which p_0 belongs is P_{x_0}, then $\mathcal{G}$ is characterised by the property that it acts as the identity on P_{x_0}.

It can be seen from the preceding discussion that the group of bundle automorphisms $Aut\,(P)$ contains the two classes of invariance possessed by the action S of a gauge invariant quantum field theory. Further, $Aut\,(P)$ provides a natural division of this invariance into the following two classes. The first class is made up of the group of gauge transformations $\widetilde{\mathcal{G}}$ or $ker\,h$— the elements of which correspond mathematically to transforming the *fibres* of P in some appropriate way. The second class is that introduced by Einstein, and is the group of coordinate transformations $Diff\,(X)$—whose elements correspond mathematically to transforming the *base space* X of P in an appropriate way.

The Lie algebra of $\mathcal{G}$ can be easily constructed. In order to do this we introduce a more explicit description of $Aut\,(P)$: Take the principal bundle P and form its associated adjoint bundle $Ad\,P$. The bundle $Ad\,P$ is the associated bundle over X where each fibre is an isomorphic copy of the group G but the action of the group is conjugation rather than right translation; we write this as $Ad\,P = P \times_{AdG} G$. We shall now show that $\mathcal{G}$ is given by all the smooth sections of $Ad\,P$ with the group product being pointwise evaluation of sections. To this end let $\alpha \in Aut\,(P)$; then, since α takes fibres of P into themselves, we can introduce a map $\widehat{\alpha} : P \to G$ which we define by

$$\alpha(p) = p\widehat{\alpha}(p) \qquad (6.19)$$

Next we impose the equivariance requirement 6.18 which then becomes

$$\begin{aligned} (pg)\widehat{\alpha}(pg) &= p\widehat{\alpha}(p)g \\ \Rightarrow \widehat{\alpha}(pg) &= g^{-1}\widehat{\alpha}(p)g \end{aligned} \qquad (6.20)$$

Now we project $\widehat{\alpha}$ onto the base as we did above; those $\widehat{\alpha}$ in the kernel of the projection will correspond to the α which are gauge transformations rather than diffeomorphisms of the base. The effect of this projection is simply to replace an $\widehat{\alpha} : P \to G$ by a map $\widehat{\alpha}_X : X \to G$. Having done this we introduce local coordinates on P and write $p = (x, g)$ and, in terms of these coordinates, we find that

$$\alpha(pg) = (x, \widehat{\alpha}_X(x)g) = (x, g^{-1}\widehat{\alpha}_X(x)g) \tag{6.21}$$

But this immediately identifies $\widehat{\alpha}_X : X \to G$ as a section of a bundle over X whose fibres are copies of G subject to the adjoint action. In other words, $\widehat{\alpha}_X$ determines a section of $Ad\,P$. Thus the group of gauge transformations $\mathcal{G}$ can be viewed as the space of sections of $Ad\,P$.

Finally we check that, under the transition from $Aut\,(P)$ to sections of $Ad\,(P)$, the group product in $\mathcal{G}$ changes from composition of automorphisms to pointwise evaluation of sections. If $\alpha, \beta \in Aut\,(P)$ then

$$\begin{aligned}
\alpha(\beta(p)) &= \alpha(p\widehat{\beta}_X(p)) = (p\widehat{\beta}_X(p))\widehat{\alpha}_X(p\widehat{\beta}_X(p)) \\
&= p\widehat{\beta}_X(p)\widehat{\beta}_X^{-1}(p)\widehat{\alpha}_X(p)\widehat{\beta}_X(p) = p\widehat{\alpha}_X(p)\widehat{\beta}_X(p)
\end{aligned} \tag{6.22}$$

That is we have shown that

$$\alpha(\beta(p)) = p\widehat{\gamma}_X(p)$$
$$\text{where} \qquad \widehat{\gamma}_X(p) = \widehat{\alpha}_X(p)\widehat{\beta}_X(p) = (\widehat{\alpha}_X \bullet \widehat{\beta}_X)(p) \tag{6.23}$$

as required.

With the alternative description of $\mathcal{G}$ in place its Lie algebra $\mathfrak{L}\mathcal{G}$ is rather naturally defined as the space of smooth sections of the corresponding *Lie algebra bundle* $ad\,P$. This latter bundle is obtained from $Ad\,P$ by replacing all the fibres G by copies of the *Lie algebra* $\mathfrak{g}$, thus $ad\,P = P \times_{AdG} \mathfrak{g}$. Summarising, we have

$$\begin{aligned}
\mathcal{G} &= \Gamma(X, Ad\,P), & \mathfrak{L}\mathcal{G} &= \Gamma(X, ad\,P) \\
Ad\,P &= P \times_{AdG} G, & ad\,P &= P \times_{AdG} \mathfrak{g}
\end{aligned} \tag{6.24}$$

Example *Current algebra*

One of the earliest occurrences of an infinite dimensional algebra in physics was the current algebra motivated by the quark model (Adler and Dashen [1] and Gell-Mann and Ne'eman [1]).

A concrete example is available in two dimensions. We take space-time to be two dimensional and periodic in space. Then we use the Kac–Moody algebra to define a a current $j^a(\theta)$ by

$$j^a(\theta) = \sum_n T_n^a \exp[in\theta] \tag{6.25}$$

The commutation relations of the currents $j^a(\theta)$ are then the equal time current algebra commutation relations

$$[j^a(\theta), j^b(\phi)] = if^{abc}j^c(\theta)\delta(\theta - \phi) + ic\delta^{ab}\delta'(\theta - \phi) \qquad (6.26)$$

Notice that the central term of the Kac–Moody algebra has emerged as the well known Schwinger term of the current algebra.

Before leaving these examples we wish to remark upon the analyticity of the groups that appear in them. All the groups are Lie groups and are smooth infinite dimensional manifolds. However, even stronger analyticity properties are possessed by some of the groups. The group $Map(X, G)$ is smooth but the target space G in $Map(X, G)$ is a finite dimensional Lie group, and is therefore *analytic*. This analyticity of G induces an analyticity for the infinite dimensional group $Map(X, G)$. The mechanism whereby this comes about is easy to pinpoint. The group $Map(X, G)$ is topologised by using the topology of uniform convergence: one says that the sequence $\alpha_n \in Map(X, G)$ converges to an element α if $\alpha_n(x)$ and *all* its derivatives converge *uniformly* to $\alpha(x)$. Thus the rôle of the analyticity of G is brought to the fore. In any case the analyticity of $Map(X, G)$ means that it has an atlas of analytic coordinate charts which can be represented by Taylor series with no remainder.

By contrast, however, the group $Diff(X)$, though smooth, has no particular reason to be analytic and indeed it is not analytic (Milnor [**2**]). The lack of analyticity of $Diff(X)$ has consequences for the exponential map from its Lie algebra to itself, i.e. the map $exp : vect(X) \to Diff(X)$. For an analytic Lie group G the exponential map $exp : \mathfrak{g} \to G$ provides, in a canonical way, a local coordinate system on G near the identity. But for $Diff(X)$ the map exp does not give canonical local coordinates near the identity; also exp is not locally a homeomorphism because one can produce examples of elements of $Diff(X)$ which, though arbitrarily close to the identity, are not on *any* one parameter subgroup. There are also elements which lie on two or more one parameter subgroups (Milnor [**2**]).

§ 2. Group extensions

In the previous section we quoted two sets of Lie algebra commutation relations in a centrally extended form. These group extensions are of considerable mathematical interest and play an extremely important part in applications to quantum field theory. Before dealing with particular examples we give a brief summary of the properties of group extensions.

Recall that extensions $\overline{G}$ of a group G by a group H are summarised by an exact sequence of group homomorphisms of the form

$$1 \longrightarrow H \xrightarrow{\alpha} \overline{G} \xrightarrow{\beta} G \longrightarrow 1 \qquad (6.27)$$

The group H is called the kernel of the extension and is a normal subgroup of the group $\overline{G}$; the quotient $\overline{G}/H$ is the original group G. If the exact sequence splits—i.e. β has a right inverse—then we say that the extension $\overline{G}$ is the semi-direct product of H with G and we write $\overline{G} = H \ltimes G$. If H is contained in the centre of $\overline{G}$ then we have a *central* extension.

A corresponding Lie algebra statement is that $\bar{\mathfrak{g}}$, considered as a vector space rather than as an algebra, is the direct sum $\mathfrak{g} \oplus \mathfrak{h}$. The content of the extension lies in the relation of the Lie bracket of $\bar{\mathfrak{g}}$ to those of $\mathfrak{g}$ and $\mathfrak{h}$. We shall write an element of $\bar{\mathfrak{g}}$ as (U, V) with $U \in \mathfrak{g}$ and $V \in \mathfrak{h}$. Then the Lie bracket can be written

$$[(U_1, V_1), (U_2, V_2)] = ([U_1, U_2], c(U_1, U_2)) \qquad (6.28)$$

where $c(U_1, U_2)$ is a G-invariant anti-symmetric form on $\mathfrak{g}$, $c : \mathfrak{g} \times \mathfrak{g} \to \mathfrak{h}$. The need for the Lie algebra $\bar{\mathfrak{g}}$ to satisfy the Jacobi identity imposes the usual cyclic permutation restriction

$$c([U, V], W) + c([V, W], U) + c([W, U], V) = 0 \qquad (6.29)$$

But this is also the condition for a G-invariant bilinear form on $\mathfrak{g}$ to determine a Lie algebra 2-cocycle. Hence Lie algebra extensions are classified by an element $c \in H^2(\mathfrak{g}; \mathfrak{h})$.

For cohomology calculations it is useful to know that the de Rham cohomology of G and $\mathfrak{g}$ coincide when G is compact. This is because if ω is a form belonging to the de Rham cohomology of G then ω can be assumed to be left invariant; but the left invariant forms on G are precisely the elements of the de Rham cohomology of the algebra $\mathfrak{g}$, hence $H^*(G; \mathbf{R}) = H^*(\mathfrak{g}; \mathbf{R})$. When G is replaced by a non-compact infinite dimensional group this is no longer generally true. Nonetheless, for LG we do still have $H^*(LG; \mathbf{R}) = H^*(Map(S^1, \mathfrak{g}); \mathbf{R})$. However, for the group $Map(X, G)$ with dim $X > 1$ the result no longer holds. This is because there are some cohomology classes in $H^*(Map(X, G); \mathbf{R})$ which cannot be made left invariant; the upshot is that $H^*(Map(X, G); \mathbf{R})$ is smaller than $H^*(Map(X, \mathfrak{g}); \mathbf{R})$.

If one has a Lie algebra extension it need not give rise to a corresponding group extension (Pressley and Segal [1]). For example, let a cocycle c specify a Lie algebra extension of $Map(S^1, \mathfrak{g})$ by $\mathbf{R}$. Represent c using de Rham cohomology as a differential form[2] $\omega/2\pi$; then the corresponding group extension $\overline{LG}$ of LG by S^1 exists if and only if $\omega/2\pi$ determines an integral cohomology class on LG. The integral class $\omega/2\pi$ can also be thought of

---

[2] The denominator factor 2π is recognisable as the standard normalisation used when representing characteristic classes as differential forms.

as the first Chern class of a $U(1)$ bundle: the point is that, topologically speaking, $\overline{LG}$ is a principal $U(1)$ bundle over LG and it is completely characterised by its Chern class which is $\omega/2\pi$. If $\omega/2\pi$ does *not* determine an integral class then $Map\,(S^1, \mathfrak{g})\oplus\mathbf{R}$ is an example of a Lie algebra without a corresponding Lie group.

To gain some insight into the quantum theoretic aspects we start with a classical theory that has a symmetry group G. Then the generators T^a of the Lie algebra of G will enter the classical theory via the Poisson bracket relation

$$\{T^a, T^b\} = f^{abc}T^c \tag{6.30}$$

where $\{T^a, T^b\}$ denote the Poisson bracket of T^a and T^b. The quantised version of this theory is obtained by replacing T^a by $i\hbar T^a$. The resulting quantum commutation relations are of the form [3]

$$[T^a, T^b] = i\hbar f^{abc}T^c \tag{6.31}$$

However, in quantum field theory we have to consider corrections to these commutation relations which are higher order in Planck's constant $\hbar$. This leads us to the equation

$$[T^a, T^b] = i\hbar f^{abc}T^c + O(\hbar^2) \tag{6.32}$$

In the case where the correction is just of order $\hbar^2$, and is proportional to the identity, the quantum corrected commutation relations are

$$[T^a, T^b] = i\hbar f^{abc}T^c + i\hbar^2 c\delta^{ab} \tag{6.33}$$

and this is a central extension of the Lie algebra of G. We would like to replace the finite dimensional algebra by the algebra of LG. This can easily be done and gives us the central extension of $Map\,(S^1, \mathfrak{g})$ that we had above

$$[T_m^a, T_n^b] = i\hbar f^{abc}T_{m+n}^c + cm\hbar^2\delta^{ab}\delta_{m,-n} \tag{6.34}$$

The centrally extended Virasoro algebra is at the heart of string theory and the central term gives rise to a conformal anomaly unless the space-time has a certain critical dimension. We shall return to this matter of a critical dimension when we discuss anomalies and strings.

Another important property of the Virasoro algebra is that it is intimately related to the affine Kac–Moody algebra. More precisely, there is a

---

[3] We normally work in units where $\hbar = 1$ but this is not sensible if one wants to compare a quantum theory to its classical limit.

Virasoro algebra naturally associated with every Kac–Moody algebra. This Virasoro algebra has its origin in the fact that the group $Diff^+(S^1)$ acts (projectively) on the loop group LG as an automorphism group. A consequence of this is that one can obtain a Virasoro algebra from a Kac–Moody algebra by the Sugawara construction (Sugawara [1], Goddard and Olive [1], Segal [2]). The idea is to form the generators of the Virasoro algebra by taking a carefully constructed expression which is bilinear in the Kac–Moody generators. Sugawara proposed that the energy momentum tensor $T_{\mu\nu}$ be bilinear in the currents and of the form (from now on $\hbar = 1$)

$$T_{\mu\nu} = j_\mu^a j_\nu^b - \frac{g_{\mu\nu}}{2} j_\alpha^a j^{a\alpha} \tag{6.35}$$

But since the currents are proportional to the T^a then this produces a quantity bilinear in the Kac–Moody generators. Sugawara's original suggestion was to work in four dimensions but two dimensional examples have the most interesting properties for us. A specific example is provided by a two dimensional sigma model with a Wess–Zumino term included (Witten [1]). In any case the generator L_n is given by

$$L_n = \frac{1}{2k + c(G)} \sum_{m \in \mathbf{Z}} {}^\circ_\circ T_m^a T_{n-m}^a {}^\circ_\circ \tag{6.36}$$

where $c(G)$ is the value of the Casimir operator for the adjoint representation of G ($c(G) = f^{abc} f^{abc} / \dim G$) and ${}^\circ_\circ T_m^a T_n^a {}^\circ_\circ$ is the usual normal ordering defined by

$$
{}^\circ_\circ T_m^a T_n^a {}^\circ_\circ = \begin{cases} T_n^a T_m^a, & \text{if } m \geq 0, \\ T_m^a T_n^a, & \text{if } m < 0. \end{cases} \tag{6.37}
$$

These L_n satisfy the Virasoro algebra with a predetermined value for the central term. We find that

$$[L_m, L_n] = (m - n)L_{m+n} + \frac{2k \dim G}{12(2k + c(G))} m(m^2 - 1)\delta_{m,-n} \tag{6.38}$$

The physical reason for the appearance of normal ordering is the familiar quantum field theoretic need to regularise operator products $A(x)B(y)$ when $x \to y$; in this case it is the energy momentum tensor which requires attention.

The mathematical origin for the normal ordering of the expression for the Virasoro generator may also be uncovered (Pressley and Segal [1]). The key observation is that a *formal* Casimir operator for the loop group algebra $Map(S^1, \mathfrak{g})$ can be written down just as in the finite dimensional case but the expression now contains a divergent infinite sum. We define the Casimir operator Δ for a finite dimensional Lie algebra with basis $\{T^a\}$ by

$$\Delta = T^a (T^a)^* \tag{6.39}$$

where $(T^a)^*$ is the adjoint of T^a with respect to the standard inner product. Now for the Kac–Moody algebra the corresponding expression is

$$\tilde{\Delta} = \sum_{n \in \mathbf{Z}} T^a_n (T^a_n)^* = \sum_{n>0} T^a_n (T^a_n)^* + \sum_{n<0} T^a_n (T^a_n)^* + T^a_0 (T^a_0)^* \tag{6.40}$$

To expose the divergence we use an orthonormal basis for $Map\,(S^1, \mathfrak{g})$ in which $(T^a_n)^* = T^a_{-n}$ (this amounts to a unitary representation of LG). Then we have

$$\sum_{n \in \mathbf{Z}} T^a_n (T^a_n)^* = \sum_{n>0} \left(T^a_n T^a_{-n} + T^a_{-n} T^a_n \right) + T^a_0 T^a_0 \tag{6.41}$$

Now the commutator can be employed to write

$$\begin{aligned}
T^a_n T^b_{-n} &= T^b_{-n} T^a_n + [T^a_n, T^b_{-n}] = T^a_{-n} T^a_n + i f^{abc} T^c_0 + k n \delta^{ab}, \quad \text{by 6.9} \\
\Rightarrow \left(T^a_n T^a_{-n} + T^a_{-n} T^a_n \right) &= 2 T^a_n T^a_{-n} + k n \dim G \\
\Rightarrow \tilde{\Delta} &= \sum_{n>0} 2 T^a_n T^a_{-n} + T^a_0 T^a_0 + k \dim G \sum_n n
\end{aligned}$$
$$\tag{6.42}$$

But the last term contains the divergent sum $\sum_n n$, hence our difficulty. If we had used normal ordering the commutator would never have appeared and this infinite term would be absent. With this motivation we define the operator $\tilde{\Delta}_0$ in which the last term is simply discarded

$$\tilde{\Delta}_0 = \sum_{n>0} 2 T^a_n T^a_{-n} + T^a_0 T^a_0 \tag{6.43}$$

This new operator $\tilde{\Delta}_0$ is not yet a Casimir operator for the Kac–Moody algebra because one can check that it does not yet commute with all the generators. A straightforward calculation of commutators shows that

$$[T^a_n, \tilde{\Delta}_0] = -n(2k + c(G)) T^a_n \tag{6.44}$$

where we have assumed for simplicity that G is simple and, as before, $c(G)$ is the value of its (finite dimensional) Casimir in the adjoint representation. The true Casimir operator is obtained by adding a multiple of the operator $(id/d\theta)$ to $\tilde{\Delta}_0$. To find this multiple we make use of the basis

$T_n^a = \exp[in\theta]T^a$ for the Kac–Moody algebra that we introduced above. We can verify that

$$\left[T_n^a, i\frac{d}{d\theta}\right] = nT_n^a \tag{6.45}$$

Thus the Casimir operator we seek is simply

$$\widetilde{\Delta}_0 + (2k + c(G))i\frac{d}{d\theta} \tag{6.46}$$

The geometric origin of this Casimir can be elucidated a little. The group $Diff^+(S^1)$ acts on LG and $\overline{LG}$, by reparametrising the loop. A subgroup of this $Diff^+(S^1)$ is the (one parameter) group of rigid rotations of the loop through some fixed angle; this subgroup is evidently isomorphic to S^1. The Lie algebra corresponding to this one parameter group is just $(id/d\theta)$. This widens the group under consideration to be the semi-direct product $S^1 \ltimes \overline{LG}$ whose Lie algebra is $\mathbf{R} \oplus \mathbf{R} \oplus Map(S^1, \mathfrak{g})$. The 'regularised' Casimir operator for this algebra is then nearly correct since it is of the form $\widetilde{\Delta}_0 + \alpha(id/d\theta)$, with α a constant; however, we cannot select the precise coefficient of $(id/d\theta)$ without further computation.

Finally the commutator of T_l^a with the operator $\widetilde{\Delta}_n$ defined by $\widetilde{\Delta}_n = \sum_{m \in \mathbf{Z}} T_m^a T_{n-m}^a$, $n \neq 0$ is

$$[T_l^a, \widetilde{\Delta}_n] = l(2k + c(G))T_{l+n}^a \tag{6.47}$$

It follows immediately that the operators L_n defined by

$$L_n = \frac{1}{2k + c(G)} \sum_{m \in \mathbf{Z}} T_m^a T_{n-m}^a, \quad n \neq 0$$

$$L_0 = \frac{1}{2k + c(G)} \left(\sum_{n>0} 2T_n^a T_{-n}^a + T_0^a T_0^a\right) \tag{6.48}$$

obey the Virasoro algebra

$$[L_m, L_n] = (m - n)L_{m+n} + \frac{2k \dim G}{12(2k + c(G))}m(m^2 - 1)\delta_{m,-n} \tag{6.49}$$

in agreement with the normal ordered Sugawara expression above.

In addition to central extensions of LG we can consider central extensions of $Map(X, G)$ for $\dim X > 1$. Actually all such extensions come from the loop group LG rather than the group $Map(X, G)$ itself (Pressley and Segal [1]). Let us explain how to construct these extensions. We work at the

Lie algebra level. Suppose to begin with we only have a central extension of $Map\,(S^1, \mathfrak{g})$ specified by $c \in H^2(Map\,(S^1, \mathfrak{g}); \mathbf{Z})$, and let f be a map

$$f : S^1 \longrightarrow X \tag{6.50}$$

Then f induces a map $\widehat{f} : Map\,(X, \mathfrak{g}) \to Map\,(S^1, \mathfrak{g})$ which we can use to pull back the cocycle c from $Map\,(S^1, \mathfrak{g})$ to $Map\,(X, \mathfrak{g})$. The definition of $\widehat{f}$ is

$$\begin{aligned} \widehat{f} : Map\,(X, \mathfrak{g}) &\longrightarrow Map\,(S^1, \mathfrak{g}) \\ \alpha &\longmapsto \alpha \circ f \end{aligned} \tag{6.51}$$

Thus $\widehat{f}^*c$ is an element of $H^2(Map\,(X, \mathfrak{g}); \mathbf{Z})$ and so determines a central extension of $Map\,(X, \mathfrak{g})$, and *all* central extensions are obtainable in this fashion.

The preceding result suggests that, when $\dim X > 1$, we should consider non-central extensions of $Map\,(X, G)$. We proceed to do this in the following way. Let $\dim X = n$ and ω be an $(n+2)$-form on the group G. We can use ω to obtain a 2-form on $Map\,(X, G)$ by using an evaluation map Ev defined by

$$\begin{aligned} Ev : X \times Map\,(X, G) &\longrightarrow G \\ (x, \alpha) &\longmapsto \alpha(x) \end{aligned} \tag{6.52}$$

The pullback $Ev^*\omega$ is an $(n+2)$-form on the product $X \times Map\,(X, G)$ and the desired 2-form c, say, on $Map\,(X, G)$ is got by integration over X. We have therefore

$$c = \int_X Ev^*\omega, \qquad [c] \in H^2(Map\,(X, G); \mathbf{R}) \tag{6.53}$$

where $[c]$ is the cohomology class determined by the 2-form c. This 2-form c determines a Lie algebra cocycle which takes values in the space $\mathcal{H}$ of functions on $Map\,(X, G)$: the Lie algebra cocycle is defined by the map $\widehat{c}$

$$\begin{aligned} \widehat{c} : Map\,(X, \mathfrak{g}) \times Map\,(X, \mathfrak{g}) &\longrightarrow \mathcal{H} \\ (\alpha, \beta) &\longmapsto c(\alpha, \beta) \end{aligned} \tag{6.54}$$

where $c(\alpha, \beta)$ denotes the value of the 2-form c on the two elements α, β of the Lie algebra $Map\,(X, \mathfrak{g})$. Note that $c(\alpha, \beta)$ is a 0-form or function on $Map\,(X, \mathfrak{g})$. Thus we have extended the Lie algebra $Map\,(X, \mathfrak{g})$ by the space $\mathcal{H}$ where $\mathcal{H} = \Omega^O(Map\,(X, \mathfrak{g}))$; the corresponding Lie algebra cocycle $\widehat{c}$ belongs to $H^2(Map\,(X, \mathfrak{g}); \mathcal{H})$.

As a specific example we take $G = SU(N)$, $N \geq 3$ and $X = S^3$. Now if P is an $SU(N)$-bundle over S^3 and A is a connection 1-form on P, then we can take $\alpha, \beta \in Map(X, \mathfrak{g})$ and write

$$c(\alpha, \beta) = C \int_{S^3} tr(d\alpha \wedge d\beta \wedge A), \qquad C \text{ a constant} \qquad (6.55)$$

Actually, in this particular example, $c(\alpha, \beta)$ takes values in the space of functions on $\mathcal{A}$ (recall that $\mathcal{A}$ denotes the connections on X) rather than in $\mathcal{H}$. Thus the Lie algebra cocycle is an element of $H^2(Map(X, \mathfrak{g}); \mathcal{E})$ where $\mathcal{E} = Map(\mathcal{A}, \mathbf{R})$. Now since the Lie algebra of S^1 is $\mathbf{R}$ the Abelian group $Map(\mathcal{A}, S^1)$ has Lie algebra $Map(\mathcal{A}; \mathbf{R})$. Thus in terms of groups we have a non-central extension—the Mickelsson–Faddeev extension—of the group $Map(X, G)$ by the group $Map(\mathcal{A}, S^1)$. These non-central extensions are responsible for the Schwinger terms in the current algebra in higher dimensions and have been considered in varying contexts by various authors (cf. Faddeev [1], Mickelsson [2], Pressley and Segal [1] and references therein). They shall also turn up in our discussion of anomalies in chapter 10.

§ 3. Representations

The representation theory of infinite dimensional groups is still under construction and the theory is nothing like as highly developed as that of finite dimensional semi-simple Lie groups. Nevertheless, for particular groups such as the loop group LG and the Virasoro group $Diff(S^1)$ there is a substantial representation theory. Much less is known about representations of $Map(X, G)$ when $\dim X > 1$; for a summary of what *is* known cf. Pressley and Segal [1]. For a systematic account of the representation theory of groups such as LG and $Diff(S^1)$ we refer the reader to the references cited in this chapter; we shall limit ourselves to some remarks on a few salient features. We begin with the Virasoro group but we shall actually describe the representations of the algebra instead of those of the group.

Example *The Virasoro algebra*

The Virasoro algebra is the centrally extended form of the Lie algebra $vect(S^1)$. In fact there are physical reasons to attach more importance to the central extensions of $vect(S^1)$ rather than to $vect(S^1)$ itself. The point is that, in applications, the spectrum of the operator L_0 is an energy so that L_0 has a lower bound. In that case we shall see below that all the non-trivial irreducible representations are representations of a *central extension* of $vect(S^1)$. In group theoretic terms the non-trivial irreducible representations are all projective representations.

Recall that if unitary operators $\{T_g\}$ form a projective representation of a group G then

$$T_g T_h = c(g, h) T_{gh} \qquad (6.56)$$

where the projective multiplier $c(g, h)$ is a complex number of unit modulus. Of course a projective representation of G is a true representation of the extension $\overline{G}$.

An irreducible representation, with L_0 bounded below, is called a *highest weight*[4] *representation* and is specified by a pair of *non-negative numbers* (c, h). The number c is the central coefficient already introduced and h is the lowest eigenvalue of L_0. The eigenvector for h is thus the 'vacuum state' and is usually denoted by $|h\rangle$. The representation has the property that

$$L_0 |h\rangle = h |h\rangle \Rightarrow L_n |h\rangle = 0, \quad n \geq 1 \qquad (6.57)$$

The reason that we must have $L_n |h\rangle = 0$, $n \geq 1$ is that otherwise $L_n |h\rangle$ would be an eigenvector of L_0 with eigenvalue $(h - n)$, and this would contradict the assumption that h is the lowest eigenvalue. On the other hand, $L_{-n} |h\rangle \neq 0$, $n \geq 1$, as may be seen by applying the commutation relations.

Only certain values of (c, h) can occur. If we restrict ourselves to unitary representations then the possible values of (c, h) were found by Friedan, Qiu and Shenker [1,2]. The allowed values are

$$c \geq 1, \qquad h \geq 0,$$

or

$$\left. \begin{array}{l} c = 1 - \dfrac{6}{(m+1)(m+2)} \\[2mm] h = \dfrac{\{(m+2)p - (m+1)q\}^2 - 1}{4(m+1)(m+2)} \end{array} \right\} \quad \begin{array}{l} m = 1, 2, \ldots \\ p = 1, 2, \ldots, m \\ q = 1, 2, \ldots, p \end{array} \qquad (6.58)$$

Representations for all these allowed values can be constructed, cf. Goddard, Kent and Olive [1,2]. We note that the unitary representations with $c < 1$ form a discrete series. Examples of these discrete series representations occur in conformally invariant statistical models, cf. chapter 11. In

---

[4] This terminology, though commonly used, is far from ideal since $|h\rangle$ is actually a lowest weight vector; the terms anti-dominant weight and positive energy representation are also used. Some difficulty with terminology seems inevitable; however, in the physical context, the interpretation of $|h\rangle$ as a lowest energy state renders it natural to use $|h\rangle$ in the description of representations.

contrast the Sugawara construction provides an example of a unitary representation of the opposite type since its c value is $2k \dim G/(2k + c(G))$ and this is never less than one because it is bounded below by the rank of G.

If the representation is a true representation rather than a projective one, then $c = 0$; actually this also implies $h = 0$ (Goddard and Olive [1]). Now we can calculate $\| L_{-n} | h \rangle \|$, $n \geq 1$ and discover that

$$
\begin{aligned}
\| L_{-n} | h \rangle \|^2 &= \langle h | L_n L_{-n} | h \rangle & \text{since } L_{-n}^* = L_n \\
&= \langle h | [L_n, L_{-n}] | h \rangle & \text{since } L_n | h \rangle = 0, \ n \geq 1 \qquad (6.59) \\
&= 0 & \text{since } c = h = 0
\end{aligned}
$$

But the vacuum $| h \rangle$ is clearly a cyclic vector; that is, the representation space is spanned by vectors of the form $L_{-1}^{\alpha_1} L_{-2}^{\alpha_2} \cdots L_{-s}^{\alpha_s} | h \rangle$, therefore the representation is the trivial one in which all the L_n's are represented by zero. Thus we see that all non-trivial highest weight representations are necessarily projective.

Example *The Kac–Moody algebra*

Moving on to the loop group LG we find a rather similar picture. It is convenient to do as we did for the Virasoro group and describe the representations of the algebra rather than the group. A similar boundedness criterion is used to select representations; we obtain this criterion as follows. In the previous section, while carrying out the Sugawara construction, we constructed a representation of the Virasoro algebra from the Kac–Moody algebra. If we use this Virasoro algebra again here we can take things a little further and work out the commutators of the L_n's with the T_n^a's. We find easily that the complete commutation relations are

$$
\begin{aligned}
[L_m, L_n] &= (m - n)L_{m+n} + \frac{2k \dim G}{12(2k + c(G))} m(m^2 - 1)\delta_{m,-n} \\
[T_m^a, T_n^b] &= i f^{abc} T_{m+n}^c + k m \delta^{ab} \delta_{m,-n} \\
[L_n, T_m^a] &= -m T_{n+m}^a, \quad [L_n, k] = [T_n^a, k] = 0
\end{aligned}
\qquad (6.60)
$$

Now we simply demand, as before, that the spectrum of L_0 be bounded from below, and this is our boundedness criterion. Let $| h \rangle$ be the vacuum state for L_0, then the commutation relations allow us to deduce that

$$
L_0 | h \rangle = h | h \rangle \implies L_n | h \rangle = 0, \ T_n^a | h \rangle = 0, \quad n \geq 1 \qquad (6.61)
$$

and this describes a highest weight representation of the Kac–Moody algebra. We can easily verify that there are no non-trivial highest weight representations of the algebra $Map(S^1, \mathfrak{g})$ itself and so we only consider the centrally extended algebra—the Kac–Moody algebra.

Now when introducing Kac–Moody algebras we remarked already that the generators $\{T_0^a\}$ form a representation of $\mathfrak{g}$. For a highest weight representation of the Kac–Moody algebra this representation of $\mathfrak{g}$ is irreducible. But, for a highest weight representation, the commutation relations show us that the vacuum is cyclic; and thus we can construct the representation space by application of polynomials in the generators to the vacuum state $|h\rangle$. This means that we can characterise irreducible highest weight representations by just two things: the central term k and this irreducible representation of the finite dimensional algebra $\mathfrak{g}$. Equivalently we can say that the representation is characterised by the central term and a highest weight λ, say, and we write this pair of objects as (k, λ). However, just as in the $c < 1$ Virasoro case, the central term k can only take certain discrete values. If l denotes the length of a long root of $\mathfrak{g}$ then these values are determined by the condition

$$N \in \mathbf{Z}, \quad \text{where } N = \frac{2k}{l^2} \tag{6.62}$$

This integer N is called the *level* of the irreducible representation. For further details of the description of the representations in terms of highest weights and levels we refer the reader to the references already cited. We now turn briefly to an alternative approach to representations which emphasises the group rather than the algebra.

Example *The Borel–Weil construction*

This alternative approach is the Borel-Weil construction of irreducible representations. In its original formulation (Borel and Weil [1], Bott [2]) it was applied to finite dimensional compact Lie groups, but it has been extended to infinite dimensions so as to include loop groups, cf. Pressley and Segal [1]. Loosely speaking the Borel–Weil theory constructs representations as Hilbert spaces of functions or sections. This is rather analogous to the way representations occur in quantum theory and this apparently rather abstract analogy can be put to use in conformal quantum field theories.

To describe the Borel–Weil construction we start with a compact finite dimensional group G and summarise the standard description of an irreducible representation. Next choose a maximal Abelian subgroup or torus T and an irreducible unitary representation of G. This representation is determined by the action of the torus T on the representation space. Since T is Abelian, and the representation is unitary, this action is one in which an element of T is represented by a diagonal matrix whose diagonal entries are complex numbers of unit modulus. Consider the action of T on each basis vector v_i of the representation; this gives us a homomorphism from T to $U(1)$, that is a map $\lambda_i : T \to U(1)$, and λ_i is simply the weight of the vector v_i. Since the representation is irreducible then one of the λ_i is a highest weight and it is this highest weight that characterises V. For simplicity, from now on, we shall denote a weight by just λ and we shall write the representation space or module as V_λ.

Now G forms a principal T-bundle over the homogeneous space G/T.

$$
\begin{array}{ccc}
T & \longrightarrow & G \\
 & & \downarrow \pi \\
 & & G/T
\end{array}
\qquad (6.63)
$$

This means that, given the homomorphism provided by choosing any weight $\lambda : T \to U(1)$, we can construct an *associated* complex line bundle L_λ over G/T. Now we want to show that L_λ is naturally a *holomorphic* line bundle. For this to be even possible the base space G/T must be a complex manifold. This is actually well known to be the case: if $G_{\mathbf{c}}$ is the complexification of G it has a subgroup B known as a Borel subgroup and we have

$$
G_{\mathbf{c}}/B = G/T \qquad (6.64)
$$

Then the fact that the LHS of 6.64 is naturally a complex manifold renders the RHS one also. A further fact about the complex manifold G/T is that it is a Kähler manifold to which the Kodaira vanishing theorem can be applied to show that the sheaf cohomology groups $H^i(G/T; \mathcal{O})$, $i \geq 1$ vanish, i.e.

$$
H^i(G/T; \mathcal{O}) = 0, \quad i \geq 1 \qquad (6.65)
$$

The significance of this is found by referring to § 3 of chapter 5 where we showed that the *difference* between holomorphic and differentiable line bundles over a complex manifold M is measured by the sheaf cohomology groups $H^{1,2}(M; \mathcal{O})$. When these groups vanish, we have $H^1(M; \mathcal{O}^*) \simeq H^2(M; \mathbf{Z})$, thus holomorphic line bundles over M are determined by their first Chern class in $H^2(M; \mathbf{Z})$ just as in the differentiable case. Since a holomorphic line bundle over M is determined by an element of $H^1(M; \mathcal{O}^*)$, then it follows at once that the isomorphism $H^1(M; \mathcal{O}^*) \simeq H^2(M; \mathbf{Z})$ implies that any line bundle over M has a *unique* holomorphic structure. Hence we see that our line bundle L_λ over G/T can be taken to holomorphic. Finally we can state the Borel–Weil theorem.

Theorem (Borel–Weil) *The space of holomorphic sections of the line bundle L_λ over G/T is non-trivial if and only if λ is the highest weight of an* *irreducible representation of G. When λ is a highest weight then this space of holomorphic sections is a realisation of the representation space V_λ.*

To make this theorem plausible we must describe the complex structure on G/T in a more explicit way. This description exploits the Lie algebras of G and T. Using the action of T we can decompose the Lie algebra $\mathfrak{g}$ in the standard fashion as

$$\mathfrak{g} = \mathfrak{t} \oplus \mathfrak{h} \tag{6.66}$$

where $\mathfrak{h}$ denotes the sum of the root spaces relative to T. To obtain complex coordinates on G/T we first complexify this decomposition, yielding

$$\mathfrak{g}_{\mathbf{c}} = \mathfrak{t}_{\mathbf{c}} \oplus \mathfrak{h}_{\mathbf{c}} \tag{6.67}$$

But we can express $\mathfrak{h}_{\mathbf{c}}$ as a sum of the spaces $\mathfrak{h}_{\mathbf{c}}^{\mp}$ of positive and negative roots (relative to some Weyl chamber) whose bases are the familiar step operators. We write

$$\mathfrak{g}_{\mathbf{c}} = \mathfrak{t}_{\mathbf{c}} \oplus \mathfrak{h}_{\mathbf{c}}^{+} \oplus \mathfrak{h}_{\mathbf{c}}^{-} \tag{6.68}$$

We give complex coordinates to the G/T in the following manner. First note that the complex space $\mathfrak{h}_{\mathbf{c}}$ can be thought of as the *complexification* of the tangent space of G/T. Now assign complex coordinates to G/T itself by identifying its tangent space as the summand $\mathfrak{h}_{\mathbf{c}}^{+}$ in the decomposition of $\mathfrak{h}_{\mathbf{c}}$. In fact the summand $\mathfrak{h}_{\mathbf{c}}^{+}$ can be thought of specifying the holomorphic directions with the other summand $\mathfrak{h}_{\mathbf{c}}^{-}$ specifying the anti-holomorphic ones. These complex coordinates on the tangent space to G/T must be shown to be integrable to G/T itself; but they are automatically so because the positive root space $\mathfrak{h}_{\mathbf{c}}^{+}$ is closed under Lie brackets.

Having provided a more explicit notion of holomorphicity on G/T we now turn briefly to the Peter–Weyl theorem for representations of compact Lie groups. This theorem decomposes the Hilbert space $L^2(G)$ under right translation $(f(g) \mapsto f(gh)$ for $f(g) \in L^2(G)$ and $h \in G)$ and expresses it as a sum of irreducible representations of G. We have

$$L^2(G) = \bigoplus_\lambda c_\lambda V_\lambda \tag{6.69}$$

where c_λ measures the multiplicity of the representation λ. However, we could also have obtained an isomorphic decomposition of $L^2(G)$ using left translation $(f(g) \mapsto f(h^{-1}g)$ for $f(g) \in L^2(G)$ and $h \in G)$ instead of right translation; identifying these two decompositions gives us the Peter–Weyl form

$$L^2(G) = \bigoplus_\lambda V_\lambda \otimes V_\lambda^* \tag{6.70}$$

where $*$ denotes dual and this form shows that the multiplicity of an irreducible representation V_λ equals its degree $\dim V_\lambda$.

Returning to the line bundle of the Borel–Weil theorem we let f be a holomorphic section of L_λ; f can be thought of as a holomorphic map $f : G/T \to \mathbf{C}$. The action of G on f is by left translation and so the space of holomorphic sections $\Gamma(G/T, \mathcal{O}(L_\lambda))$ forms a representation. Now f can be pulled back from G/T to G using the projection $\pi : G \to G/T$. On G we denote f by $\tilde{f}$ and $\tilde{f} = \pi^* f$, note too that $\tilde{f}$ is an element of $L^2(G)$. But $\tilde{f}$ is a pullback of a T-invariant quantity, so under left translation by a Lie algebra element $Y \in \mathfrak{t}$ it is just multiplied by a constant, i.e. $\tilde{f} \mapsto \lambda(Y)\tilde{f}$ with $\lambda(Y) \in \mathbf{C}$. Also f is holomorphic so it is annihilated by all $W \in \mathfrak{h}_{\mathbf{c}}^-$; pulling this statement back to G we find that such W annihilate $\tilde{f}$. However, these two properties mean that $\tilde{f} \in L^2(G)$ is a lowest weight vector for an irreducible representation. Referring to the Peter–Weyl decomposition permits us to identify $\Gamma(G/T, \mathcal{O}(L_\lambda))$ as V_λ.

Finally we mention the case of the loop group. In the extension of the Borel–Weil construction to LG the homogeneous space G/T of the finite dimensional case is replaced by the space LG/T. The highest weight representations of the central extension $\overline{LG}$ of LG are realised as spaces of holomorphic sections of line bundles over LG/T.

CHAPTER VII

Morse Theory

§ 1. The topology of critical points

The aim in Morse theory is to study the relation between critical points and topology. More specifically one extracts topological information from a study of the critical points of a smooth real valued function

$$f : M \longrightarrow \mathbf{R} \qquad (7.1)$$

where M is a compact manifold usually without boundary. For a suitably behaved class of functions f there exists quite a tight relationship between the number and type of critical points of f and topological invariants of M such as the Euler–Poincaré characteristic, the Betti numbers and other cohomological data. This relationship can then be used in two ways: one can take certain special functions whose critical points are easy to find and use this information to derive results about the topology of M; on the other hand, if the topology of M is well understood, one can use this topology to infer the existence of critical points of f in cases where f is too complex, or too abstractly defined, to allow a direct calculation.

Successful applications of Morse theory in mathematics are impressive and widespread; a few notable examples are the proof by Morse [1] that there exist infinitely many geodesics joining a pair of points on a sphere S^n endowed with any Riemannian metric, Bott's [1,3] proof of his celebrated periodicity theorems on the homotopy of Lie groups, Milnor's construction [1] of the first exotic spheres, and the proof by Smale [1] of the Poincaré conjecture for dim $M \geq 5$. An excellent mathematical reference is the classic of Milnor [3]. Morse theory has also found a variety of applications in physics; this is not too surprising in view of the central position occupied by the variational principle in both classical and quantum physics. Some of these are described in Nash and Sen [1] which contains a short introduction to the subject. In this book we shall be interested in some of the more recent applications involving quantum mechanics and Yang–Mills fields.

We begin with some fundamental results of the theory. The setting is very simply described. One takes the smooth function $f : M \to \mathbf{R}$ of 7.1 and solves the equation for its critical points, namely

$$df = 0 \tag{7.2}$$

We *assume* that all the critical points p of f are non-degenerate; this means that the Hessian matrix Hf of second derivatives is invertible at p, or

$$det\, Hf(p) \neq 0 \quad \text{where} \quad Hf(p) = \left[\partial^2 f / \partial x^i \partial x^j \big|_p \right]_{n \times n} \tag{7.3}$$

Each critical point p has an index λ_p which is defined to be the number of *negative* eigenvalues of $Hf(p)$. In a neighbourhood of a non-degenerate critical point p of index λ_p we can represent f as

$$f(x) = f(p) - \overbrace{x_1^2 - x_2^2 - \cdots - x_{\lambda_p}^2}^{\lambda_p \text{ terms}} + \underbrace{x_{\lambda_p+1}^2 + \cdots + x_n^2}_{n-\lambda_p \text{ terms}} \tag{7.4}$$

for suitable coordinates $(x_1, \ldots, x_n)$. The nullity of p is defined to be the integer $\dim M - \operatorname{rank} Hf(p)$. Clearly all non-degenerate critical points have nullity zero.

For the present we shall only deal with functions whose critical points are all non-degenerate; these are called Morse functions. We can then associate to the function f and its critical points p the Morse series $M_t(f)$ defined by

$$M_t(f) = \sum_{\text{all } p} t^{\lambda_p} = \sum_i m_i t^i \tag{7.5}$$

The sum will always converge since it only contains a finite number of terms; this is because the non-degeneracy makes the critical points all discrete and the compactness of M permits only a finite number of such discrete points. The topology of M now enters via $P_t(M)$: the Poincaré series of M. This is the following polynomial constructed out of the Betti numbers of M; we have

$$P_t(M) = \sum_{i=0}^n \dim H^i(M; \mathbf{R}) t^i = \sum_{i=0}^n b_i t^i \tag{7.6}$$

The fundamental result of Morse theory is the statement that

$$M_t(f) \geq P_t(M) \tag{7.7}$$

In fact we can state this result using cohomology with coefficients in a general field $\mathbf{F}$, giving us

$$M_t(f) \geq P_t(M; \mathbf{F}) \quad \text{where} \quad P_t(M; \mathbf{F}) = \sum_{i=0}^{n} \dim H^i(M; \mathbf{F}) t^i \quad (7.8)$$

If we substitute the relevant polynomials into this inequality then we deduce that

$$M_t(f) \geq P_t(M)$$

$$\Rightarrow \sum_i m_i t^i \geq \sum_{i=0}^{n} b_i t^i \quad (7.9)$$

$$\Rightarrow m_i \geq b_i, \ i = 0, \ldots, n$$

Actually Morse obtained a strong result about what happens when one subtracts the smaller polynomial from the larger: setting $\Delta(t) = M_t(f) - P_t(M)$ Morse found that $\Delta(t)$ always contains the factor $(1+t)$ and that the remaining polynomial $R(t)$ has only non-negative coefficients. Summarising, we have

$$M_t(f) - P_t(M) = (1 + t)R(t)$$

$$\text{where} \quad R(t) = \sum_i r_i t^i \quad \text{with} \quad r_i \geq 0, \ i = 0, \ldots, n \quad (7.10)$$

This is also true for any coefficient field $\mathbf{F}$. With this stronger statement we obtain a more powerful set of inequalities. We use $(1+t)^{-1} = 1 - t + t^2 - \cdots$, and write

$$\frac{(M_t(f) - P_t(M))}{(1 + t)} = R(t)$$

$$\Rightarrow \sum_{i=0}^{n}(m_i - b_i)t^i \sum_{j=0}^{\infty}(-1)^j t^j \geq 0$$

$$\Rightarrow \sum_{i,j}(-1)^j(m_i - b_i)t^{i+j} \geq 0 \quad (7.11)$$

$$\Rightarrow (-1)^j(m_0 - b_0) + (-1)^{j-1}(m_1 - b_1) + \cdots + (m_j - b_j) \geq 0$$

$$\Rightarrow (m_j - m_{j-1} + m_{j-2} - \cdots \mp m_0) \geq (b_j - b_{j-1} + \cdots \mp b_0)$$

These results 7.7–7.11 are known as the Morse inequalities. Note that the number m_i is just the number of critical points of f of index i; adding these all together we find that

$$M_t(f)|_{t=1} = \text{the total number of critical points of } f \quad (7.12)$$

Thus $M_1(f)$ is a strongly f-dependent quantity. On the other hand, if we set $t = -1$ in 7.10 then we find that

$$M_{-1}(f) = P_{-1}(M) = \sum_{i=0}^{n}(-1)^i b_i \tag{7.13}$$
$$= \chi(M)$$

Hence $M_1(f)$, being the Euler–Poincaré characteristic of M, is completely independent of f. This last result shows some of the power of the stronger version 7.11 of the Morse inequalities; further evidence of their power is provided by the next result which is

Theorem (Morse's lacunary principle) *If $M_t(f)$ has no consecutive powers of t then, whatever the coefficient field $\mathbf{F}$, the polynomial $R(t)$ is identically zero so that*

$$M_t(f) = P_t(M; \mathbf{F}) \tag{7.14}$$

If we use 7.11 this is easy to prove: Let t^i be the first non-zero power in $M_t(f))$, then, because $m_i \geq b_i$, t^i is also the first non-zero power in the difference $M_t(f) - P_t(M)$. Thus t^i occurs in the product $(1 + t)R(t)$. But $(1 + t)R(t) = \sum r_j(t^j + t^{j+1})$ thus if $R(t) \neq 0$ then t^{i+1} also occurs in the product $(1 + t)R(t)$, and so too in $M_t(f) - P_t(M)$. However this last fact is impossible because t^{i+1} does not occur in $M_t(f)$ by our lacunary assumption and it cannot occur in $P_t(M)$ either because this would violate $m_{i+1} \geq b_{i+1}$. Hence $R(t)$ must vanish.

A function f for which $M_t(f) = P_t(M; \mathbf{F})$ for every $\mathbf{F}$ is called a *perfect Morse function*. It follows that if $M_t(f) = P_t(M; \mathbf{F})$ for all $\mathbf{F}$ then the cohomology ring $H^*(M; \mathbf{Z})$ must be free of torsion. Clearly it is no use looking for perfect Morse functions on any manifold we know to possess torsion.

A proof of the Morse inequalities, such as that in Nash and Sen [1], usually uses the level surfaces of the function f—the level surfaces of f are just a higher dimensional analogue of the level curves of chapter 2; they are the sets $\{x \in M : f(x) = c\}$. Rather than give another proof we shall just explain the part that they play in determining the topology of M. In Morse theory one constructs a half space M_c out of level surfaces where

$$M_c = \{x \in M : f(x) \leq c\} \tag{7.15}$$

Some of the topology of M begins to emerge when we consider M_c as a function of c. What happens is that, as c varies, the topology of M_c is unchanged until c passes through a critical point, when it either acquires or sheds a cell of dimension λ where λ is the index of the critical point. More precisely we have

Theorem (Bott–Morse–Smale) M_a *is diffeomorphic to* M_b *if there is no critical point in the interval* $[a, b]$. *Alternatively, if* (a, b) *contains just one critical point of index* λ *then* $M_b \simeq M_a \cup e_\lambda$.

The notation $M_a \cup e_\lambda$ means that a cell of dimension λ has been attached to M_a; also $M_b \simeq M_a \cup e_\lambda$ means that the two spaces have the same homotopy type. Thus, as far as the homotopy type of M is concerned (and this will be sufficient for computing the cohomology of M) one can think of M as being 'decomposed' into a set of cells

$$M = \bigcup_\lambda e_\lambda \qquad (7.16)$$

The number of these cells being equal to the number of critical points and the dimension of the cells being given by the index of the critical points. This decomposition is known as a *stratification* of M.

Example *The height function on* T^2

A well known example in Morse theory is given by taking M to be the two dimensional torus T^2. Then, if T^2 is regarded as a doughnut standing on its end, $f : T^2 \to \mathbf{R}$ is defined by setting $f(x)$ equal to the height of the point x off the ground. It is easy to see that f is a perfectly good function with 4 non-degenerate critical points: a maximum and minimum and two saddle points given by the two distinguished points on the inner circle. The indices of saddle points are both 1 and those of the maximum and the minimum are clearly 2 and 0 respectively. Therefore $M_t(f)$ is given by

$$M_t(f) = 1 + 2t + t^2 \qquad (7.17)$$

We can immediately verify that

$$\begin{aligned} M_{-1}(f) &= 1 - 2 + 1 = 0 \\ &= \chi(T^2) \end{aligned} \qquad (7.18)$$

as it should. We move on to another example.

Example *The cohomology of* $\mathbf{CP}^n$

The lacunary principle is often employed to compute the cohomology of $\mathbf{CP}^n$. To do this we use the fact that $\mathbf{CP}^n$ is the quotient of the odd sphere

S^{2n+1} by an action of S^1. Let us recall how this works. We use the complex numbers $(z_1, z_2, \ldots, z_{n+1})$ so that the equation of S^{2n+1} is

$$z_1 \bar{z}_1 + z_2 \bar{z}_2 + \cdots + z_{n+1} \bar{z}_{n+1} = 1 \qquad (7.19)$$

If $e^{i\theta}$ is an element of S^1, the S^1 action is given by defining

$$e^{i\theta} \cdot (z_1, \ldots, z_{n+1}) = (e^{i\theta} z_1, \ldots, e^{i\theta} z_{n+1}) \qquad (7.20)$$

Since $(z_1, \ldots, z_{n+1})$ are also homogeneous coordinates on $\mathbf{CP}^n$ we see that the orbits of this action are the space $\mathbf{CP}^n$. Now we select $(n + 1)$ real numbers $\{a_i\}$ such that $a_1 < a_2 < \cdots < a_{n+1}$ and define an S^{2n+1}-function $\widetilde{f}$ by

$$\widetilde{f} = a_1 z_1 \bar{z}_1 + \cdots + a_{n+1} z_{n+1} \bar{z}_{n+1} \qquad (7.21)$$

The function $\widetilde{f}$ is invariant under the S^1 action and so gives a function on $\mathbf{CP}^n$ which is our desired function f. There are clearly $(n + 1)$ critical points of f, one for each direction z_i. We must not forget that, despite our use of complex coordinates, f is a real valued function; when the Hessian is calculated, and viewed as a real matrix, each negative eigenvalue occurs twice corresponding to the two real variables in z_i. Now the a_i get bigger as i increases and it is a straightforward piece of calculus to check that the ith critical point has index $2i$, giving

$$M_t(f) = 1 + t^2 + t^4 + \cdots + t^{2n} \qquad (7.22)$$

The lacunary principle immediately implies that

$$P_t(\mathbf{CP}^n) = 1 + t^2 + t^4 + \cdots + t^{2n} \qquad (7.23)$$

and so we know the Betti numbers of $\mathbf{CP}^n$. This also illustrates the point that we can use Morse theory on complex manifolds even though f is a real valued function.

Returning to the height function we can generalise the situation by replacing T^2 with a Riemann surface Σ_p of genus p with $p > 1$. Thus we have

Example *The height function on Σ_p*

In this case we stick p doughnuts together, place the resulting surface on its end and define the same height functions as for T^2. It is clear that there are now $2p + 2$ critical points comprised of $2p$ saddle points, a maximum and a minimum. The Morse series is

$$M_t(f) = 1 + 2pt + t^2 \qquad (7.24)$$

allowing us to correctly deduce that $\chi(\Sigma_p) = 2 - 2p$.

So far we have insisted that our critical points be non-degenerate; but we would also like to investigate some of the properties of the degenerate case. The Morse inequalities show that, in the non-degenerate case, a function on Σ_p has at least $(2p + 2)$ critical points. However, even when $p > 1$, it is possible to reduce this number to just four, of which two are saddle points and the remaining two are a maximum and a minimum. When this is done the function f can still be taken to be the height function but the surface Σ_p has to be appropriately deformed. Of course some of the critical points must now be degenerate. The two saddle points are degenerate while the maximum and the minimum are not. Near a saddle point, whose (x, y) coordinates are $(0, 0)$, f can be represented by the function $Re\,(x + iy)^{p+1} = Re\,z^{p+1}$.

This dramatic reduction in the number of critical points raises the question of whether there is any other topological lower limit on the number of critical points of a general smooth f. The answer is in the affirmative: for each manifold M there is a positive integer $cat\,(M)$ known as the Lusternik–Schnirelmann category of M; in brief, if $\{M_1, \ldots, M_k\}$ are a collection of k closed contractible subsets of M such that $M = M_1 \cup \cdots \cup M_k$, then the smallest integer k for which this is true is $cat\,(M)$. Every smooth f has at least $cat\,(M)$ distinct critical points and, in fact, $cat\,(M)$ is a topological invariant of M. It is not particularly easy to compute $cat\,(M)$ but in two dimensions we have the result that

$$cat\,(S^2) = 2 \quad \text{and} \quad cat\,(\Sigma_p) = 3, \text{ for } p \geq 1 \qquad (7.25)$$

When $M = S^2$ we can see at once that the lower bound of $cat\,(S^2) = 2$ critical points is actually realised by the height function. When $M = \Sigma_p$ with $p \geq 1$, the lower bound, which is now 3 critical points, can also be attained but not by any embedding of Σ_p in three dimensions.

We return to our degenerate height function with its four critical points. The degeneracy is easy to remove by applying a small perturbation to f. Under this perturbation z^{p+1} changes to the product $(z - \epsilon_1) \cdots (z - \epsilon_{p+1})$ where the ϵ_i are all distinct. This perturbation achieves a bifurcation of the p-fold degenerate $z = 0$ critical point into p non-degenerate critical points. In this manner we can deform f into a good Morse function with non-degenerate critical points.

However, there are circumstances where a simple perturbation will not remove degenerate critical points. Examples of this can occur when the critical points of f are no longer discrete but form a set of positive dimension. We turn to this topic in the next two sections.

§ 2. Critical sub-manifolds

The easiest way to begin the discussion is by giving an example which illustrates the nature of the problem.

Example *The height function on T^2 revisited*

We choose f to be the height function and M to be T^2. But, this time, T^2 is oriented at an angle of $\pi/2$ relative to its position in the example on p. 196; regarded as a doughnut it sits as it would on a dinner plate. The height function f now has two extrema but they are now circles instead of points. These two circles are parallel to each other and one is a maximum, the other a minimum; there are no more extrema. For Morse theory to deal with this case we have to be able to treat functions with critical sub-manifolds rather than just critical points.

The quantity to focus on is the Morse series $M_t(f)$. In the ordinary case each critical point p contributes a term t^{λ_p} to $M_t(f)$. If f is critical on a whole sub-manifold C, say, then one feels that C must make a more substantial contribution to $M_t(f)$. It turns out (Bott [4]) that C contributes a whole polynomial to $M_t(f)$ instead of a monomial t^{λ_p}. Proceeding to the details we pass from the notion of a non-degenerate critical point to that of a non-degenerate critical *manifold:* A (connected oriented) sub-manifold $C \subset M$ is a non-degenerate critical manifold of M if $df = 0$ everywhere on C and the Hessian Hf is non-degenerate in the directions normal to C. In linear algebraic terms this says that $\det Hf(p)$ is non-vanishing when restricted to the normal directions. In terms of the normal bundle $N(C)$ of C, non-degeneracy is the statement that $\det Hf(p)$ is non-vanishing on $N(C)$.

We also have to define an *index* for a critical sub-manifold. To do this we give C a Riemannian metric and then, using the eigenvalues of Hf, we make an orthogonal decomposition of $N(C)$ into positive and negative parts. We write

$$N(C) = N^+(C) \oplus N^-(C) \tag{7.26}$$

where $N^{\mp}(C)$ correspond to the spans of the positive and negative eigenvalues of Hf. This achieved, the index of the critical manifold C is the integer λ_C where λ_C is the rank of the negative bundle $N^-(C)$.

Lastly we must give the contribution made by C to the Morse series $M_t(f)$. This contribution is the polynomial

$$t^{\lambda_C} P_t(C) \tag{7.27}$$

where $P_t(C)$ is the Poincaré polynomial of C; when C collapses to a point p this polynomial reduces to the monomial t^{λ_p} as it should.

We can now write down the full Morse series $M_t(f)$ and apply it to our current example. The new definition of $M_t(f)$ is just

$$M_t(f) = \sum_{\text{all } C} t^{\lambda_C} P_t(C) \tag{7.28}$$

We have seen already that there are two identical critical sub-manifolds both of which are circles. Let us call these C and C' where C is the maximum and C' is the minimum. It is then immediate that

$$P_t(C) = P_t(C') = 1 + t$$
$$\lambda_C = 1 \quad \text{and} \quad \lambda_{C'} = 0$$
$$\Rightarrow M_t(f) = (1+t) + t(1+t) \tag{7.29}$$
$$= (1 + 2t + t^2)$$
$$= P_t(T^2)$$

Thus, when we apply the correctly formulated Morse theory for critical sub-manifolds, the height function is restored to perfection. Finally we should say that the Morse inequalities hold for the extended case, that is

$$M_t(f) \geq P_t(M; \mathbf{F}) \quad \text{where} \quad M_t(f) = \sum_{\text{all } C} t^{\lambda_C} P_t(C; \mathbf{F}) \tag{7.30}$$

We have not yet verified the $(1+t)R(t)$ structure in the Morse inequalities. This requires us to select an example with a non-perfect function. We do this now.

Example *The square of the height function on S^2*

Let

$$M = S^2 = \{(x, y, z) : x^2 + y^2 + z^2 = 1\} \quad \text{and} \quad f(x, y) = z^2 \tag{7.31}$$

Notice that f is the square of the height function on S^2. Evidently f has three extrema. These have the following structure: there are two critical *points* which are the maxima at $z = \mp 1$ and so have index 2, then there is a minimum consisting of the circle of critical points given by $z = 0$ and this has index 0. According to our extended Morse theory the two points each contribute t^2 to the Morse series while the circle contributes $t^0 P_t(S^1) = (1 + t)$. The Morse series is therefore

$$M_t(f) = 1 + t + 2t^2 \tag{7.32}$$

But the Poincaré series of S^2 is $P_t(S^2) = 1 + t^2$ and so f is not perfect. If we calculate the difference we find that

$$M_t(f) - P_t(S^2) = t + t^2$$
$$= (1 + t)t \qquad (7.33)$$
$$\Rightarrow R(t) = t$$

Thus we have verified that, $M_t(f) - P_t(S^2)$ is divisible by $(1 + t)$, and that $R(t)$ has non-negative coefficients. In our next example we investigate the situation where f is invariant under some Lie group G which acts on M.

§ 3. Equivariant Morse theory

Example *The height function on S^2*

Let us stay with the manifold S^2 but take f to be simply the height function on S^2, that is we choose

$$M = S^2 = \{(x, y, z) : x^2 + y^2 + z^2 = 1\} \quad \text{and} \quad f(x, y) = z \qquad (7.34)$$

Now f has two critical points given by the north and south poles of the sphere. It is also easy to check that f is perfect. However, f is invariant under the group $U(1)$ realised by the rotations about the z axis. This gives us hope that we might be able to use Morse theory to study the cohomology of the quotient $S^2/U(1)$. This hope is dashed because the quotient is

$$S^2/U(1) \simeq [0, 1] \qquad (7.35)$$

and the cohomology of the interval $[0, 1]$ is trivial, i.e. $P_t(S^2/U(1)) = 1$. We shall see below that the reason for this failure is that the $U(1)$ action on S^2 is not free. The fixed points in this case are actually the two critical points $z = \mp 1$. The remedy for our difficulty is that we have not used the right cohomology theory; we need a cohomology theory which takes proper account of the group action. This is *equivariant cohomology*.

When group actions are not free we have seen (cf. chapter 5) that the resulting orbit spaces are not smooth manifolds; this is actually the source of our difficulty and the use of equivariant cohomology can get round it. First let us be clear that if f is a G-invariant function $f : M \to \mathbf{R}$ on M and G acts smoothly and *freely* on M then there is no problem. The invariance of G means that it induces a well defined smooth function f_G on the necessarily smooth space of orbits M/G. The Morse theory of the function $f_G : M/G \rightarrow \mathbf{R}$ on the manifold M/G has then no reason to be pathological and is the appropriate one to use in such a case. With this remark understood, let M be a manifold with a non-free action of a Lie group G. In equivariant cohomology we manage to replace M by a closely related space on which the action of G *is* free; having done this we can pass to the quotient and calculate as usual. This is done in the following way (Atiyah and Bott [4]).

Let BG be the classifying space for G-bundles. We know from chapter 1 that BG is the base space of a universal bundle whose total space is E_G and G must act freely on E_G because of the principal fibring over BG. Now we have two spaces on which G acts, namely M and E_G, thus G also acts on the Cartesian product $M \times E_G$ via the diagonal action

$$
\begin{aligned}
G \times (M \times E_G) &\longrightarrow M \times E_G \\
(g, x, e) &\longmapsto (g \cdot x, g \cdot e)
\end{aligned}
\tag{7.36}
$$

Happily the action on the product is also free and so it is sensible to form the quotient M_G where

$$
M_G = (M \times E_G)/G
\tag{7.37}
$$

which is a well behaved space called the *homotopy quotient* of M by G. Incidentally, when the action of G is free, M_G has the same homotopy type as M/G and so we automatically revert to the situation described in the previous paragraph. Another fact used in calculations is that M_G is fibred over BG with fibre M itself, that is we have

$$
\begin{array}{ccc}
M & \longrightarrow & M_G \\
& & \downarrow{\scriptstyle \pi} \\
& & M
\end{array}
\tag{7.38}
$$

The equivariant cohomology of M is simply the ordinary cohomology of M_G. We write the equivariant cohomology as $H_G^*(M)$ so we have

$$
H_G^*(M; \mathbf{F}) = H^*(M_G; \mathbf{F})
\tag{7.39}
$$

Equivariant cohomology is highly non-trivial even when M itself is topologically trivial. For example, if M is the space $\{p\}$ consisting of a single point then we have

$$
M_G = (\{p\} \times E_G)/G \simeq E_G/G = BG
\tag{7.40}
$$

Hence

$$
H_G^*(\{p\}) = H^*(BG)
\tag{7.41}
$$

That is, the equivariant cohomology of a point is the ordinary cohomology of the classifying space BG and this latter is highly non-trivial.

We can now give the equivariant version of the Morse inequalities between $M_t(f)$ and $P_t(M)$. The equivariant Poincaré series $P_t^G(M)$ is obtained by substituting equivariant cohomology for ordinary cohomology. Thus we define

$$
\begin{aligned}
P_t^G(M) &= \sum_i \dim H_G^i(M) t^i \\
\Rightarrow P_t^G(M) &= \sum_i \dim H^i(M_G) t^i
\end{aligned}
\tag{7.42}
$$

The equivariant Morse series for f is the ordinary Morse series for the function that f induces on M_G; we call this function f_G. If f has critical manifolds then this must be taken account of just as in the ordinary case. If N is a non-degenerate critical manifold then f_G will be critical on its homotopy quotient N_G; the index of f_G on N_G will be the *same* as the index of f on N. The equivariant Morse series in this general situation is thus

$$
\begin{aligned}
M_t^G(f) &= \sum_N P_t^G(N) t^{\lambda_N} \\
&= \sum_N P_t(N_G) t^{\lambda_N}
\end{aligned}
\tag{7.43}
$$

and the equivariant Morse inequalities are

$$
M_t^G(f) - P_t^G(M) = (1 + t) R(t)
\tag{7.44}
$$

and these hold for Morse functions $f : M \to \mathbf{R}$ which are G-invariant.

Now we return to the height function f of our example above. First we must compute the equivariant Poincaré series of S^2, that is, the quantity

$$
P_t^G(S^2) = P_t(S_G^2) \quad \text{where} \quad G = U(1)
\tag{7.45}
$$

To carry out this computation we observe that the fibration M_G of 7.38 is that of an S^2 bundle over BG; this in turn means that S_G^2 comes from a rank 3 vector bundle over BG. Since an odd dimensional vector bundle V_{2k+1}, say, has vanishing Euler class, the Leray–Hirsch theorem (Bott and Tu [1]) asserts that the cohomology of the associated sphere bundle $S(V_{2k+1})$ factorises into a product of the cohomology of its fibre S^{2k} and that of its base M. In short, we can write

$$
H^*(S(V_{2k+1})) = H^*(S^{2k}) \otimes H^*(M)
\tag{7.46}
$$

Applying this to the present case where $k = 1$ gives

$$H^*(S^2_{U(1)}) = H^*(S^2) \otimes H^*(BU(1))$$
$$\Rightarrow P_t(S^2_{U(1)}) = P_t(S^2)P_t(BU(1))$$
$$= (1 + t^2)(1 + t^2 + t^4 + \cdots) \qquad (7.47)$$
$$= \frac{1 + t^2}{1 - t^2}$$

where we have used $P_t(BU(1)) = (1/(1 - t^2))$, and we digress briefly to prove this fact: Return to our example on p. 197 where we used the lacunary principle to compute the cohomology of $\mathbf{CP}^n$. If we pass to the limit $n \to \infty$ we obtain the result

$$P_t(\mathbf{CP}^\infty) = 1 + t^2 + t^4 + \cdots = \frac{1}{(1 - t^2)} \qquad (7.48)$$

But in this example $\mathbf{CP}^n$ is expressed as the quotient of S^{2n+1} by $U(1)$; moreover, the action of $U(1)$ on S^{2n+1} is free. However, S^{2n+1} only has non-trivial cohomology in dimension $(2n + 1)$, so that, in the limit $n \to \infty$, all its Betti numbers are zero. In fact S^∞ is contractible and hence the $U(1)$-fibration of S^∞ is universal and so its base is a classifying space. Therefore

$$BU(1) = \mathbf{CP}^\infty \qquad (7.49)$$

as we require. We now return to the function f.

To compute the Morse series for f we recall that it has two critical points, each of which contribute a Poincaré series term of the form $P_t(BG)$ to $M_t^G(f)$; adding to this the information that the indices of these points are zero and two we find that

$$M_t^G(f) = \sum_N P_t^G(N)t^{\lambda_N}$$
$$= P_t^G(\{p\}) + t^2 P_t^G(\{p\})$$
$$= (1 + t^2)P_t(BG) \qquad (7.50)$$
$$= \frac{1 + t^2}{1 - t^2}$$
$$= P_t^G(S^2)$$

Thus the height function is *equivariantly perfect*.

Before leaving this topic we may as well look at the equivariant situation for the other function that we studied on S^2, namely $f = z^2$. The Poincaré series is unchanged so we only have to compute the Morse series. We found above that the three extrema on S^2 two are the points $z = \mp 1$ and the circle $z = 0$. The two points, being maxima, have index 2 and so contribute, to $M_t^G(f)$, the amount

$$2t^2 P_t(BU(1)) = \frac{2t^2}{1 - t^2} \qquad (7.51)$$

However things are different for the circle. The circle is a sub-manifold of S^2 on which $U(1)$ acts freely; thus, using our remark above that $M_G \simeq M/G$ when G acts freely, we see that

$$S^1_{U(1)} \simeq S^1/U(1) \simeq \{p\} \qquad (7.52)$$

In other words the $z = 0$ extremum, whose index is 0 since it is a minimum, only contributes the constant term $t^0 = 1$ to $M_t^G(f)$. The complete equivariant Morse series is therefore given by

$$M_t^G(f) = 1 + \frac{2t^2}{1 - t^2} = \frac{1 + t^2}{1 - t^2} \qquad (7.53)$$

Thus $f = z^2$ is equivariantly perfect; we learn, too, that equivariant perfection does not imply perfection in the ordinary sense.

In all our examples the manifold M has been finite dimensional, equivalently our functions $f : M \to \mathbf{R}$ have depended on finitely many variables. Actually the original geodesic problem of Morse is one where M is *infinite dimensional*. Since in subsequent chapters we encounter applications of Morse theory to Yang–Mills theories, all of which involve infinite dimensional M, we close this section by commenting briefly on the infinite dimensionality in the geodesic problem.

The function used by Morse to study geodesics is the energy functional E defined by

$$
\begin{aligned}
E(\gamma) &= \int_0^1 \left| \frac{d\gamma(t)}{dt} \right|^2 dt \\
&\equiv \int_0^1 g_{ij} \frac{d\gamma^i(t)}{dt} \frac{d\gamma^j(t)}{dt} dt
\end{aligned}
\qquad (7.54)
$$

where $\gamma(t)$ is a parametrised path on M with end points p and q labelled by 0 and 1, and g_{ij} is the Riemannian metric on M. We keep the metric g_{ij} and the manifold M fixed; this makes E a function or functional of $\gamma(t)$. Hence E is a positive real valued function on the space $PM(p, q)$ of paths on M from p to q. Slightly more formally we can represent E as

$$
\begin{aligned}
E : PM(p, q) &\longrightarrow \mathbf{R} \\
\gamma &\longmapsto E(\gamma)
\end{aligned}
\qquad (7.55)
$$

The space $PM(p,q)$ is of course infinite dimensional. The extrema of E are easily seen to be the geodesics joining p to q. They have the usual equation

$$\frac{d^2\gamma^i}{dt^2} + \Gamma^i_{jk}\frac{dx^j}{dt}\frac{dx^k}{dt} = 0 \tag{7.56}$$

where Γ^i_{jk} are the components of the Christoffel symbol for the metric g_{ij}. When $E(\gamma)$ is a minimum it coincides with the square of the integral $L(\gamma)$ giving the length of the geodesic joining p to q; that is, we have

$$\gamma \text{ a geodesic } \Rightarrow E(\gamma) = L^2(\gamma)$$
$$\text{where} \qquad L(\gamma) = \int_0^1 \left|\frac{d\gamma(t)}{dt}\right| dt \tag{7.57}$$

For a general γ we have the inequality

$$E(\gamma) \geq L^2(\gamma) \tag{7.58}$$

To consider *closed* geodesics we simply require γ to be a closed path, or loop, on M; this means that we take elements of $Map(S^1, M)$ instead of $PM(p,q)$. Now we regard E as a functional of the form

$$E : Map(S^1, M) \longrightarrow \mathbf{R} \tag{7.59}$$

We note, in passing, that $U(1)$ acts on $Map(S^1, M)$ by rotating the S^1; thus the equivariant theory with $G = U(1)$ should have some relevance here. To tackle the infinite dimensionality of $Map(S^1, M)$, in the case where M is a sphere S^k, Morse approximated the loops by geodesic polygons with n vertices $p_1, \ldots, p_n$. This makes $E(\gamma)$ a function of the n variables $p_1, \ldots, p_n$ instead of γ, i.e. $E = E(p_1, \ldots, p_n)$; these variables are also subject to some geodesic criterion. If we denote the space of these $\{p_i\}$ by $Map_n(S^1, S^k)$ then $Map_n(S^1, S^k)$ is to be viewed as a finite dimensional subset of the infinite dimensional $Map(S^1, S^k)$. The idea then is to compute the topology (e.g. to compute $P_t(Map_n(S^1, S^k))$) of $Map_n(S^1, S^k)$ and to understand its dependence on n. This allows the passage to the limit $n \to \infty$ where one eventually deduces results such as the existence of an infinite number of closed geodesics on S^k and that E is a perfect Morse function. This considerable piece of work is described in much greater detail and generality in Klingenberg [1], where further references are also to be found.

We already know that the topology of the space $Map(S^1, M)$ is closely related to that of the space ΩM of based loops. Granted the success of Morse theory in linking the critical points of E to the topologies of $Map(S^1, S^k)$

and ΩS^k, it is possible to imagine replacing S^k by a Lie group G and trying to use Morse theory to discover topological information about ΩG. This strategy was very successfully employed by Bott in the proof of his periodicity theorems.

§ 4. Supersymmetric quantum mechanics and Morse theory

We end this chapter by giving a quantum mechanical account of Morse theory due to Witten [2]. Witten's paper provides a new way of looking at Morse theory which has proved very influential. It also provides a point of departure for the Floer theory discussed in chapter 12. We shall see that an important extra feature of the quantum mechanical proof of the Morse inequalities is that, as well as establishing a connection between critical points and the Betti numbers of a manifold M, the cohomology of M is also explicitly constructed.

We begin by introducing the supersymmetry needed in the present context. Supersymmetric quantum theories are theories which are possess symmetry properties under the exchange of Fermions and Bosons. When Fermions are present in a quantum theory there is always an anti-symmetric structure. In the case of supersymmetry one has N supersymmetry generators $Q_1, \ldots, Q_N$ which all anti-commute with each other, giving

$$\{Q_i, Q_j\} = Q_i Q_j + Q_j Q_i = 0, \quad \text{for} \quad i \neq j \tag{7.60}$$

The above algebra does not yet constrain the squares of the generators so this has be done next. We consider a two dimensional theory with space and time both being one dimensional. We can then take $N = 2$; and, with this choice, the squares of the two supersymmetry generators are expressed in terms of the momentum P and the Hamiltonian $\mathcal{H}$ by the equations

$$Q_1^2 = \mathcal{H} + P, \qquad Q_2^2 = \mathcal{H} - P \tag{7.61}$$

If we combine this with the anti-commutation relations above we discover that

$$[Q_i, \mathcal{H}] = 0 \quad \text{and} \quad \mathcal{H} = \frac{1}{2}(Q_1^2 + Q_2^2) \tag{7.62}$$

showing, on the one hand, that the Q_i are bona fide symmetry operators, and on the other that the Hamiltonian $\mathcal{H}$ is non-negative.

To distinguish Fermions from Bosons we can introduce the 'mod 2' counting operator $(-1)^F$. It has the usual properties when it acts on particle states, these are

$$(-1)^F |P\rangle = \begin{cases} -|P\rangle & \text{if } P \text{ is a Fermion} \\ |P\rangle & \text{if } P \text{ is a Boson} \end{cases} \tag{7.63}$$

More explicitly we have the following. If H_1 denotes the Hilbert space of one particle states we can decompose H_1 into a direct sum of a Bosonic and Fermionic part, giving

$$H_1 = H_1^B \oplus H_1^F \tag{7.64}$$

The counting operator $(-1)^F$ will be block diagonal with respect to a basis of H_1^F and H_1^B and we will have

$$(-1)^F = \begin{pmatrix} I & 0 \\ 0 & -I \end{pmatrix} \tag{7.65}$$

Supersymmetry generators are created with the express intention of interchanging Fermions and Bosons; thus they must anti-commute with $(-1)^F$ and so they are block off diagonal with respect to this one particle basis. We have

$$Q_i = \begin{pmatrix} 0 & Q_i^{BF} \\ Q_i^{FB} & 0 \end{pmatrix}, \quad i = 1, 2 \tag{7.66}$$

The operators Q_i^{BF} and Q_i^{FB} transform one type of particle into the other. Their action is summarised in the maps below

$$Q_i^{BF} : H_1^F \longrightarrow H_1^B \quad \text{and} \quad Q_i^{FB} : H_1^B \longrightarrow H_1^F, \quad i = 1, 2 \tag{7.67}$$

This completes the supersymmetry information that we require.

To make contact with Morse theory the next step is to construct an example of this supersymmetry algebra making use only of data provided by some manifold M. It turns out that M should also be Riemannian. Having selected a Riemannian M all one needs is the exterior derivative d and its adjoint d^*. The supersymmetry quantities introduced above are now very simply constructed via the following definitions

$$Q_1 = (d + d^*) \quad Q_2 = i(d - d^*)$$
$$H_1^B = \bigoplus_{p \geq 0} \Omega^{2p}(M) \quad H_1^F = \bigoplus_{p \geq 0} \Omega^{2p+1}(M) \tag{7.68}$$

We see that a particle is a p-form and the parity of p determines whether it is a Boson or a Fermion. The supersymmetry algebra 7.60 is obeyed because the identities $d^2 = (d^*)^2 = 0$ imply that

$$Q_1 Q_2 = i(d + d^*)(d - d^*) = i(d^2 + d^*d - dd^* - d^*d^*)$$
$$= i(d^*d - dd^*)$$
$$\text{and} \quad Q_2 Q_1 = i(d - d^*)(d + d^*) = i(d^2 - d^*d + dd^* - d^*d^*) \tag{7.69}$$
$$= -i(d^*d - dd^*)$$
$$= -Q_1 Q_2$$

The Hodge Laplacian provides us with the Hamiltonian since we have

$$\mathcal{H} = \frac{Q_1^2 + Q_2^2}{2} = dd^* + d^*d = \bigoplus_{p \geq 0} \Delta_p \tag{7.70}$$

Actually one can also verify that $Q_1^2 = Q_2^2$, so that in this non-relativistic model the momentum P is zero. Thus far we have introduced a Riemannian manifold, but to study critical points we must have a function as well.

A Morse function f can be incorporated into this model without changing the supersymmetry algebra. This was done in Witten [2]. The technique is to replace d by d_t where

$$d_t = e^{-ft} d e^{ft} \tag{7.71}$$

With this simple but effective change we get a family of supersymmetry algebras parametrised by t; if we note that $d_t^* = e^{ft} d^* e^{-ft}$ then the new generators are given by

$$Q_1(t) = (d_t + d_t^*), \qquad Q_2(t) = i(d_t - d_t^*) \tag{7.72}$$

It is a routine matter to verify that this conjugation of d by e^{ft} leaves the algebra unchanged. The proof of the Morse inequalities rests on an analysis of the spectrum of the associated Hamiltonian, which is now

$$\mathcal{H}_t = d_t d_t^* + d_t^* d_t = \bigoplus_{p \geq 0} \Delta_p(t) \tag{7.73}$$

The easiest part of the spectrum of $\mathcal{H}_t$ to analyse is the null space. Hodge theory tells us that the Betti numbers of M are given by

$$b_p = \dim \ker \Delta_p \tag{7.74}$$

Now, though conjugation by e^{ft} means that the spectrum of $\mathcal{H}_t$ depends on t, the null spaces of $\mathcal{H}_t$ and $\mathcal{H}$ clearly coincide. Thus we still have a Betti number formula, namely

$$b_p = \dim \ker \Delta_p(t) \tag{7.75}$$

To understand the nature of the rest of the spectrum we expand $\mathcal{H}_t$ about $t = 0$ and this necessitates a few technical preliminaries about differential forms. Let $\{e_i\}$ and $\{e_i^*\}$ be bases for the tangent space and cotangent spaces $T_x M$ and $T_x^* M$; now the operations of exterior multiplication and interior multiplication are dual to one another, so, if a_i denotes interior multiplication of a form by e_i, the corresponding exterior multiplication by e_i^* is given by a_i^*. In quantum field theory a_i^* and a_i usually denote creation and annihilation operators; this notation is chosen here (Witten [2]) because, in this geometric realisation of supersymmetry, there is a formal equivalence between exterior and interior multiplication and Fermion creation and annihilation. We also need the second order covariant derivative of f, which we write as $D^2 f / Dx^i Dx^j$; the index of a critical point is given by the number of its negative eigenvalues. Having dealt with the preliminaries we return to $\mathcal{H}_t$. Because $(dd^* + d^*d)$ is a second order operator the expansion terminates at the t^2 term, providing us with the formula

$$
\begin{aligned}
\mathcal{H}_t = d_t d_t^* + d_t^* d_t &= e^{-ft}(dd^* + d^*d)e^{ft} \\
&= dd^* + d^*d + t\frac{D^2 f}{Dx^i Dx^j}[a_i^*, a_j] + t^2(g^{ij}\frac{\partial f}{\partial x^i}\frac{\partial f}{\partial x^j}) \\
&\equiv \Delta + V
\end{aligned}
\tag{7.76}
$$

where the last line of 7.76 is intended to suggest that $\mathcal{H}_t$ be viewed as an operator of Schrödinger type with the obvious potential term V. If this view is taken then the crucial observation for the Morse theory of f is that one should investigate $\mathcal{H}_t$ for large t. When t is large the 'potential' is dominated by its t^2 term, whose coefficient, we note, is $(grad\, f)^2$; that is, we have

$$
V \longrightarrow t^2(grad\, f)^2
\tag{7.77}
$$

If we think of the term $V\psi$ in the Schrödinger equation then we see that this enormous growth of the potential energy, if not stopped, will force the eigenfunctions $\mathcal{H}_t$ to be zero. However, at a critical point, $grad\, f$ vanishes and so the coefficient of t^2 is zero. Thus the actual state of affairs is that, for large t, the supports of the eigenfunctions of $\mathcal{H}_t$ are concentrated at the critical points of f. Hence this analysis suggests that to obtain Morse-theoretic results one should approximate quantities by taking t large and expanding about the critical points.

Following Witten we expand the metric about the critical point to make it flat to order two in the coordinates. Then we use our standard expansion of the form 7.4 for f in the neighbourhood of a critical point. This gives

$$
f(x) = f(0) + \frac{(\lambda_1 x_1^2 + \cdots + \lambda_n x_n^2)}{2}
\tag{7.78}
$$

the factor 2 in the denominator is just chosen for convenience; also the λ_i are all ∓ 1 and so only serve to determine the index of the critical point.

The approximated Hamiltonian $\widetilde{\mathcal{H}}_t$ is then expressible in terms of the simple harmonic oscillator Hamiltonian plus a finite correction. We obtain

$$\widetilde{\mathcal{H}}_t = \sum_i \left\{ -\frac{\partial^2}{\partial x_i^2} + t^2 x_i^2 + t\lambda_i [a_i^*, a_i] \right\} \tag{7.79}$$

the correction being the linear term in t. But the harmonic oscillator contribution to $\mathcal{H}_t$ commutes with the correction terms $\lambda_i [a_i^*, a_i]$ and we know that these latter terms have the simple spectrum ∓ 1. Thus we can write down the spectrum of $\mathcal{H}_t$. Each harmonic oscillator piece contributes a term of the form $t(1 + 2n_i)$, with n_i a non-negative integer and so the spectrum of $\widetilde{\mathcal{H}}_t$ is

$$t \sum_i \{ (1 + 2n_i) + \lambda_i \epsilon_i \}, \quad \text{where } \epsilon_i = \mp 1 \quad \text{for each } i \tag{7.80}$$

Now we want to make contact with Betti numbers, so we restrict $\widetilde{\mathcal{H}}_t$ to p-forms and look for the zero eigenvalues of this restriction. The first objective is achieved by restricting to only those terms which have precisely p positive ϵ_i's; and, since the n_i are the only terms which can be bigger than one, a zero eigenvalue is only obtained if all the n_i are zero. If we insert this information into 7.80 it is not difficult to check that, to obtain a zero eigenvalue, we are forced to set $\epsilon_i = +1$ precisely when $\lambda_i = -1$. This argument has provided us with a single zero eigenvalue corresponding to a p-form where p is the index of the critical point. Thus the total number of these zero p-forms is equal to the number of critical points of index p; also, if our approximation were exact we would know that the number of these zero p-forms is equal to the Betti number b_p. Taking into account the approximation means that some of the zero p forms may disappear in an exact calculation. In other words, we have shown that

$$b_p \leq m_p$$

which is the weak form of the Morse inequalities. This is as far as one gets with the present asymptotic analysis in t. Our next goal is to prove the stronger form of the Morse inequalities.

To attain this goal requires some additional mathematics and some additional physics; we explain the mathematical part first. The additional mathematics is a result which says that if one can form a cohomology complex from the set of critical points of f then this gives the cohomology of M and also implies the strong form of the Morse inequalities. In more detail the result is the following. Let

$$C^p = \{ \text{The set of all critical points of index } p \} \tag{7.81}$$

Included in C^p are integral linear combinations of critical points. Now if we can find a coboundary operator $\delta : C^p \rightarrow C^{p+1}$ which also obeys $\delta^2 = 0$ then we can form the cohomology complex C^* below

$$\cdots \xrightarrow{\delta} C^p \xrightarrow{\delta} C^{p+1} \xrightarrow{\delta} \cdots \qquad (7.82)$$

Having succeeded in defining δ, the cohomology of C^* coincides with that of M and the Morse inequalities are also valid. Thus, as far as the mathematics is concerned, we just have to construct δ.

The additional physics turns out to be just what is needed to construct δ. It all comes from the consideration of quantum mechanical tunnelling between critical points, that is tunnelling between the minima of the potential $t^2(grad\,f)^2$. The advantage of tunnelling is that, unlike our spectral analysis, it is not limited to working in the neighbourhood of only one critical point.

For the remainder of the argument we want to consider all the critical points; the method of steepest descent provides a framework within which one can do this. To construct the paths of steepest descent on M we first consider $grad\,f$ as a vector field on M. The integral curves of this vector field are paths (of steepest descent) $\gamma(s)$ on M which are solutions of the differential equation

$$\frac{d\gamma(s)}{ds} = -grad\,f(\gamma(s)) \qquad (7.83)$$

We could also call this the descending gradient flow of f on M. But when $grad\,f$ vanishes we have a critical point and so there are critical points at the ends of the paths. We shall see below that such paths are precisely the instantons which tunnel between the minima of the potential. In any case, in our desire to define a map between C^p and C^{p+1}, let us consider just those paths which connect pairs of critical points whose degree differs by one. Now we have to give the definition of δ.

First let γ_{ab} be a steepest descent path between the critical points a and b whose indices are p and $(p+1)$ respectively. Note that this means that $f(b) > f(a)$, so the flow is *from* b to a. We need to define a crucial sign $\epsilon_{\gamma_{ab}}$ associated with this flow. This sign is determined by comparing the orientation of appropriate tangent spaces. Let $T_b^- M$ be the $(p+1)$-dimensional space corresponding to the $(p+1)$ negative eigenvalues of the Hessian Hf at b, and let $T_b\gamma_{ab}$ be the tangent space to the path γ_{ab} at b. We can decompose $T_b^- M$ into $T_b\gamma_{ab}$ plus an orthogonal complement O_b, say, giving

$$T_b^- M = T_b\gamma_{ab} \oplus O_b \qquad (7.84)$$

Now, using the critical point b thought of as a $(p+1)$-form to orient $T_b^- M$, we can induce an orientation O_b by interior multiplication with any tangent vector in $T_b \gamma_{ab}$. Next we flow the p-dimensional O_b to a and compare its orientation with the orientation of the p-dimensional space $T_a^- M$. When these orientations agree we define $\epsilon_{\gamma_{ab}}$ to be $+1$ and when they disagree we define it to be -1. In general there will be more than one path of steepest descent from b to a; the (finite) sum over all these paths gives us our final number $\epsilon(a, b)$, which we define by

$$\epsilon(a, b) = \sum_{\text{all } \gamma_{ab}} \epsilon_{\gamma_{ab}} \tag{7.85}$$

All our signs are now determined and we can finish the definition: if $a \in C^p$ and b is a typical basis element of C^{p+1} then we define

$$\delta(a) = \sum_{b \in C^{p+1}} \epsilon(a, b) b \tag{7.86}$$

Finally we must make sure that $\delta^2 = 0$. A direct calculation with the present combinatorial definition is difficult. The way round this difficulty is to return to the quantum mechanical framework of instantons and tunnelling. The point is that, up to some factors which will be calculated below, we achieve the property $\delta^2 = 0$ by defining δ to be the large t limit of an operator whose square is zero, namely d_t. In doing this we discover that δ has a physical origin.

Bearing in mind the remark just made we must turn our attention to the operator d_t. This operator will map a critical point of index p to one of index $(p+1)$ and also satisfies $d_t^2 = 0$. The relevant physical quantity to calculate is the amplitude

$$\langle b | d_t | a \rangle \tag{7.87}$$

We must resort to some approximation method to calculate this amplitude and what we do is to take t large and sum over the instanton or tunnelling configurations connecting a and b. The equation for these paths is obtained by taking a Euclidean action for our quantum mechanical system and finding the minima. The action is S where

$$S = \frac{1}{2} \int \left\{ g_{ij} \frac{dx^i}{ds} \frac{dx^j}{ds} + t^2 g^{ij} \frac{\partial f}{\partial x^i} \frac{\partial f}{x^j} \right.$$
$$\left. + i g_{ij} \bar{\psi}^i \frac{D \bar{\psi}^j}{Ds} + \frac{1}{4} R_{ijkl} \bar{\psi}^i \psi^k \bar{\psi}^j \psi^l + t \frac{D^2 f}{Dx^i Dx^j} \bar{\psi}^i \psi^j \right\} ds \tag{7.88}$$

with x^i now counting as a Boson field and R_{ijkl} the curvature tensor of the Euclidean metric g_{ij}. The instantons of this action are got by just varying the Bosons and so we only have to minimise the Bosonic action, which we write in inner product form as

$$S_B = \frac{1}{2} \left\| \frac{dx}{ds} \right\|^2 + \frac{t^2}{2} \left\| \frac{\partial f}{\partial x} \right\|^2$$
$$= \frac{1}{2} \left\{ \left\| \frac{dx}{ds} + t\frac{\partial f}{\partial x} \right\|^2 - t\left\langle \frac{dx}{ds}, \frac{\partial f}{\partial x} \right\rangle \right\}$$

(7.89)

By changing f to $-f$, if necessary, we can assume that the inner product $-\langle (dx/ds), (\partial f/\partial x) \rangle$ is positive; this then provides a lower bound for S_B and the minimum is attained when

$$\left\| \frac{dx}{ds} + t\frac{\partial f}{\partial x} \right\| = 0$$
$$\Rightarrow \frac{dx}{ds} = -t\frac{\partial f}{\partial x}$$

(7.90)

which, after identifying x with γ_{ab} and absorbing t in a redefinition of s, is our equation of steepest descent. We have thus found the connection between instantons and the paths of steepest descent. At a minimum the action for a path from b to a is therefore

$$-t\left\langle \frac{dx}{ds}, \frac{\partial f}{\partial x} \right\rangle = -t \int ds \frac{dx^i}{ds} \frac{\partial f}{\partial x^i} = t(f(b) - f(a))$$

(7.91)

The standard WKB semi-classical approximation now gives the contribution of this path to the amplitude $\langle b| d_t |a \rangle$ as

$$det\, F_{ab} \exp[-t(f(b) - f(a))]$$

(7.92)

where $det\, F_{ab}$ is the determinant coming from the Gaussian integration over the fluctuations about the instanton. However, in this supersymmetric theory F_{ab} contains Fermionic and Bosonic contributions which mutually cancel; all that is left is the sign of $det\, F_{ab}$, the WKB calculation determines this sign and it is actually equal to $\epsilon_{\gamma_{ab}}$. Having dealt with $det\, F_{ab}$, and especially its sign, we can add together all the contributions to the amplitude $\langle b| d_t |a \rangle$. Our final expression for the large t behaviour of the amplitude is therefore

$$\langle b| d_t |a \rangle = \sum_{\gamma_{ab}} \epsilon_{\gamma_{ab}} \exp[-t(f(b) - f(a))]$$

(7.93)

This would be consistent with the action of d_t on $|a\rangle$ being given by

$$d_t |a\rangle = \sum_{\gamma_{ab}} \exp[-t(f(b) - f(a))]\epsilon_{\gamma_{ab}} |b\rangle \qquad \text{for } t \text{ large} \qquad (7.94)$$

It would now appear natural to define δ by setting $\delta(a) = d_t |a\rangle$, but Witten [2] points out that the exponential factors are inessential and can be discarded. This then dictates the choice of definition

$$\delta(a) = \sum_b \epsilon(a, b)b \qquad (7.95)$$

which is just what we had in 7.86. Thus we see that the vanishing of the square of δ is really a consequence of $d_t^2 = 0$ and that the origin of δ itself, which is the key ingredient of the whole construction, is intimately tied to the physics of instantons.

Thus with δ defined we have both constructed the cohomology of M and established the Morse inequalities. The instanton ideas employed for this purpose will be useful again in the somewhat analogous infinite dimensional Floer theory discussed in chapter 12.

CHAPTER VIII

Instantons and Monopoles

§ 1. The topology of gauge fields

In this chapter we wish to discuss instantons and monopoles. Our interest, in the main, is in geometrical and topological properties of the fields and the various solution spaces; however, many explicit solutions are known, cf. for example Nash and Sen [1] and references therein. We begin with an account of some of the general topological properties of the space of connections.

The space of all connections $\mathcal{A}$ is an affine space, i.e. it is a vector space once it has had an origin chosen: Think, for example, of expressing an arbitrary connection A in terms of a fixed connection A_0 according to $A = A_0 + a$, a is then a vector referred to the origin A_0. Alternatively, note that if A_1 and A_2 are two connections then so is the combination

$$(1 - t)A_1 + tA_2 \tag{8.1}$$

i.e. one can draw straight lines through $\mathcal{A}$. A simple, but important, consequence of this is that $\mathcal{A}$ can be contracted to a point and is thus homotopically trivial.

Any connection is acted on by the group of gauge transformations. Actually, in chapter 6, we distinguished between two gauge transformation groups $\mathcal{G}$ and $\tilde{\mathcal{G}}$ but favoured the former since it acts freely on the space $\mathcal{A}$. We can now prove this assertion. The proof follows fairly easily from the fact that the operations of parallel transport and gauge transformation commute with one another.

We begin by describing the formulation of parallel transport that is most suitable for our present argument. Let P be a principal G-bundle over M endowed with a connection A. Next we take a pair of points a and b on the base M. These points are joined by a smooth path

$$\gamma_{ab}(t) : [0, 1] \longrightarrow M \qquad \text{with} \qquad \gamma_{ab}(0) = a, \ \gamma_{ab}(1) = b \tag{8.2}$$

Parallel transport associates the path $\gamma_{ab}(t)$ to a lifted path $\widehat{\gamma}_{ab}(t)$ on P. We describe this using local coordinates. If (x, g) are local coordinates on P then, with respect to these coordinates, the lifted path is expressed as

$$\widehat{\gamma}_{ab}(t) = (\gamma_{ab}(t), g(t)) \qquad (8.3)$$

where $g(t)$ is required to satisfy the standard parallel transport equation (Nash and Sen [1])

$$\frac{dg(t)}{dt} + A\dot{\gamma}_{ab}g(t) = 0 \qquad (8.4)$$

The initial value $g(0)$ of the solution to this differential equation is arbitrary but, since the equation is first order, once $g(0)$ is chosen the final point $g(1)$ is determined. This means that the initial point of the lifted path can be any point in the fibre of P above a. In local coordinates, we have

$$\widehat{\gamma}_{ab}(0) = (a, g(0)) \qquad (8.5)$$

Passing to the final point we similarly have

$$\widehat{\gamma}_{ab}(1) = (b, g(1)) \qquad (8.6)$$

where $g(1)$ is computed from the solution to the parallel transport equation. The foregoing description makes it evident that, if we exercise our freedom to vary $g(0)$, parallel transport provides us with a map from the fibre P_a above a to the fibre P_b above b. If we denote this map by $PT_{\gamma_{ab}}(A)$ then we can write

$$\begin{aligned} PT_{\gamma_{ab}}(A) : P_a &\longrightarrow P_b \\ g(0) &\longmapsto g(1) \end{aligned} \qquad (8.7)$$

Next we bring in the group of gauge transformations $\mathcal{G}$. Recall from chapter 6 p. 176 that $\mathcal{G}$ acts as the identity on the fibre above some point m_0 of M. Now we choose $\alpha \in \mathcal{G}$ and consider its restriction to a fibre P_a. Denoting this restriction by α_a, the commutativity of gauge transformations and parallel transport becomes the statement

$$\alpha_b \circ PT_{\gamma_{ab}}(A) = PT_{\gamma_{ab}}(\alpha \cdot A) \circ \alpha_a \qquad (8.8)$$

where $\alpha \cdot A$ is the gauge transform of the connection A. To show that $\mathcal{A}$ acts freely on $\mathcal{A}$, let A be a fixed point of $\mathcal{A}$ and let $a = m_0$. Then we know that

$$\begin{aligned} \alpha \cdot A = A \qquad &\text{and} \qquad \alpha_a = I \\ \Rightarrow \alpha_b \circ PT_{\gamma_{ab}}(A) &= PT_{\gamma_{ab}}(A) \end{aligned} \qquad (8.9)$$

In other words, α_b is the identity map on P_b. But if M is path connected (which we now assume) then all points b can be joined to the base point a by some path γ_{ab} and α is the identity everywhere on M. Thus $\mathcal{G}$ acts on $\mathcal{A}$ without fixed points and the action is free.

Closely related to this discussion is the notion of the holonomy group of the connection A. This is obtained by joining the ends of the *path* γ_{ab} so as to produce a *loop* γ_a, say, beginning and ending at a. The parallel transport map will still be non-trivial but is now a map from P_a to itself

$$PT_{\gamma_a}(A) : P_a \longrightarrow P_a \tag{8.10}$$

As γ_a varies over all the loops based at a the set of all the $PT_{\gamma_a}(A)$ form a group known as the holonomy group of A at a; for a path connected M the holonomy groups at different points a will be isomorphic. By its definition the holonomy group of A is clearly a subgroup of the structure group G. If this subgroup actually coincides with G itself then the connection A is called *irreducible;* alternatively, if the holonomy group is a proper subgroup of G then A is called *reducible.*

Having established that $\mathcal{G}$ acts freely on $\mathcal{A}$ we pass to the quotient $\mathcal{A}/\mathcal{G}$ and thereby obtain a bundle

$$
\begin{array}{ccc}
\mathcal{G} & \longrightarrow & \mathcal{A} \\
 & & \downarrow \\
 & & \mathcal{A}/\mathcal{G}
\end{array}
\tag{8.11}
$$

Since $\mathcal{A}$ is also contractible it follows that $\mathcal{A}$ is the total space of a universal $\mathcal{G}$-bundle; hence $\mathcal{A}/\mathcal{G}$ is the classifying space for $\mathcal{G}$ bundles, or

$$\mathcal{A}/\mathcal{G} = B\mathcal{G} \tag{8.12}$$

$\mathcal{A}/\mathcal{G}$ is also the space of gauge orbits and is the space of physically measurable fields: the configuration space. In quantum field theory the partition function Z and its associated correlation functions are expressed as functional integrals over $\mathcal{A}$. In the absence of anomalies, which we discuss in chapter 10, the gauge invariance of the theory means that these integrals project onto the quotient $\mathcal{A}/\mathcal{G}$. This process requires gauge fixing and the introduction of the familiar Faddeev–Popov ghost Jacobian. The bundle $\mathcal{A} \to \mathcal{A}/\mathcal{G}$ is in general non-trivial (Singer [1]). To see this we suppose $\mathcal{A}$ is trivial, then

$$\mathcal{A} = \mathcal{G} \times (\mathcal{A}/\mathcal{G}) \tag{8.13}$$

and hence

$$\pi_j(\mathcal{A}) = \pi_j(\mathcal{G}) + \pi_j(\mathcal{A}/\mathcal{G}) \tag{8.14}$$

But the LHS of 8.14 is zero since $\mathcal{A}$ is contractible, while on the RHS, reference to 8.23 below shows that $\mathcal{G}$ possesses non-zero homotopy groups. We conclude that $\mathcal{A}$ is non-trivial and has, therefore, no global continuous section; the equivalent physical statement is that no continuous choice of gauge exists; this is known as the Gribov ambiguity, cf. Gribov [1], Singer [1] and Mitter and Viallet [1]. The topology of the configuration space plays an important part in the theory of these integrals.

To calculate the topology of $B\mathcal{G}$ we choose $M = S^n$ and return to the principal bundle P. The bundle P is completely characterised by the degree k of a homotopy class in $\pi_{n-1}(G)$ (Nash and Sen [1]). Further, if n is even, k is given by evaluating the Chern class $c_{n/2}$ on S^n. This k-dependence of P applies also to $\mathcal{A}$, which splits into components $\mathcal{A}_k$; displaying the various k-dependencies together we have

$$
\begin{array}{ccc}
G \longrightarrow P_k & \qquad & \mathcal{G}_k \longrightarrow \mathcal{A}_k \\
\downarrow & & \downarrow \\
S^n & & B\mathcal{G}_k \\
c_{n/2}(P_k) = k & & \mathcal{G}_k \subset Aut\,(P_k)
\end{array}
\qquad (8.15)
$$

Fortunately, as far as our topological calculations are concerned, this dependence of P_k on k does *not* live on in $B\mathcal{G}$. The reason for this is that the homotopy type of $B\mathcal{G}_k$ is *independent* of k (Atiyah and Jones [1]). Thus we can even take $k = 0$ so that P_k is trivial, yielding

$$
\begin{aligned}
P_0 &= G \times S^n \\
\Rightarrow \mathcal{G}_0 &= Map_0(S^n, G) \equiv \Omega^n G
\end{aligned}
\qquad (8.16)
$$

It is then immediate that, up to homotopy, we have

$$
\mathcal{G}_k \simeq \mathcal{G}_{k-1} \simeq \Omega^n G
\qquad (8.17)
$$

Since we only want to calculate the homotopy and cohomology of $\mathcal{G}_k$ and $B\mathcal{G}_k$ we can drop the k from our notation. Thus we write

$$
\mathcal{G} = \Omega^n G
\qquad (8.18)
$$

However, if we use the inverse relationship between the operands Ω and B (cf. p. 73), we find that

$$
\begin{aligned}
B\mathcal{G} = B\Omega^n G &\simeq B\Omega\Omega^{n-1}G \\
&\simeq \Omega^{n-1}G
\end{aligned}
\qquad (8.19)
$$

We can also see that

$$
\begin{aligned}
\Omega^{n-1}G &\simeq \Omega^{n-1}\Omega BG \\
&\simeq \Omega^n BG \\
&\simeq Map_0(S^n, BG) \\
\Rightarrow BG &\simeq Map_0(S^n, BG)
\end{aligned}
\tag{8.20}
$$

Continuing in this fashion we have

$$
\begin{aligned}
\Omega^{m+n}BG &\simeq \Omega^m\Omega^n BG \simeq Map_0(S^m, BG) \\
\Rightarrow Map_0(S^n, Map_0(S^m, BG)) &\simeq Map_0(S^m, BG)
\end{aligned}
\tag{8.21}
$$

Bearing in mind the fact that elements of $Map_0(X, BH)$ correspond to H-bundles over X, this last result suggests that a G bundle over a sphere S^m is rather like an n-parameter family of G-bundles over S^m. Precisely this property is of importance when studying anomalies—cf. § 6 of chapter 10.

Summarising the situation for G and BG we have shown that

$$
G = \Omega^n G \qquad \text{and} \qquad BG = \Omega^{n-1}G
\tag{8.22}
$$

The first topological consequence of 8.22 is that the homotopy groups of G and BG are now easily determined by expressing them in terms of those of G. We have

$$
\begin{aligned}
\pi_m(G) &= \pi_m(\Omega^n G) = \pi_{m+n}(G) \\
\pi_m(BG) &= \pi_m(\Omega^{n-1}G) = \pi_{m+n-1}G
\end{aligned}
\tag{8.23}
$$

and
$$
\pi_m(BG) = \pi_{m-1}(G)
$$

Many of these homotopy groups are non-zero: for example, if $M = S^4$ and $G = U(N)$ then

$$
\pi_1(BG) = \pi_4(G) = \begin{cases} 0, & N = 1 \\ \mathbf{Z}_2, & N = 2 \\ 0, & N \geq 3 \end{cases}
\tag{8.24}
$$

$$
\pi_2(BG) = \pi_5(G) = \begin{cases} 0, & N = 1 \\ \mathbf{Z}_2, & N = 2 \\ \mathbf{Z}, & N \geq 3 \end{cases}
$$

with the same results applying to $\pi_0(G)$ and $\pi_1(G)$.

Next we wish to calculate *cohomology* rather than its homotopy; working with G this time we turn our attention to the cohomology ring $H^*(G)$.

Let us also assume that n is even. In this chapter we shall simplify matters by only calculating the free cohomology; the more difficult task of detecting the torsion in $H^*(\mathcal{G})$ is addressed when we deal with anomalies. With this simplification understood, we draw on the property possessed by Lie groups G which says that, up to *rational homotopy*, G is a product of odd dimensional spheres. When $G = U(N)$ we obtain

$$G \sim S^1 \times S^3 \times \cdots \times S^{2N-1} \tag{8.25}$$

where $\sim$ denotes rational homotopy equivalence. This information is enough to enable us to calculate $H^*_{de\,Rham}(G)$. In fact we can straightaway deduce that

$$H^*_{de\,Rham}(U(N)) = H^*_{de\,Rham}(S^1) \otimes H^*_{de\,Rham}(S^3) \otimes \cdots \otimes H^*_{de\,Rham}(S^{2N-1}) \tag{8.26}$$

and so we can pick off the generators of the LHS from those on the RHS—for each sphere S^i we shall have just one generator ω_i in dimension i. To use this information to calculate the cohomology of $\mathcal{G}$ we bring in the evaluation map Ev (cf. 6.52). With $x \in S^n$ and $\alpha \in \mathcal{G}$ we define

$$\begin{aligned} Ev : S^n \times \mathcal{G} &\longrightarrow G \\ (x, \alpha) &\longmapsto \alpha(x) \end{aligned} \tag{8.27}$$

The last step is to use Ev to pull back the cohomology of G to that of $S^n \times \mathcal{G}$, and then one 'divides out' the cohomology of S^n leaving behind the cohomology of $\mathcal{G}$. More precisely, let ω_i be a closed i-form on G representing a generator of $H^i_{de\,Rham}(G)$, then

$$Ev^*\omega_i \in H^i_{de\,Rham}(S^n \times \mathcal{G}) \tag{8.28}$$

Now we integrate the i-form $Ev^*\omega_i$ over S^n and obtain a closed $(i-n)$-form on $\mathcal{G}$, we denote this form by μ_{i-n}. Hence μ_{i-n} represents a cohomology class in $\mathcal{G}$ and we have

$$\mu_{i-n} = \int_{S^n} Ev^*\omega_i \in H^{i-n}_{de\,Rham}(\mathcal{G}) \tag{8.29}$$

where for simplicity we use the same notation for the cohomology class as the form. Each generator ω_i can be treated in this way and the collection of μ_{i-n} so created generate an exterior algebra. The following argument shows that this exterior algebra coincides with the cohomology ring $H^*_{de\,Rham}(\mathcal{G})$.

We saw in chapter 1 that a topological group $\widehat{\mathcal{G}}$ is a H-space; this leads to a standard result of Hopf that $H^*(\widehat{\mathcal{G}}; \mathbf{Q})$ is determined by the rational homotopy groups $\pi_i(\widehat{\mathcal{G}}) \otimes \mathbf{Q}$. $H^*(\widehat{\mathcal{G}}; \mathbf{Q})$ is then an example of a Hopf algebra and, when $\widehat{\mathcal{G}}$ is connected, this Hopf algebra can be expressed (Whitehead [1]) as the product of a symmetric algebra $S(A)$ and an exterior algebra $\bigwedge(A)$. If $A^i = \pi_i(\widehat{\mathcal{G}}) \otimes \mathbf{Q}$ one has

$$H^*(\widehat{\mathcal{G}}; \mathbf{Q}) = S(A) \bigotimes \bigwedge(A) \qquad \text{with} \qquad \begin{cases} S(A) = \bigotimes_i S(A^{2i}) \\ \bigwedge(A) = \bigotimes_i \bigwedge(A^{2i-1}) \end{cases}$$

(8.30)

In our case, where $\widehat{\mathcal{G}} = \mathcal{G}$, we know that

$$\mathcal{G} \simeq \Omega^n G \qquad \text{and} \qquad G \sim S^1 \times S^3 \times \cdots \times S^{2N-1} \qquad (8.31)$$

therefore

$$\pi_i(\mathcal{G}) \otimes \mathbf{Q} = \bigoplus_{j=1}^{N} \pi_{n+i}(S^{2j-1}) \otimes \mathbf{Q} \simeq \bigoplus_{j=1}^{N} H^{n+i}(S^{2j-1}; \mathbf{Q}) \qquad (8.32)$$

However, since n is even it is immediate that $\pi_i(\mathcal{G}) \otimes \mathbf{Q}$ is only non-zero when i is odd and given by the non-negative values in the sequence

$$i = 2j - n - 1, \quad j = 1, \dots, N \qquad (8.33)$$

Referring to 8.30 we see that there is no symmetric algebra piece and that $H^*(\mathcal{G}; \mathbf{Q})$ is an exterior algebra generated by odd dimensional elements of dimension i with i belonging to the sequence 8.33 above. These elements are the ones that we constructed using the evaluation map and we can now conclude that the μ_{i-n} generate $H^*_{de\,Rham}(\mathcal{G})$.

In a similar way we can apply this technique to calculate the cohomology of $B\mathcal{G}$ using the homotopy equivalence

$$B\mathcal{G} \simeq Map_0(S^n, BG) \qquad (8.34)$$

We use the same notation as before and define another evaluation map Ev by

$$Ev : S^n \times Map_0(S^n, BG) \longrightarrow BG$$
$$(x, \alpha) \longmapsto \alpha(x) \qquad (8.35)$$

Luckily the generators of the cohomology of BG are well known: they are the universal Chern classes $c_i \in H^{2i}_{de\,Rham}(BG)$. Thus the generators of $H^*_{de\,Rham}(B\mathcal{G})$ are represented by the forms

$$\nu_{2i-n} = \int_{S^n} Ev^* c_i \in H^{2i-n}_{de\,Rham}(B\mathcal{G}) \qquad (8.36)$$

If we set $n = 2m$ we can write

$$\mu_{i-2m} \in H_{de\,Rham}^{i-2m}(\mathcal{G}), \qquad \nu_{2i-2m} \in H_{de\,Rham}^{2i-2m}(\mathcal{BG}) \qquad (8.37)$$

In contrast to our discovery above that the generators of $H_{de\,Rham}^{i-2m}(\mathcal{G})$ are all odd dimensional we note that the generators of $H_{de\,Rham}^{*}(\mathcal{BG})$ are all even dimensional. Let us now observe how $H_{de\,Rham}^{*}(\mathcal{G})$ varies as the N in the $U(N)$ increases. For definiteness choose $m = 2$ so that $n = 4$. Then the need to maintain $i - 2m$ positive shows that:

$$
\begin{array}{lll}
U(1) & \text{has} & H_{de\,Rham}^{i}(\mathcal{G}) \quad \text{all zero} \\[4pt]
U(2) & \text{has} & H_{de\,Rham}^{i}(\mathcal{G}) \quad \text{all zero} \\[4pt]
U(3) & \text{has just} & H_{de\,Rham}^{1}(\mathcal{G}) \neq 0 \\[4pt]
U(4) & \text{has just} & H_{de\,Rham}^{1}(\mathcal{G}) \neq 0, \quad H_{de\,Rham}^{3}(\mathcal{G}) \neq 0
\end{array}
\qquad (8.38)
$$

and so on. For $H_{de\,Rham}^{*}(\mathcal{BG})$ it is clear that $U(3)$ also provides the first non-trivial case; notice, too, that there is no sensitivity to the difference between $U(N)$ and $SU(N)$ here.

§ 2. Secondary characteristic classes

We can obtain explicit formulae for the cohomology classes μ_{2i-1} and ν_{2i} using the theory of secondary characteristic classes.

We must begin with a brief explanation of the construction of secondary characteristic classes—for a full account, cf. Chern [1]. The central point is that the vanishing of a standard characteristic class gives rise to the existence of a new characteristic class known as a Chern–Simons secondary characteristic class. Suppose that P is a bundle over the manifold M.

$$
\begin{array}{c}
P \\
\downarrow \pi \\
M
\end{array}
\qquad (8.39)
$$

Let $c_j(F)$ be a $2j$-dimensional characteristic class expressed in the usual way an invariant polynomial in the curvature F of a connection A. It will be useful below to write this in the form

$$c_j(F) = P_j \underbrace{(F, F, \ldots, F)}_{j\ \text{entries}} \qquad (8.40)$$

where $P_j(F, F, \ldots, F)$ denotes the invariant polynomial. If we use π to pullback $c_j(F)$ from M to P then, viewed as a $2j$-form on P rather than

M, it becomes exact (Nash and Sen [1]). There is a standard formula (Chern [1]) for this pull back which is obtained by using the curvature F_t of a family of connections A_t where

$$A_t = tA$$
$$F_t = tdA + t^2 A \wedge A$$

(8.41)

The formula that one obtains is

$$\pi^* c_j(F) = dTP_j(A), \qquad \text{with} \qquad TP_j(A) \in \Omega^{2j-1}(P)$$

and $\qquad TP_j(A) = j \int_0^1 dt \, P_j(A, F_t, \dots, F_t)$

(8.42)

where d denotes the exterior derivative acting on P and the notation $TP_j(A)$ is used to signify that T is the transgression operator acting on P. Now we consider what happens if $c_j(F)$ vanishes. In that event 8.42 asserts that

$$dTP_j(A) = 0$$

(8.43)

In other words, $TP_j(A)$ is closed and thus it determines a cohomology class on P. Since $TP_j(A)$ is a $(2j-1)$-form it is an *odd dimensional* cohomology class, that is

$$TP_j(A) \in H^{2j-1}(P; \mathbf{R})$$

(8.44)

It is $TP_j(A)$ that is known as a secondary characteristic class. It should be noted that $TP_j(A)$ is only a real cohomology class on P, unlike $c_j(F)$ which is an integral class on M. However, if we project $TP_j(A)$ onto the base M it becomes another cohomology class $\widehat{T}P_j(A)$, say, on M but with $\mathbf{R}/\mathbf{Z}$ coefficients. Summarising this cohomology information we have

$$c_j(F) \in H^{2j}(M; \mathbf{Z}), \ \ TP_j(A) \in H^{2j-1}(P; \mathbf{R}), \ \ \widehat{T}P_j(A) \in H^{2j-1}(M; \mathbf{R}/\mathbf{Z})$$

(8.45)

where in each case we have simplified matters by using the same notation to denote the differential form and the cohomology class it represents. In practice the $\mathbf{R}/\mathbf{Z}$-valued nature of the class $\widehat{T}P_j(A)$ can be seen if one calculates $\widehat{T}P_j(A)$ in two different gauges: unlike the situation for the ordinary characteristic class $c_j(F)$, the two classes will differ but their difference will be an integral class which projects to zero when we impose $\mathbf{R}/\mathbf{Z}$ coefficients. We shall encounter examples of this phenomenon in 8.62 below and in chapter 12.

We are now ready to apply the preceding construction to $H^*(\mathcal{G})$. We start with another use of the evaluation map, applied, this time, to the product $\mathcal{G} \times P$, giving

$$
\begin{array}{ccc}
\mathcal{G} \times P & \xrightarrow{Ev} & P \\
(\alpha, p) & \longmapsto & \alpha(p) \\
& & \downarrow{\scriptstyle \pi} \\
& & S^n
\end{array}
\tag{8.46}
$$

Next let $TP_j(A)$ be the secondary characteristic class on P; we pull this back to $\mathcal{G} \times P$ where it becomes

$$
Ev^*TP_j(A)
\tag{8.47}
$$

Then we restrict $Ev^*TP_j(A)$ to $\mathcal{G} \times S^n \subset \mathcal{G} \times P$; denoting this restriction by $\widehat{Ev}^*TP_j(A)$, we integrate over S^n and obtain a closed $(2j - 1 - n)$-form μ_{2j-1-n} on $\mathcal{G}$ which represents a cohomology class. Using the same notation for the cohomology class as the form, we write

$$
\mu_{2j-1-n} \in H^{2j-1-n}(\mathcal{G}; \mathbf{R}), \qquad \mu_{2j-1-n} = \int_{S^n} \widehat{Ev}^*TP_j(A)
\tag{8.48}
$$

We have not yet checked that the characteristic class $c_j(F)$ vanishes but it turns out that this is automatic. To see this we simply observe that, for μ_{2j-1-n} to be non-trivial, $2j - 1 - n$ has to be non-negative. Thus we must require

$$
2j - 1 - n \geq 0 \Rightarrow 2j \geq n + 1
\tag{8.49}
$$

But

$$
c_j(F) \in H^{2j}(S^n; \mathbf{Z})
\tag{8.50}
$$

and so $c_j(F)$ does indeed vanish when $2j > n$. Moving on to the ν_{2j}, which represents cohomology classes on $B\mathcal{G}$, we can relate the ν_{2j} to the μ_{2j-1} by using the fibration

$$
\begin{array}{ccc}
\mathcal{G} & \longrightarrow & \mathcal{A} \\
& & \downarrow{\scriptstyle \pi} \\
& & B\mathcal{G}
\end{array}
\tag{8.51}
$$

To do this we pull back ν_{2j} from $B\mathcal{G}$ to $\mathcal{A}$ where it is exact. This gives us

$$
\pi^*\nu_{2j} = df_j, \quad \text{for some} \quad f_j \in \Omega^{2j-1}(\mathcal{A})
\tag{8.52}
$$

But the fibre of $\mathcal{A}$ is $\mathcal{G}$ and so if we restrict f_j to a fibre—i.e. an orbit $\mathcal{G} \cdot A$ with $A \in \mathcal{A}$—we obtain a closed $(2j - 1)$-form on $\mathcal{G}$ which one can easily check coincides cohomologically with μ_{2j-1}. Hence, adjusting indices a little, we have

$$\mu_{2j-1} = \int_{S^n} \widehat{Ev}^* TP_{j+(n/2)}(A)$$
$$= f_j|_{\mathcal{G}\cdot A} \tag{8.53}$$
$$\text{with} \qquad \pi^* \nu_{2j} = df_j$$

Let us now set j to a specific value and display the resulting formula for μ_{2j-1}. In our work on anomalies in chapter 10 the cohomology group $H^1(\mathcal{G}; \mathbf{R})$ is of central importance, for this reason we calculate μ_1. First we must commit ourselves to a particular Lie group G and a value of n; we choose $G = SU(N)$ and $n = 4$. With this data in mind we refer to 8.48 which says that

$$\mu_{2j-1-n} = \int_{S^4} \widehat{Ev}^* TP_j(A) \tag{8.54}$$

Thus, for μ_1, we require

$$2j - 1 - n = 1$$
$$\Rightarrow j = \frac{2+n}{2} = 3 \tag{8.55}$$

With $j = 3$ we have

$$TP_3(A) = 3 \int_0^1 dt P_3(A, F_t, F_t) \tag{8.56}$$

But for $SU(N)$ the invariant polynomial $P_3(F, F, F)$ is given by

$$P_3(F, F, F) = -\frac{i}{24\pi^3} tr(F \wedge F \wedge F) \tag{8.57}$$

Putting all this together yields the following formula for μ_1

$$\mu_1 = -\frac{i}{(2\pi)^3} \int_0^1 dt \int_{S^4} tr(A \wedge F_t \wedge F_t), \quad \text{with} \quad F_t = tdA + t^2 A \wedge A \tag{8.58}$$

where A and F now denote the connection and curvature on S^4. We see that combining connection and curvature with secondary characteristic classes provides us with an explicit formula for the cohomology of $\mathcal{G}$. It is also clear that if μ_1 is exact the same is true for ν_2, conversely, using our formulae it is straightforward to verify that if ν_2 is exact then so is μ_1.

Another interesting example where a secondary characteristic class appears is the 3-dimensional gauge theory of Deser, Jackiw and Templeton [1] and Schonfeld [1]. In 3-dimensions $\mathcal{G}$ is disconnected: we have $\mathcal{G} \simeq \Omega^3 G$ and the number of connected components is counted by

$$\pi_0(\mathcal{G}) = \pi_3(G) = \mathbf{Z} \tag{8.59}$$

where G is a compact connected simple Lie group. The action S for this 3-dimensional gauge theory contains the usual $\|F\|^2$ term as well as a secondary characteristic class contribution of the form $\alpha T P_2(A)$. If M is the 3-dimensional manifold on which the theory is defined, the actual expression for S is

$$S = \|F\|^2 + \alpha T P_2(A) \tag{8.60}$$

If we take $G = SU(N)$, for which we know that $P_2(F) = (-1/8\pi^2)tr(F \wedge F)$, we can easily compute $T P_2(A)$ which is known as the Chern–Simons 3-form. Using 8.42 we find that

$$\begin{aligned}
T P_2(A) &= 2 \int_0^1 dt \, P_2(A, F_t) \\
&= -\frac{2}{8\pi^2} \int_0^1 dt \, tr \left\{ A \wedge (tA \wedge dA + t^2 A \wedge A \wedge A) \right\} \tag{8.61} \\
&= -\frac{1}{8\pi^2} tr(A \wedge dA + \frac{2}{3} A \wedge A \wedge A)
\end{aligned}$$

Having calculated the Chern–Simons term, the detailed expression for S is

$$S = -tr \int_M \left\{ (F \wedge *F) + \frac{\alpha}{8\pi^2} (dA \wedge A + \frac{2}{3} A \wedge A \wedge A) \right\} \tag{8.62}$$

However, S is not gauge invariant because of the lack of gauge invariance of the Chern–Simons form. Nevertheless, because $T P_2(A)$ changes by an integer under a gauge transformation, the physically measurable quantity $\exp[iS]$ will be gauge invariant provided α satisfies the quantisation condition $\alpha = 2\pi n$ where n is an integer. This condition has an important physical interpretation since α has the dimension of a mass: we obtain a 3-dimensional gauge theory with a quantised mass. Also $\mathcal{G}$ is disconnected for all the larger odd values of $\dim M$.

§ 3. Instantons and their moduli

In this section we want to calculate the dimension of the moduli space of instantons. The setting is a very specific one: we have a non-Abelian gauge theory with G a compact simple Lie group and action

$$S \equiv S(A) = \|F\|^2 = -\int_M tr(F \wedge *F) \tag{8.63}$$

with M a closed four dimensional orientable Riemannian manifold and $*$ the Hodge dual with respect to the Riemannian metric. Instantons are those A which correspond to critical points of S. To fix our conventions, and for the sake of completeness, we review briefly some basic properties of S relevant for a study of instantons—for additional background, cf. Nash and Sen [1].

First we should obtain the Euler–Lagrange equations of motion, i.e. the extremum condition. Let A be an arbitrary connection through which passes the family of connections

$$A_t = A + ta \tag{8.64}$$

The curvature and action for this family are

$$F(A_t) = dA_t + A_t \wedge A_t$$
$$S(A_t) = \|F(A_t)\|^2 = < F(A_t), F(A_t) > \tag{8.65}$$

Expanding in the vicinity of $t = 0$ gives

$$S(A_t) = < F(A), F(A) > +t\frac{d}{dt} < F(A_t), F(A_t) >|_{t=0} + \cdots$$

and

$$F(A_t) = F(A) + t(da + A \wedge a + a \wedge A) + t^2 a \wedge a$$
$$= F(A) + t d_A a + t^2 a \wedge a$$
$$\Rightarrow S(A_t) = \|F(A)\|^2 + t\{< d_A a, F(A) > + < F(A), d_A a >\} + \cdots$$
$$= S(A) + 2t < F(A), d_A a > + \cdots \tag{8.66}$$

A is a critical point if

$$\frac{dS(A_t)}{dt}\bigg|_{t=0} = 0 \tag{8.67}$$

That is, if

$$< F(A), d_A a > = 0$$
$$\Rightarrow < d_A^* F(A), a > = 0 \tag{8.68}$$

But a is arbitrary so our equation for a critical point is

$$d_A^* F(A) = 0 \tag{8.69}$$

However, $F(A) = d_A A$ also satisfies the Bianchi identity $d_A F(A) = 0$ and so we have the pair of equations

$$d_A F(A) = 0, \qquad d_A^* F(A) = 0 \tag{8.70}$$

This is similar to the condition 2.50 for a form ω to be harmonic, which is

$$d\omega = 0, \qquad d^*\omega = 0 \tag{8.71}$$

It should be emphasised, though, that the Yang–Mills equations are not linear; thus they really express a kind of non-linear harmonic condition. The most distinguished class of solutions to the Yang–Mills equations $d_A^* F(A) = 0$ is that consisting of those connections whose curvature is self-dual or anti-self-dual. To see how such solutions originate we first prove that, on 2-forms, d_A^* has the property that

$$d_A^* = - * d_A * \tag{8.72}$$

To prove this we take a p-form ω and a $(p-1)$-form η, both forms being matrix valued. Using ω and η we construct the $(n-1)$-form $tr(\omega \wedge * \eta)$ and observe that, since M is closed, we have

$$\int_M d\, tr(\omega \wedge * \eta) = 0$$

$$\Rightarrow \int_M tr(d_A\omega \wedge * \eta) + (-1)^p \int_M tr(\omega \wedge d_A * \eta) = 0 \tag{8.73}$$

But, from 2.43, we know that, on p-forms, $** = (-1)^{p(n-p)}$, and so

$$\int_M tr(d_A\omega \wedge * \eta) + (-1)^{p-(4-p)p} \int_M tr(\omega \wedge ** d_A * \eta) = 0$$

$$\Rightarrow <d_A\omega, \eta> + (-1)^{p-(4-p)p} <\omega, *d_A * \eta> = 0 \tag{8.74}$$

$$\Rightarrow <\omega, d_A^*\eta> + (-1)^{p-(4-p)p} <\omega, *d_A * \eta> = 0$$

$$\Rightarrow d_A^* = (-1)^{1+p-(4-p)p} * d_A *$$

On setting $(p-1) = 2$ this is the property that we require; when we make use of it the Yang–Mills equations become

$$d_A * F(A) = 0 \tag{8.75}$$

Thus if $F = \mp * F$ the Bianchi identities immediately show that we have a solution to the Yang–Mills equations—we have managed to solve a non-linear second order equation by solving a non-linear first order equation. We shall now show that these critical points are all *minima* of the action S.

First we decompose F into its self-dual and anti-self-dual parts F^+ and F^-, giving

$$F = \frac{1}{2}(F + *F) + \frac{1}{2}(F - *F)$$

$$= F^+ + F^-$$

$$\Rightarrow S = \|(F^+ + F^-)\|^2 \tag{8.76}$$

$$= \|F^+\|^2 + \|F^-\|^2$$

where the crossed terms in the norm contribute zero. The instanton number k is minus $c_2(F)$ so that, for $SU(N)$,

$$
\begin{aligned}
k &= \frac{1}{8\pi^2} \int_M tr(F \wedge F) \\
&= \frac{1}{8\pi^2} \int_M tr\{(F^+ + F^-) \wedge (F^+ + F^-)\} \qquad (8.77) \\
&= \frac{\|F^+\|^2 - \|F^-\|^2}{8\pi^2}
\end{aligned}
$$

The inequality $(a^2 + b^2) \geq |a^2 - b^2|$ shows that, for each k, the absolute minima of S are attained when

$$
S = 8\pi^2 |k| \qquad (8.78)
$$

and this corresponds to $F^{\mp} = 0$ or equivalently

$$
F = \mp * F \qquad (8.79)
$$

and we have the usual self-dual and anti-self-dual conditions. Changing the orientation of M has the effect of changing the sign of the $*$ operation and so interchanges F^+ with F^-. Hence there is no essential distinction between self-duality and anti-self-duality. For convenience we deem all our minima to be self-dual instantons.

If we think of the Yang–Mills equations as a non-linear generalisation of Hodge theory we expect that these equations are elliptic as is the case for Laplace's equation. Actually the Yang–Mills equations 8.69 are not yet elliptic as they stand. The reason for this is that they are the equations for the critical points of a function which possesses an invariance group. From a physical perspective this is very easy to understand: it is just the statement that the action $S(A)$ is invariant under gauge transformations. The non-ellipticity is easy to demonstrate: gauge invariance gives the Yang–Mills operator $d_A^* d_A$ a non-invertible symbol, whose kernel is parametrised by the space of gauge transformations. Further, suppose A is a smooth solution to

$$
d_A^* d_A A \equiv d_A^* F(A) = 0
$$

then so is $A_g = g^{-1} A g + g^{-1} dg$. But A_g need not be smooth if g is not sufficiently smooth. To obtain an elliptic problem we have to choose a (local) gauge. We shall do this by imposing a gauge condition. To find this gauge condition we have to look at connections in a neighbourhood of A.

Consider a small perturbation of A in a gauge direction; that is, subject A to an infinitesimal gauge transformation g by writing

$$g = \exp[tf] \sim I + tf, \qquad \text{for } t \text{ small}$$

so that
$$
\begin{aligned}
A_g &= g^{-1}Ag + g^{-1}dg \\
&\sim (I - tf)A(I + tf) + (I - tf)d(I + tf) \qquad (8.80) \\
&\sim A + t[A, f] + tdf \\
&= A + td_A f
\end{aligned}
$$

Taking the derivative at $t = 0$ shows that the tangent to $\mathcal{A}$ at A is just $d_A f$. However, this is a tangent in a gauge direction and, using standard linear algebra, we can decompose the tangent space according to

$$T_A \mathcal{A} = Im\, d_A \oplus ker\, d_A^* \qquad (8.81)$$

This makes it clear that the non-gauge directions are given by $ker\, d_A^*$: the orthogonal complement to $Im\, d_A$. This analysis suggests a natural gauge condition to impose on perturbations of A to a nearby connection $A + a$, namely the condition

$$d_A^* a = 0 \qquad (8.82)$$

Also if $\mathcal{G} \cdot A$ is the orbit under $\mathcal{G}$ of the connection A then it is clear that

$$T_{\mathcal{G} \cdot A} \mathcal{A} \simeq Im\, d_A \qquad (8.83)$$

and so we can also write

$$T_A \mathcal{A} = T_{\mathcal{G} \cdot A} \mathcal{A} \oplus ker\, d_A^* \qquad (8.84)$$

Having provided $\mathcal{A}$ with a gauge condition we can say that Yang–Mills equations are elliptic when restricted to the non-gauge directions $ker\, d_A^*$. More concretely, the equations

$$d_A^* F(A) = 0, \qquad d_A^* a = 0 \qquad (8.85)$$

form an elliptic system of non-linear partial differential equations with only smooth solutions. This completes our review of basic properties of the Yang–Mills action and we can now turn to the matter of the moduli space.

Let A be a self-dual connection of instanton number k. When gauge equivalence is taken properly into account, the space of such A form a finite dimensional space

$$\mathcal{M}_k \qquad (8.86)$$

which we call the instanton moduli space. The moduli space should be viewed as a finite dimensional subspace of the infinite dimensional configuration space $\mathcal{A}_k/\mathcal{G}_k$

$$\mathcal{M}_k \subset \mathcal{A}_k/\mathcal{G}_k \tag{8.87}$$

However, we have not quite dealt with all the gauge invariance present in the theory (cf. Atiyah and Jones [1]). There is a remaining invariance present because a purely physical quantity, such as the Yang–Mills action $S(A)$, does not depend on our choice of base point $p_0 \in P_a$; recall that p_0 is the base point used in the definition of $\mathcal{G}$. Now if $p_0' \in P_a$ is another base point, then $p_0' = p_0 g$ for some $g \in G$. Thus by acting with G we can move from one base point to another. This action extends to $\mathcal{A}_k/\mathcal{G}_k$ but unfortunately it is not a free action.

To obtain a good moduli space we have to cut down both the configuration space and the group G slightly: G is simple and so has a discrete centre $Z(G)$ and all the elements of $Z(G)$ leave the points of $\mathcal{A}_k/\mathcal{G}_k$ fixed: this is because, if $g \in Z(G)$, then $dg = 0$ and $g^{-1}Ag = g^{-1}gA = A$, thus $g^{-1}(A + d)g = A$. To get round this problem we factor out the centre by replacing G by its adjoint $Ad\,G = G/Z(G)$. To cut down $\mathcal{A}_k/\mathcal{G}_k$ we restrict to the subspace of *irreducible* G-connections. The reason for this is that a given connection A may be a fixed point for some elements of $Ad\,G$ but such an A is necessarily reducible (cf. next paragraph). Hence $Ad\,G$ acts without fixed points on the irreducible connections and so our new configuration space is therefore the quotient

$$(\mathcal{A}_k^{irred}/\mathcal{G}_k)/(Ad\,G) \tag{8.88}$$

Before continuing with our investigation of moduli we elucidate the relation between the action of $Ad\,G$ and reducible connections. We return to the parallel transport formula 8.8, which is

$$\alpha_b \circ PT_{\gamma_{ab}}(A) = PT_{\gamma_{ab}}(\alpha \cdot A) \circ \alpha_a \tag{8.89}$$

Reducibility has to do with holonomy so we set $a = b$ and replace $PT_{\gamma_{ab}}(A)$ by $PT_{\gamma_a}(A)$; if we further suppose that $\alpha \in Ad\,G$ and that A is a fixed point of α then the formula becomes

$$\alpha \circ PT_{\gamma_a}(A) = PT_{\gamma_a}(A) \circ \alpha \tag{8.90}$$

Thus those α which fix A—the stability group of A—commute with the elements of the holonomy group of A and so belong to its centraliser. Now if A is irreducible its holonomy group is G and the centraliser is just $Z(G)$, the centre of G; but $Z(G)$ is discrete and does not belong to $Ad\,G$, hence the stability group is trivial. Thus a connection with a non-trivial stability group must be reducible. A simple means of detecting the existence of a non-trivial stability group for A is to see whether there are any covariantly constant sections f of $ad\,P$, that is, any f satisfying

$$d_A f = 0 \qquad (8.91)$$

Given such an f, the gauge transformation $g = \exp[tf]$ leaves A fixed since

$$
\begin{aligned}
A_g &= \exp[-tf](A + d)\exp[tf] \\
&= A + td_A f + \frac{t^2}{2}[d_A f, f] + \cdots \qquad (8.92) \\
&= A
\end{aligned}
$$

If we were to allow reducible self-dual connections then the moduli space would be singular at the reducible points. We shall continue with our study of $\mathcal{M}_k$, which we take to be the space of equivalence classes of irreducible self-dual connections. If we do not divide by $Ad\,G$, then, instead of $\mathcal{M}_k$, we obtain a space $\mathcal{M}'_k$, and this space is fibred over $\mathcal{M}_k$ with fibre $Ad\,G$. Our next task is to find the dimension of $\mathcal{M}_k$.

We employ a similar technique to that used in chapter 5 when calculating the dimension of the Riemann moduli space $\mathcal{M}_p$. The main idea is to work infinitesimally, by which we mean to work with the tangent space to $\mathcal{M}_k$. The advantage of doing this is that the dimension of the tangent space can be calculated using the index theorem. Let $A + ta$ be a one parameter family of smooth connections. Hence, by construction, a is a tangent to this family at $t = 0$, so that

$$a \in T_A \mathcal{A}_k \qquad (8.93)$$

We wish to produce, from a, a corresponding tangent to the moduli space. If we use $[A]$ to denote the point in the moduli space to which A corresponds then we wish to produce an element of

$$T_{[A]}\mathcal{M}_k \qquad (8.94)$$

To do this we must do two things: firstly we must render this family self-dual, and secondly we must project out those a which correspond to gauge directions; in geometrical language this latter condition means those a which belong to the tangent space in the orbital directions, i.e. to $T_{g\cdot A}\mathcal{A}$. To achieve our *first* goal we introduce π_-, which is the operator which projects a 2-form ω onto its anti-self-dual part, i.e.

$$\pi_- \omega = \frac{1}{2}(\omega - *\omega) \qquad (8.95)$$

Our self-duality equation is now

$$\pi_- F(A) = 0 \qquad (8.96)$$

and substituting $A + ta$ into this equation gives

$$\pi_- F(A + ta) = 0 \qquad (8.97)$$

But

$$\begin{aligned}
F(A + a) &= d(A + ta) + (A + ta) \wedge (A + ta) \\
&= F(A) + t(da + a \wedge A + A \wedge a) + t^2 a \wedge a \qquad (8.98) \\
&= F(A) + td_A a + t^2 a \wedge a
\end{aligned}$$

The connection $A + ta$ is self-dual, so

$$\begin{aligned}
\pi_- (F(A) + td_A a + t^2 a \wedge a) = 0 \\
\Rightarrow \pi_- (F(A)) + t\pi_- (d_A a + ta \wedge a) = 0
\end{aligned} \qquad (8.99)$$

Thus, at the point $t = 0$, the tangent a satisfies

$$\pi_- (d_A a) = 0 \qquad (8.100)$$

For convenience we define the operator d_A^- by $d_A^- = \pi_- \circ d_A$ and then 8.100 becomes

$$d_A^- a = 0 \qquad (8.101)$$

To achieve our *second* goal we must identify those a which differ by an element of $T_{g \cdot A} \mathcal{A}$. But, since $T_{g \cdot A} \mathcal{A} \simeq Im \, d_A$, the two requirements we demand of a are met if

$$a \in \frac{ker \, d_A^-}{Im \, d_A} \qquad (8.102)$$

This has an obvious cohomological interpretation, an interpretation which we now exploit so that we can use the index theorem.

A more detailed description of the Lie algebra valued 1-forms a is that they are sections of the bundle

$$ad \, P \otimes \wedge^1 T^* M \qquad (8.103)$$

The first factor in 8.103 (which was defined in 6.24) ensures that a has the correct behaviour under gauge transformations, namely $a \mapsto g^{-1} a g$; the second factor is simply because it is a 1-form. Let us use the notation

$\Omega^i(M, ad\, P)$ to denote the various spaces of sections where $\Omega^i(M, ad\, P) = \Gamma(M, ad\, P \otimes \wedge^i T^*M)$; the natural elliptic complex to use for our index calculation is

$$0 \xrightarrow{\;\;i\;\;} \Omega^0(M, ad\, P) \xrightarrow{\;\;d_A\;\;} \Omega^1(M, ad\, P) \xrightarrow{\;\;d_A^-\;\;} \Omega_-^2(M, ad\, P) \xrightarrow{\;\;\pi_+\;\;} 0$$

$$(8.104)$$

It is necessary to check that 8.104 is a complex in the sense that

$$d_A^- \circ d_A = 0 \qquad (8.105)$$

To do this let $f \in \Omega^0(M, ad\, P)$, and so we must show that

$$\pi_- d_A d_A f = 0 \qquad (8.106)$$

This is guaranteed to be true since we can reason as follows: Recalling our identification in 8.80 of $d_A f$ as the change in the connection A under an infinitesimal gauge transformation, we see that $d_A d_A f$ is the corresponding change in the curvature $F(A)$ under this gauge transformation; but the self-duality equation $F = *F$ is clearly gauge invariant, thus the gauge transformed curvature must also be self-dual. In other words, we have

$$\pi_-(F(A) + d_A d_A f) = 0$$
$$\Rightarrow \pi_-(F(A)) + \pi_- d_A d_A f = 0 \qquad (8.107)$$
$$\Rightarrow \pi_- d_A d_A f = 0$$

as required. Thus we do have an elliptic complex. It is instructive to check this explicitly. To do so we write [1]

$$
\begin{aligned}
d_A f &= df + [A, f] \\
\Rightarrow d_A d_A f &= d(df + [A, f]) + A \wedge (df + [A, f]) + (df + [A, f]) \wedge A \\
&= dAf - A \wedge df - df \wedge A - fdA + A \wedge df + A \wedge Af \\
&\quad - A \wedge fA + df \wedge A + Af \wedge A - fA \wedge A \\
&= F(A)f - fF(A), \\
\Rightarrow \pi_- d_A d_A f &= \pi_-(F(A))f - f\pi_-(F(A)) \\
&= 0
\end{aligned}
$$

$$(8.108)$$

and so our check has been successful.

---

[1] For this, and similar calculations, it is useful to recall the formula $d_A \omega = d\omega + A \wedge \omega + (-1)^{p+1}\omega \wedge A$ where ω is an $ad\, P$-valued p-form.

Let us use the notation $\Omega_-(M, ad\, P)$ to denote the complex. The co-homology data for this complex is

$$H^0(\Omega_-(M, ad\, P)) = ker\, d_A^{(0)}, \qquad \dim H^0(\Omega_-(M, ad\, P)) = h_0$$

$$H^1(\Omega_-(M, ad\, P)) = \frac{ker\, d_A^-}{Im\, d_A^{(0)}}, \qquad \dim H^1(\Omega_-(M, ad\, P)) = h_1 \qquad (8.109)$$

$$H^2(\Omega_-(M, ad\, P)) = \frac{ker\, \pi_+}{Im\, d_A^-}, \qquad \dim H^2(\Omega_-(M, ad\, P)) = h_2$$

where the notation $d_A^{(p)}$ denotes the exterior covariant derivative acting on $ad\, P \otimes \wedge^p T^* M$. Only one of these dimensions is required for our moduli space calculation, this being h_1; in other words, we wish to compute

$$h_1 = \dim T_{[A]}\mathcal{M}_k = \dim \mathcal{M}_k \qquad (8.110)$$

However, the index of the complex is the alternating sum

$$h_0 - h_1 + h_2 \qquad (8.111)$$

Nevertheless, we shall still be able to calculate h_1 because it turns out that $h_0 = h_2 = 0$, a result we now prove before using the index theorem itself. h_0, coming from a cohomology group in dimension zero, is the dimension of a space of sections. More precisely it is the dimension of the space of sections of $ad\, P$ which are covariantly constant. But from our earlier discussion of reducible connections we know that the irreducibility of the connection A means that this space is empty, hence $h_0 = 0$. To deal with h_2 requires a vanishing theorem. This is done by a Weitzenböck positivity argument: one proves that $H^2(\Omega_-(M, ad\, P))$ is trivial by showing that the associated Laplacian

$$\Delta_A^{(2)} = d_A^-(d_A^-)^* + \pi_+\pi_+^* \qquad (8.112)$$

is positive definite and hence has no kernel; the term $\pi_+\pi_+^*$ is just zero and computing the remaining term in local coordinates shows that

$$\Delta_A^{(2)} = \frac{1}{2}d_A^{(1)}(d_A^{(1)})^* + \frac{R}{6} - W_- \qquad (8.113)$$

where R is the scalar curvature of M and W_- the anti-self-dual part of its Weyl tensor. Positivity will result if we assume that W_- is zero; that is, M is what is known as a self-dual manifold, and the scalar curvature is positive. Following Atiyah, Hitchin and Singer [1] we now make these two assumptions.

As usual in such calculations we can also compute the index by using the elliptic operator

$$D : \Omega^0(M, ad\, P) \oplus \Omega^2_-(M, ad\, P) \longrightarrow \Omega^1(M, ad\, P)$$
$$\text{where} \quad D = \{d_A^{(0)} + (d_A^-)^*\}$$

(8.114)

Since $h_0 = h_2 = 0$ we have

$$index\, D = -h_1 = -\dim \mathcal{M}_k$$

(8.115)

After complexification we can use our index formula 4.48, which is

$$index\, D = (-1)^{n/2} \frac{ch\left(\sum_p (-1)^p [E^p]\right)}{e(M)} \cdot td\,(T M_{\mathbf{c}})[M]$$

(8.116)

In the present case, $n = 4$, and the E^p are given by

$$E^0 = ad_{\mathbf{c}} P \otimes \wedge^0 T^* M_{\mathbf{c}}, \quad E^1 = ad_{\mathbf{c}} P \otimes \wedge^1 T^* M_{\mathbf{c}}, \quad E^2 = ad_{\mathbf{c}} P \otimes \wedge^2_- T^* M_{\mathbf{c}}$$

(8.117)

with $ad_{\mathbf{c}} P$ the complexification of the adjoint bundle $ad\, P$. This rather formidable looking formula can be dealt with fairly easily: Substituting in the expressions for the E^p, and using the multiplicative property of the Chern character, gives

$$index\, D = ch\,(ad_{\mathbf{c}} P)\, \bar{E}\,(M)[M]$$
$$\text{where} \quad \bar{E}\,(M) = \frac{ch\left(\sum_p (-1)^p [\bar{E}^p]\right)}{e(M)} \cdot td\,(T M_{\mathbf{c}})$$
$$\text{and} \quad \bar{E}^0 = \wedge^0 T^* M_{\mathbf{c}}, \quad \bar{E}^1 = \wedge^1 T^* M_{\mathbf{c}}, \quad \bar{E}^2 = \wedge^2_- T^* M_{\mathbf{c}}$$

(8.118)

Since M is only four dimensional the Chern character $ch\,(ad_{\mathbf{c}} P)$ need only be expanded to its first three terms, consequently we have

$$ch\,(ad_{\mathbf{c}} P) = rk\,(ad_{\mathbf{c}} P) + c_1(ad_{\mathbf{c}} P) + \frac{1}{2}(c_1^2(ad_{\mathbf{c}} P) - 2c_2(ad_{\mathbf{c}} P)) \quad (8.119)$$

But, drawing on § 5 of chapter 3, we can reason as follows. Because $ad_{\mathbf{c}} P$ is the complexification of a real bundle, it is self-conjugate and has only even dimensional Chern classes; also it is clear that $rk\,(ad_{\mathbf{c}} P) = \dim G$. Finally we can employ Pontrjagin classes to write $p_1(ad_{\mathbf{c}} P) = -2c_2(ad_{\mathbf{c}} P)$. This gives the result that

$$ch\,(ad_{\mathbf{c}} P) = \dim G + \frac{1}{2} p_1(ad_{\mathbf{c}} P)$$

(8.120)

To deal with the rest of the formula we note that inspection of 8.118 shows that the complex which defines $\bar{E}(M)$ is got from a certain truncation of the de Rham complex. The de Rham complex has been truncated in two ways: its last two terms are missing and the middle dimensional term is $\wedge^2_- T^* M_{\mathbf{c}}$ instead of just $\wedge^2 T^* M_{\mathbf{c}}$. But the missing terms in the de Rham complex contribute the same amount as the first two by Poincaré duality, while the difference between the present middle dimensional term and its usual de Rham form is measured by the signature $\tau(M)$. Indeed, if b_i are the Betti numbers of M, we note that the index of this complex is

$$b_0 - b_1 + b_2^- \tag{8.121}$$

However, the Euler characteristic and the signature of M are given by

$$\chi(M) = b_0 - b_1 + b_2 - b_3 + b_4$$
$$\tau(M) = b_2^+ - b_2^- \tag{8.122}$$

Thus we have

$$\begin{aligned}
\chi(M) &= b_0 - b_1 + b_2 - b_3 + b_4 \\
&= 2b_0 - 2b_1 + b_2 \qquad \text{by Poincaré duality} \\
&= 2b_0 - 2b_1 + 2b_2^- + (b_2 - 2b_2^-) \\
&= 2b_0 - 2b_1 + 2b_2^- + (b_2^+ + b_2^- - 2b_2^-) \\
&= 2b_0 - 2b_1 + 2b_2^- + \tau(M),
\end{aligned} \tag{8.123}$$

$$\Rightarrow b_0 - b_1 + b_2^- = \frac{1}{2}(\chi(M) - \tau(M))$$

and this gives the four dimensional contribution from $\bar{E}(M)$. Putting this together with the contribution for $ad_{\mathbf{c}}P$ we find that the dimension of $\mathcal{M}_k$ is given by the four dimensional part of the expression

$$\left\{ \dim G + \frac{1}{2} p_1(ad_{\mathbf{c}}P) \right\} \left\{ 2 - \frac{1}{2}(\chi(M) - \tau(M)) \right\} \tag{8.124}$$

$$\Rightarrow \dim \mathcal{M}_k = p_1(ad_{\mathbf{c}}P) - \frac{\dim G}{2}(\chi(M) - \tau(M))$$

This formula is valid when M is self-dual with positive scalar curvature R, and when G is compact and simple; in Taubes [1] the formula for $\dim \mathcal{M}_k$ is also established for a class of non-self-dual manifolds. It is now possible to make specific choices for G and M and work out the details.

Example $SU(2)$ *instantons on* S^4

$SU(2)$ is compact and simple and S^4 both has positive scalar curvature and is conformally flat, so we can apply our results. The Euler characteristic and signature for S^4 are given by

$$\chi(S^4) = 2, \qquad \tau(S^4) = 0 \tag{8.125}$$

and so

$$\dim \mathcal{M}_k = p_1(ad_{\mathbf{c}}P) - 3 \tag{8.126}$$

It remains to calculate $p_1(ad_{\mathbf{c}}P)$. Let E be the $SU(2)$ bundle which gives the instanton number k, i.e.

$$k = -c_2(E) \tag{8.127}$$

Then E carries the fundamental two dimensional representation of $SU(2)$ and, if we tensor E with itself, there is a natural decomposition of this tensor product bundle into three dimensional and one dimensional parts corresponding to the three dimensional and one dimensional representations. However, because $E \otimes E$ is quadratic in E, it is clear that the elements $\pm g$ of the fundamental representation are both mapped onto the same element in the tensor product; hence the representations carried by $E \otimes E$ are representations of $SU(2)/\mathbf{Z}_2 = Ad\, SU(2)$. Thus the bundle $ad_{\mathbf{c}}P$ is the appropriate three dimensional part of the tensor product $E \otimes E$. This three dimensional part is easy to find: We decompose $E \otimes E$ into the sum of a symmetric and an anti-symmetric piece

$$E \otimes E = S^2 E \oplus \wedge^2 E \tag{8.128}$$

Here $S^2 E$ denotes the symmetric square of E. Clearly $\wedge^2 E$ has rank $\binom{2}{2}$ and is a line bundle while $S^2 E$ has rank $(2.3/2) = 3$ and is, in fact, $ad_{\mathbf{c}}P$. Now we are just left with a little manipulation of characteristic classes.

Applying the Chern character to 8.128 gives

$$ch\,(E)\,ch\,(E) = ch\,(ad_{\mathbf{c}}P) + ch\,(\wedge^2 E) \tag{8.129}$$

If we expand both sides we get

$$(2 + c_1(E) + \frac{1}{2}(c_1^2(E) - 2c_2(E)))^2 = 3 + \frac{1}{2}p_1(ad_{\mathbf{c}}P) + 1 + c_1(E) \tag{8.130}$$

But on S^4 we need only keep cohomology in dimension 4 so that

$$4 - 4c_2(E) = 4 + \frac{1}{2}p_1(ad_{\mathbf{c}}P)$$
$$\Rightarrow p_1(ad_{\mathbf{c}}P) = -8c_2(E) = 8k \tag{8.131}$$

Thus the dimension of the $SU(2)$ moduli space on S^4 is given by

$$\dim \mathcal{M}_k = 8k - 3 \qquad (8.132)$$

If we keep the group $SU(2)$ but replace M by a simply connected manifold with positive definite intersection form, then we will still have an $(8k - 3)$-dimensional moduli space. To prove this we only have to refer to 8.123, where we had

$$b_0 - b_1 + b_2^- = \frac{1}{2}(\chi(M) - \tau(M)) \qquad (8.133)$$

But $b_0 = 1$, and the simply connectedness and positive definiteness give $b_1 = 0$ and $b_2^- = 0$ respectively, so

$$\frac{1}{2}(\chi(M) - \tau(M)) = 1$$
$$\Rightarrow \dim \mathcal{M}_k = 8k - 3 \qquad (8.134)$$

as claimed. Referring to chapter 1 we see that simply connected M with positive definite intersection form are precisely the class of M to which Donaldson's theorem applies.

We have a final remark about reducible connections for $G = SU(2)$. If an $SU(2)$ connection is reducible then all the bundle E can do is to decompose into the sum of two line bundles

$$E = L \oplus L^{-1} \qquad (8.135)$$

the choice of L and L^{-1} being necessary to maintain $c_1(E) = 0$. Since $c_1(E) = 0$ and $c_2(E) = -k$ we have

$$c(E) = c(L \oplus L^{-1})$$
$$\Rightarrow (1 + c_2(E)) = (1 + c_1(L))(1 + c_1(L^{-1})) \qquad (8.136)$$
$$\Rightarrow c_2(E) = -c_1(L) \cup c_1(L) = -q(c_1(L), c_1(L)) = -k$$

where q is the intersection form. Thus for E and L to be non-trivial we require $H^2(M; \mathbf{Z}) \neq 0$. It now follows easily that there are no non-trivial reducible connections on S^4, but $\mathbf{CP}^2$—which is simply connected and has an intersection form that can be taken to be positive definite—is not similarly excluded.

We return to the matter of moduli spaces in general. As we have made manifest above, the Atiyah–Singer index calculation is an infinitesimal one.

It assumes that we have an instanton A and then provides us with the dimension of the tangent space to $\mathcal{M}_k$ at $[A]$. Before we can be sure that we have a global moduli space, three conditions must be met: at least one instanton must be found; the infinitesimal coordinates at the *point* $[A] \in \mathcal{M}_k$ must be extended to a *neighbourhood* of $[A]$; finally the resulting system of local coordinates must be shown to fit together to form a global moduli space. The first condition is easily met as many instantons are known. The last two conditions are analytic rather than topological in character. The second condition is an integrability question similar those that arise in complex analysis when passing from an almost complex structure to a full complex structure. What is needed here is to prove that every element in the 'deformation space' $H^1(ad\,P \otimes \wedge_- T^* M)$ comes from a one parameter family $A + tA$; this is done in Atiyah, Hitchin and Singer [1], who also show that the third condition is met, ensuring that $\mathcal{M}_k$ exists as a global (Hausdorff) manifold.

We can also consider groups other than $SU(2)$ and the corresponding computation of dim $\mathcal{M}_k$ can be done with the addition of some group theory, cf. Bernard, Christ, Guth and Weinberg [1] and Atiyah, Hitchin and Singer [1]. In this context a point worth noting is that the need to make sure that the instanton is irreducible becomes more pressing. This is because, unlike $SU(2)$, for G large there may be many subgroups H to which a self-dual connection might reduce. For the convenience of the reader we provide a table of moduli space information for various Lie groups G and $M = S^4$

G	dim $\mathcal{M}_k$	Irreducibility restrictions
$SU(N)$	$4Nk - N^2 + 1$	$k \geq N/2$
$Spin(N)$	$4(N-2)k - N(N-1)/2$	$k \geq N/4, N \geq 7$
$Sp(N)$	$4(N+1)k - N(2N+1)$	$k \geq N$
E_6	$48k - 78$	$k \geq 3$
E_7	$72k - 133$	$k \geq 3$
E_8	$120k - 248$	$k \geq 3$
F_4	$36k - 52$	$k \geq 3$
G_2	$16k - 14$	$k \geq 2$

$$(8.137)$$

When $M = S^4$ the $SU(2)$ moduli space $\mathcal{M}_1$ is five dimensional; this is the space used in the proof of Donaldson's theorem. $\mathcal{M}_1$ is a non-compact space with boundary S^4; in fact $\mathcal{M}_1$ is the five dimensional hyperbolic space H^5. It is also true that the boundary of $\mathcal{M}_1$ is M for $SU(2)$ instantons on the other M to which Donaldson's theorem applies. Donaldson's theorem can be viewed as a kind of analogue to Teichmüller theory. In the latter the moduli of a linear elliptic operator—the $\bar{\partial}$-operator—are used to deduce topological information about Riemann surfaces, the main result being that dim $\mathcal{M}_p$ gives the genus of the Riemann surface. In Donaldson's theorem the moduli of a non-linear elliptic operator—the operator $d_A^* d_A$—are used to deduce differential-topological information about 4-manifolds; for example, we have the smoothability results described in chapter 1.

The instanton moduli spaces also carry interesting topological information about the configuration space $\mathcal{A}/\mathcal{G}$. This is more easily described using the moduli space $\mathcal{M}_k'$ introduced on p. 233. For $SU(2)$ instantons on S^4 the homology of $\mathcal{M}_k'$ approximates that of $\mathcal{A}/\mathcal{G}$ more and more closely as k increases. Atiyah and Jones [1] show that the inclusion map

$$i(k) : \mathcal{M}_k' \longrightarrow \mathcal{A}/\mathcal{G} \qquad (8.138)$$

has the property that, for k large enough, $i(k)$ induces a projection of the homology of $\mathcal{M}_k'$ onto part of the homology of $\mathcal{A}/\mathcal{G}$. It then follows that, since $\mathcal{M}_k'$ is growing in dimension as k increases, more and more of the homology of $\mathcal{A}/\mathcal{G}$ is captured by that of $\mathcal{M}_k'$. Thus, homologically speaking, in the large k limit instantons give a kind of approximation to the physical configuration space $\mathcal{A}/\mathcal{G}$.

§ 4. Monopoles and symmetries of instantons

The Yang–Mills equations $d_A^* F = 0$ change considerably when the dimension of the manifold M is changed. In this section M is three dimensional and our interest is in magnetic monopoles. Monopoles are static, finite energy, objects which give the critical points of the energy of an appropriate system of fields defined on a three dimensional M. Like instantons they possess a discrete topological invariant k which can be normalised to be an integer; unlike instantons monopoles are trivial if M is closed and compact. In fact the usual choice for M is the non-compact space $\mathbf{R}^3$. Analysis on a non-compact M introduces some new technical difficulties but these have not proved insurmountable.

The physical system studied consists of Yang–Mills G-connection A and a Higgs scalar field ϕ transforming according to a representation of G. We make the usual choice and take this to be the adjoint representation of G. From the point of view of quantum field theory, a natural physical system with these fields is characterised by specifying that their energy E is given by the expression

$$E = \frac{1}{2} \int_M \left\{ -tr(F \wedge *F) - tr(d_A \phi \wedge *d_A \phi) + \lambda * (|\phi|^2 - C^2)^2 \right\} \qquad (8.139)$$

where $|\phi|^2 = -2tr(\phi^2)$. However, the field equations for the critical points of this system are the second order system

$$d_A^* F = [d_A\phi, \phi], \qquad d_A^* d_A\phi = 4\lambda\phi(|\phi|^2 - C^2) \qquad (8.140)$$

and these equations are difficult to solve explicitly, cf. Jaffe and Taubes [1]. The limit in which many explicit solutions are available is the Prasad–Sommerfield limit where the scalar potential term vanishes. This is achieved by setting $\lambda = 0$. The energy is now

$$E \equiv E(A, \phi) = -\frac{1}{2} \int_M \{tr(F \wedge *F) + tr(d_A\phi \wedge *d_A\phi)\}$$
$$= \frac{1}{2}\{\|F\|^2 + \|d_A\phi\|^2\} \qquad (8.141)$$

and the field equations are

$$d_A^* F = [d_A\phi, \phi], \qquad d_A^* d_A\phi = 0 \qquad (8.142)$$

These equations can be solved by imposing the first order Bogomolny equation (Bogomolny [1])

$$F = *d_A\phi \qquad (8.143)$$

To see this we substitute into the LHS of both field equations and note that the second equation just becomes the Bianchi identity

$$d_A F = 0 \qquad (8.144)$$

while the LHS of the first equation becomes

$$\begin{aligned}
d_A^* * d_A\phi &= *d_A(d\phi + [A, \phi]) \\
&= *\{d(d\phi + [A, \phi]) + A \wedge (d\phi + [A, \phi]) + (d\phi + [A, \phi]) \wedge A\} \\
&= *\{dA\phi - A \wedge d\phi - d\phi \wedge A - \phi dA + A \wedge d\phi + A \wedge A\phi \\
&\quad - A \wedge \phi A + d\phi \wedge A + A\phi \wedge A - \phi A \wedge A\} \\
&= *\{F\phi - \phi F\} = \{d_A\phi\phi - \phi d_A\phi\} \\
&= [d_A, \phi]
\end{aligned}$$
$$(8.145)$$

and thus we do indeed have a solution to the field equations. Actually it is much easier to see this by completing the square in the expression for the energy by writing

$$E = \frac{1}{2}\{< F \mp *d_A\phi, F \mp *d_A\phi > \pm 2 < F, *d_A\phi >\}$$
$$= \frac{1}{2}\{\|F \mp *d_A\phi\|^2 \pm 2 < F, *d_A\phi >\} \qquad (8.146)$$

This shows that the absolute minima of E are attained when the pair (A, ϕ) satisfy

$$F = \mp * d_A \phi \tag{8.147}$$

which is the Bogomolny equation; we see, too, that there is a solution whatever the sign in front of the term $*d_A\phi$. The expression

$$< F, *d_A\phi > = - \int_M tr(F \wedge d_A\phi) \tag{8.148}$$

is the absolute minimum and looks like a topological charge. However, we must be careful because

$$\begin{aligned} d\, tr(F\phi) &= tr(d_A F \wedge \phi) + tr(F \wedge d_A\phi) \\ &= tr(F \wedge d_A\phi) \end{aligned} \tag{8.149}$$

Thus if M is compact and closed Stokes' theorem gives

$$< F, *d_A\phi > = - \int_M d\, tr(F\phi) = 0 \tag{8.150}$$

from which it follows that the energy E is zero. Hence we have

$$\begin{aligned} E &= \frac{1}{2}\{\|F\|^2 + \|d_A\phi\|^2\} = 0 \\ &\Rightarrow F = d_A\phi = 0 \end{aligned} \tag{8.151}$$

and this means that A is a flat connection; its only departure from triviality lies in its holonomy, which relies on $\pi_1(M)$ to be non-vanishing. Thus if M is simply connected the ground state of our system is entirely trivial. We can avoid this state of affairs by choosing M to be non-compact or to have a boundary. We shall consider two situations where the Yang–Mills–Higgs monopole system has non-trivial ground states: the first is when $M = \mathbf{R}^3$ and the second when $M = H^3$, where H^3 is three dimensional hyperbolic space.

We begin with the case $M = \mathbf{R}^3$ furnished with the Euclidean metric and also set $G = SU(2)$. The first consequence of choosing $M = \mathbf{R}^3$ is that we must specify boundary conditions at infinity. This is necessary to make the integrals for the energy converge and to make the field equation problem well posed. A standard boundary condition for ϕ is

$$\lim_{r \to \infty} |\phi| \longrightarrow (C + O(r^{-2})) \tag{8.152}$$

where r is the distance from the origin in $\mathbf{R}^3$. This behaviour of ϕ at infinity is linked to that of F via the Bogomolny equation; the result is that it is sufficient to make the energy integrals converge and to render the Bogomolny equation well posed. The integral 8.148 can now be non-zero and it does have a topological interpretation which forces it to take discrete values. More precisely, if k is an integer, then

$$\frac{1}{4\pi C} < F, *d_A\phi >= k \tag{8.153}$$

This integer is the magnetic charge and can be thought of as the Chern class of a $U(1)$ bundle over a two sphere which is S^2_∞, the two sphere at infinity; setting $C = 1$ and using Stokes' theorem, we have

$$k = \frac{1}{4\pi} \int_{\mathbf{R}^3} tr(F \wedge d_A\phi) = \frac{1}{4\pi} \int_{S^2_\infty} tr(F\phi) \tag{8.154}$$

The condition $|\phi| = 1$ on the boundary defines an S^2 inside the Lie algebra $su(2)$ and $(F\phi)$, when evaluated at infinity, becomes the $U(1)$-curvature of a bundle over S^2_∞ and k is its Chern class. Alternatively, one can write k as the winding number, or degree, of a map $\hat{\phi} : S^2_\infty \to S^2_{su(2)}$, giving

$$\hat{\phi} : S^2_\infty \longrightarrow S^2_{su(2)}$$

$$x \longmapsto \hat{\phi}(x) = \frac{\phi}{|\phi|} \tag{8.155}$$

$$\text{and} \qquad k = -\frac{1}{2\pi} \int_{S^2_\infty} tr(\hat{\phi}d\hat{\phi} \wedge d\hat{\phi})$$

This boundary integer k is the only topological invariant associated with the monopole system; the $SU(2)$ bundle over $\mathbf{R}^3$ is topologically trivial since $\mathbf{R}^3$ is contractible.

There is an important formulation of the Bogomolny equation as a variant of the four dimensional self-duality equation. In fact the Bogomolny equation is simply the static version of the equation $F = *F$ on $\mathbf{R}^4$. This is not difficult to see. Let $\widehat{A}$ be a connection on $\mathbf{R}^4$ which we decompose into its components

$$\widehat{A} = \widehat{A}_0 dx^0 + \widehat{A}_1 dx^1 + \widehat{A}_2 dx^2 + \widehat{A}_3 dx^3 \tag{8.156}$$

Then, with respect to the Euclidean metric on $\mathbf{R}^4$, the self-duality equations for the curvature $\widehat{F}$ of $\widehat{A}$, when written in component form, are

$$\widehat{F}_{\mu\nu} = \frac{1}{2}\epsilon_{\mu\nu\alpha\beta}\widehat{F}_{\alpha\beta} \tag{8.157}$$

Using Latin indices for the last three components of $\mathbf{R}^4$ we can separate this into the pair of equations

$$\widehat{F}_{ij} = -\epsilon_{ijk}\widehat{F}_{k0}, \qquad \widehat{F}_{i0} = -\frac{1}{2}\epsilon_{ijk}\widehat{F}_{jk} \qquad (8.158)$$

Actually both these equations are the same and, settling on the first one, and recalling our imposition of x_0 independence, we find that

$$\begin{aligned}\widehat{F}_{ij} &= -\epsilon_{ijk}(\partial_k\widehat{A}_0 - \partial_0\widehat{A}_k + [\widehat{A}_k, A_0]) \\ &= -\epsilon_{ijk}(\partial_k\widehat{A}_0 + [\widehat{A}_k, \widehat{A}_0])\end{aligned} \qquad (8.159)$$

Now we define the Yang–Mills–Higgs pair (A, ϕ) on $\mathbf{R}^3$ by

$$A = \widehat{A}_1 dx^1 + \widehat{A}_2 dx^2 + \widehat{A}_3 dx^3, \qquad \phi = A_0 \qquad (8.160)$$

With F_{ij} denoting the curvature of A, the time independent self-duality equations become

$$F_{ij} = -\epsilon_{ijk}(\partial_k\phi + [\widehat{A}_k, \phi]) \qquad (8.161)$$

and this is recognisable as the component form of the Bogomolny equation

$$F = -*d_A\phi \qquad (8.162)$$

Thus instantons which are invariant under time translation are monopoles; of course we are using the term instanton in a loose sense since time independent instantons necessarily have infinite action, which we usually exclude.

Time translation invariance is not the only invariance possible for the self-duality equations. We can also consider rotational invariance, and following this line of investigation leads to monopoles on H^3 or hyperbolic monopoles, cf. Atiyah [3,4].

To obtain hyperbolic monopoles we look for axially symmetric solutions of the self-duality equations. Now consider the flat line element in $\mathbf{R}^4$:

$$ds^2 = dx_0^2 + dx_1^2 + dx_2^2 + dx_3^2$$

If we choose the rotations to be in the $x_0 - x_1$ plane and let r, θ be polar coordinates in that plane, we have

$$\begin{aligned}ds^2 &= dr^2 + r^2 d\theta^2 + dx_2^2 + dx_3^2 \\ &= r^2\left\{d\theta^2 + \frac{dr^2 + dx_2^2 + dx_3^2}{r^2}\right\}\end{aligned} \qquad (8.163)$$

Now because of the conformal invariance in $\mathbf{R}^4$ of the self-dual equations $F = *F$ a conformal change of metric leaves these equations invariant. Thus we can as well take

$$ds^2 = d\theta^2 + \frac{dr^2 + dx_2^2 + dx_3^2}{r^2} \qquad (8.164)$$

This is the line element for the space $S^1 \times H^3$ where H^3 is the 3-dimensional hyperbolic space

$$0 < r < \infty, \qquad -\infty < x_2 < \infty, \qquad -\infty < x_3 < \infty$$

with metric determined by 8.164. It is important to note that the H^3 metric is singular at $r = 0$ and that $r = 0$ is both a plane $\mathbf{R}^2$ in $\mathbf{R}^4$ and the 'axis' of the rotation in $\mathbf{R}^4$. Therefore, in passing from 8.163 to 8.164, we have made use of the conformal equivalence

$$\mathbf{R}^4 - \mathbf{R}^2 \simeq S^1 \times H^3 \qquad (8.165)$$

For the present we turn to the Bogomolny equation in the hyperbolic space H^3. The equation is

$$d_A \phi = - * F \qquad (8.166)$$

and the $*$ operation is with respect to the hyperbolic metric on H^3 defined by 8.164. We work throughout with gauge group $SU(2)$. As boundary condition we require $|\phi|$ to be constant on the boundary of H^3; we denote this constant value by $|\phi|_{as}$. The boundary of H^3 is, of course, the axis $\mathbf{R}^2$. At this point we find it useful to represent H^3 as the interior of a three dimensional ball of radius 2. This spherical model of hyperbolic space is completely equivalent to the 'upper-half-plane model' 8.165.The metric of the spherical model is defined by

$$ds^2 = \frac{(dx^2 + dy^2 + dz^2)}{(1 - r^2/4)^2}, \qquad r = \sqrt{x^2 + y^2 + z^2} \qquad (8.167)$$

One can now verify that the spherical model is actually isometric to the upper-half-plane model; the transformation from the former to the latter is given by T where

$$T = \alpha^{-1} \beta \alpha \qquad (8.168)$$

where α is stereographic projection from $\mathbf{R}^3$ onto $S^3 - \{\text{north pole}\}$ and β is a certain $\pi/2$ rotation of S^3 into itself, cf. Nash [1]. It follows that, under T, the boundary $r = 2$ of the ball is mapped onto the 'axis' $r = 0$ in $\mathbf{R}^4$.

An important difference between hyperbolic monopoles and monopoles on $\mathbf{R}^3$ is that the axially symmetric instanton associated to the hyperbolic monopole has finite action. This being the case it is not surprising that there is a relation between the instanton number and the magnetic charge. To find this relation suppose that A is an axially symmetric $SU(2)$-instanton on S^4. Let V be the associated rank two vector bundle whose Chern class $c_2(V)$ gives the instanton number. Now consider the action of a rotation through an angle θ in the $x_0 - x_1$ plane. Such a rotation also acts on the bundle V and its base; however, if we restrict to the axis of the rotation then the rotation only acts on the fibres of the corresponding restricted bundle V' say. But V' also has rank two and the $U(1)$ action on its fibres gives a two dimensional representation of $U(1)$. Hence if M_θ denotes the $SU(2)$-matrix representing rotation through the angle θ, then we must have

$$M_\theta = \begin{pmatrix} e^{in\theta} & 0 \\ 0 & e^{-in\theta} \end{pmatrix}, \qquad n \in \mathbf{Z} \qquad (8.169)$$

The integer n represents $\partial/\partial\theta$, the infinitesimal action of the rotation on the fibres but, since we are on the boundary, the covariant derivative in the θ direction vanishes and so the Higgs field, evaluated on the boundary, generates the rotations. Thus, appropriately normalised, we have $|\phi|_{as} = n$ and $|\phi|_{as}$ is quantised. In the hyperbolic picture V' is a complex vector bundle over the boundary S^2 of H^3. But, from the $U(1)$ action, we see that V' is a sum of two line bundles, in fact

$$V' = L \oplus L^{-1} \qquad (8.170)$$

where L is a line bundle over $\partial H^3 \simeq S^2$. This line bundle is characterised by its Chern class $c_1(L)$, which is the magnetic charge k. The topological data of the hyperbolic monopole is contained in the three integers k, $c_2(V)$ and n; moreover, if we integrate out the redundant variable θ in the instanton action, we can verify that they are related according to

$$c_2(V) = 2nk \qquad (8.171)$$

A further matter of geometrical interest is the rôle played by the curvature of the hyperbolic space H^3. We can import a parameter R into the metric ds^2 if we write

$$ds^2 = \frac{r^2}{R^2}\left\{ R^2 d\theta^2 + \frac{R^2}{r^2}\left(dr^2 + dx_1^2 + dx_2^2 \right) \right\}$$

When we delete the conformal factor this corresponds to working on $S^1(R) \times H^3(R)$ where $S^1(R)$ is a circle of radius R, and $H^3(R)$ is a hyperbolic space of scalar curvature $-6/R^2$. The spherical model for $H^3(R)$ is now the interior of a ball of radius $2R$. On $S^1(R) \times H^3(R)$ the Bogomolny equation in component form is

$$D_k \phi^a = -\frac{R}{2\sqrt{f_R}} \epsilon_{ijk} F^a_{jk} \qquad (8.172)$$

where $f_R = (1 - r^2/4R^2)^{-2}$. But in the limit $R \to \infty$, $H^3(R)$ becomes the flat space $\mathbf{R}^3$. Thus we ought to be able to produce $\mathbf{R}^3$ monopoles by an appropriate limiting procedure applied to hyperbolic monopoles and, thinking of hyperbolic monopoles as axially symmetric instantons, this provides us with a picture of an $\mathbf{R}^3$ monopole as limits of instantons. This can indeed be done, cf. Chakrabarti [1] and Nash [1].

Now we return to hyperbolic space H^3 and would like to examine the question of the location of the monopole. This is really the same as discussing the zeros of the Higgs field ϕ. The existence of a zero of ϕ is forced by the non-vanishing of the magnetic charge k: we know that the magnetic charge is the winding number of the map

$$\hat{\phi} : \partial H^3 \longrightarrow S^2_{su(2)}$$

Also, a standard homotopy argument says that $\hat{\phi}$ only extends to the interior of H^3 in a singularity free manner if the winding number k is zero. Hence for $k \neq 0$, ϕ has a zero, and in general k counts the zeros of ϕ. For the simplest hyperbolic monopole with $k = 1$ there is a single zero of ϕ at the origin of the ball H^3 (Nash [1]). Then, because of the dual interpretation of a hyperbolic monopole as an instanton, it is of interest to ask for something in the instanton picture to which the zero of ϕ corresponds. First of all the origin of the ball is transformed, under T, to the point $r = 2$, $x_2 = 0$, $x_3 = 0$ in the upper-half-plane model. This in turn corresponds to a circle of radius 2 in $\mathbf{R}^4$ centred at the origin and lying in the $x_0 - x_1$ plane. If this correspondence is pursued for $R \to \infty$ and for larger values of k, one can obtain a configuration where a ring of instantons corresponds to an $\mathbf{R}^3$ monopole (Chakrabarti [2]).

It is natural to consider the properties of a hyperbolic monopole for a general value of $|\phi|_{as}$. We have seen how axially symmetric instantons correspond to hyperbolic monopoles with quantised $|\phi|_{as}$. For forbidden values of $|\phi|_{as}$ the hyperbolic monopole (ϕ, A) may exist but the corresponding axially symmetric instanton A', say, does not. Indeed were A' to exist it could have non-integral c_2 and would possess a singularity located on its axis;

for an example, see Forgács, Horváth and Palla [1]. Thus the notion of a hyperbolic monopole is wider than that of an axially symmetric instanton.

It is worth noting that the conformal equivalence of 8.165 just a special case of the more general equivalence

$$\mathbf{R}^n - \mathbf{R}^m \simeq S^{n-m-1} \times H^{m+1}, \quad m < n \tag{8.173}$$

To obtain this more general form we ask that a connection in $\mathbf{R}^n$ be invariant under rotations specified by l angles $l < n$. Then for the line element ds^2 in $\mathbf{R}^n$ we can write ($r = \sqrt{x_1^2 + \cdots + x_{l+1}^2}$, Ω_l represents the solid angle in $\mathbf{R}^n$)

$$
\begin{aligned}
ds^2 &= dx_1^2 + \cdots + dx_{l+1}^2 + dx_{l+2}^2 + \cdots + dx_n^2 \\
&= dr^2 + r^2 d\Omega_l^2 + dx_{l+2}^2 + \cdots + dx_n^2 \\
&= r^2 \left\{ d\Omega_l^2 + \frac{dr^2 + dx_{l+2}^2 + \cdots + dx_n^2}{r^2} \right\}
\end{aligned}
\tag{8.174}
$$

Deleting the conformal factor r^2 amounts to the conformal correspondence $\mathbf{R}^n - \mathbf{R}^{n-l-1} \simeq S^l \times H^{n-l}$, which implies 8.173. This correspondence would be relevant if one were to study $SO(3)$-symmetric instantons. One would then choose $n = 4$ and $m = 1$, giving

$$\mathbf{R}^4 - \mathbf{R}^1 \simeq S^2 \times H^2 \tag{8.175}$$

It also turns out that the dimensional reduction of the $SU(2)$ Yang-Mills theory from S^4 to H^2 corresponds to a Yang-Mills-Higgs system which is Abelian but has a non-zero, predetermined, value for the coupling constant λ (Jaffe and Taubes [1]).

§ 5. Monopole moduli and monopole scattering

In this section we begin by considering the moduli space for monopoles and then make some remarks about monopole scattering; the monopoles are all standard $\mathbf{R}^3$ monopoles with group $SU(2)$.

The dimension of the monopole moduli space can be calculated using index theory methods supplemented by extra analysis to deal with the technicalities of the non-compactness of $\mathbf{R}^3$. We denote the monopole moduli space by M_k'; parameter counting gives its dimension as $4k$ (Weinberg [1]). Actually it will emerge below that the true parameter space of gauge inequivalent monopoles is a space M_k whose dimension is $4k - 1$. The index theory part of the calculation proceeds in an analogous fashion to the instanton case. Let the pair (A, ϕ) be a solution to the Bogomolny equation, which we perturb to the family of solutions $(A+ta, \phi+t\eta)$. The Bogomolny equation gives

$$F(A + ta) = *d_{(A+ta)}(\phi + t\eta)$$
$$\Rightarrow F(A) + td_A a + t^2 a \wedge a = *\{d(\phi + t\eta) + [A + ta, \phi + t\eta]\}$$
$$= *\{d\phi + [A, \phi] + t(d\eta + [A, \eta] + [a, \phi])$$
$$+ t^2[a, \eta]\}$$
(8.176)

and passing to the tangent at $t = 0$ yields the equation

$$d_A a = *\{d_A \eta + [a, \phi]\} \tag{8.177}$$

We must also project out those (a, η) which correspond to an infinitesimal gauge transformation: if $g = \exp[tf]$ then the infinitesimal effect on (A, ϕ) is that

$$(A, \phi) \longmapsto (A, \phi) + (d_A f, [f, \phi]) \tag{8.178}$$

The preceding pair of equations suggests the definition of the two elliptic operators D_0 and D_1

$$D_0 : \Omega^0(\mathbf{R}^3, ad\, P) \longrightarrow \Omega^1(\mathbf{R}^3, ad\, P) \oplus \Omega^0(\mathbf{R}^3, ad\, P)$$
$$f \longmapsto (d_A f, [f, \phi])$$
(8.179)

$$D_1 : \Omega^1(\mathbf{R}^3, ad\, P) \oplus \Omega^0(\mathbf{R}^3, ad\, P) \longrightarrow \Omega^2(\mathbf{R}^3, ad\, P)$$
$$(a, \eta) \longmapsto d_A a - *d_A \eta - *[a, \phi]$$

If m denotes the equivalence class of the monopole (A, ϕ) then the tangent space to the moduli space at m is then expressed as the cohomology group

$$\frac{ker\, D_1}{Im\, D_0} = T_m M_k' \tag{8.180}$$

The associated elliptic complex is then

$$0 \xrightarrow{i} \Omega^0(\mathbf{R}^3, ad\, P) \xrightarrow{D_0} \Omega^1(\mathbf{R}^3, ad\, P) \oplus \Omega^0(\mathbf{R}^3, ad\, P) \xrightarrow{D_1} \Omega^2(\mathbf{R}^3, ad\, P) \xrightarrow{\pi} 0$$
(8.181)

Let the cohomology of the complex be given by

$$H^0(\Omega(\mathbf{R}^3, ad\, P)) = ker\, D_0, \qquad \dim H^0(\Omega(\mathbf{R}^3, ad\, P)) = \hat{h}_0$$

$$H^1(\Omega(\mathbf{R}^3, ad\, P)) = \frac{ker\, D_1}{Im\, D_0}, \qquad \dim H^1(\Omega(\mathbf{R}^3, ad\, P)) = \hat{h}_1 \tag{8.182}$$

$$H^2(\Omega(\mathbf{R}^3, ad\, P)) = \frac{ker\, \pi}{Im\, D_1}, \qquad \dim H^2(\Omega(\mathbf{R}^3, ad\, P)) = \hat{h}_2$$

The index of this complex is that of the operator $D_0 + D_1^*$ and hence we know that

$$index\,(D_0 + D_1^*) = \dim ker\,(D_0 + D_1^*) - \dim ker\,(D_0^* + D_1)$$
$$= \hat{h}_0 - \hat{h}_1 + \hat{h}_2 \qquad (8.183)$$
$$= \hat{h}_0 + \hat{h}_2 - \dim M_k'$$

In Taubes [2] it is shown that $index\,(D_0 + D_1^*) = -4k$ and a Weitzenböck argument is used to prove that

$$ker\,(D_0^* + D_1) = \phi \qquad (8.184)$$

Thus $\hat{h}_0 = \hat{h}_2 = 0$ and we obtain

$$\dim M_k' = 4k \qquad (8.185)$$

This index calculation, involving the non-compact space $\mathbf{R}^3$, requires considerable extra analysis. One needs to show that the cohomology above can be constructed using only square integrable sections and this entails showing that D_0 and D_1 are Fredholm so that they have finite dimensional kernels. We can now see why the true moduli space M_k only has dimension $4k - 1$. The reason is that the Higgs field itself generates a gauge transformation, namely $g = \exp[t\phi]$; but the boundary condition obeyed by ϕ means that it is not square integrable, hence ϕ is not included in the space $\Omega^0(\mathbf{R}^3, ad\,P)$. However, the image of ϕ under the operator D_0 is just the one dimensional vector $(d_A\phi, [\phi, \phi]) = (d_A\phi, 0)$; this vector is square integrable, despite its origin, and belongs to $\Omega^1(\mathbf{R}^3, ad\,P) \oplus \Omega^0(\mathbf{R}^3, ad\,P)$. Projecting out this one dimensional subspace gives the moduli space M_k, for which we then have

$$\dim M_k = 4k - 1 \qquad (8.186)$$

It is also common to fix the centre of the monopole and thus get rid of the translational invariance of M_k'; when this is done the final space M_k^0 is $(4k - 4)$-dimensional. We can check this for $k = 1$: if $k = 1$ there is just a single monopole possible, the Prasad–Sommerfield monopole. It is spherically symmetric with its Higgs field vanishing at the origin, which is its centre. In terms of the spaces above we have the simple situation $M_1' = \mathbf{R}^3 \times S^1$, $M_1 = \mathbf{R}^3$ and $M_1^0 = \{p\}$. Thus, after we have assigned the centre, there is only one $k = 1$ monopole. More generally, in an analogous manner to the instanton situation, the space M_k' is fibred over M_k with fibre S^1; this fibre is the one parameter family of gauge transformations generated by the Higgs field ϕ. The main properties of the three monopole space are therefore

$$
\begin{array}{c}
S^1 \longrightarrow M'_k \\
\downarrow \\
M_k
\end{array}
\qquad
M_k = \frac{M'_k}{S^1}, \qquad
M^0_k = \frac{M'_k}{S^1 \times \mathbf{R}^3}
\qquad (8.187)
$$

$$
\dim M'_k = 4k, \qquad \dim M_k = 4k - 1, \qquad \dim M^0_k = 4k - 4
$$

The topology of the extended moduli space M'_k also has an interesting structure; it is diffeomorphic to the space of rational functions of degree k which vanish at infinity (Donaldson [2]). These rational functions can be written in the form

$$
f = \frac{a_{k-1} z^{k-1} + \cdots + a_1 z + a_0}{z^k + b_{k-1} z^{k-1} + \cdots + b_1 z + b_0} = \frac{\hat{a}_1}{z - \hat{b}_1} + \cdots + \frac{\hat{a}_k}{z - \hat{b}_k} \qquad (8.188)
$$

where the numerator and denominator in the first expression are relatively prime. We note that these functions are specified by $4k$ real parameters; the second expression is useful when trying to approximate a monopole of charge k as a linear superposition of k monopoles of charge 1. If $k = 1$ then

$$
f = \frac{a_0}{z + b_0} \qquad (8.189)
$$

and, if $a_0 = 1$ and $b_0 = 0$, this represents the Prasad–Sommerfield monopole centred at the origin. For larger k we can make the simple choice

$$
f = \frac{1}{z^k} \qquad (8.190)
$$

which corresponds to an axially symmetric monopole of charge k. If we want to pass to the true moduli space M_k we must identify the functions f and $\exp[i\theta]f$.

The space M'_k has also arisen as a key element in the scattering of monopoles on each other. So far we have considered static monopoles and deduced that the energy of a monopole of charge k is given by

$$
E = 4\pi k \qquad (8.191)
$$

In particular, E is linear in k. Let us consider those k-monopole states which correspond to k single, well separated, monopoles. The absence of any dependence of E on distance can be interpreted as the constancy of the potential energy and means that there are no forces between these

monopoles. If we bring the monopoles close together this particle interpretation breaks down due to the existence of monopoles whose charge is
greater than one but are localised in their support. This intuition raises the
possibility of describing the coming together of the k single monopoles as
some kind of interaction; if some of the monopoles come apart again then
we have scattering.

In Manton [1] a precise idea of this type is developed: static monopoles
are given velocities so that they come together and, for low velocities, the
subsequent dynamics is assumed to be determined only by their kinetic
energy. Now the kinetic energy T of a monopole (A, ϕ) is given by an
expression of the usual quadratic type using a metric on the relevant coordinate space which, in this case, is the Yang–Mills–Higgs space $\mathcal{A}_{YMH}$ of
all pairs (A, ϕ). More precisely metrics are used to form inner products of
tangent vectors to the coordinate space. There is a natural metric available
on $\mathcal{A}_{YMH}$ and it is defined in the following way. Let $m = (A, \phi)$ represent
the monopole and $T_m = m^I \partial m / \partial X^I$ represent a formal tangent vector to
$\mathcal{A}_{YMH}$ at m with X^I some formal infinite dimensional system of local coordinates. The metric g_{IJ} can then be defined by setting $A_t = A + ta$ and
$\phi_t = \phi + t\eta$ and writing

$$
\begin{aligned}
< T_m, T_m > &\equiv g_{IJ} m^I m^J \\
&= < a, a > + < \eta, \eta > \\
&= - \int_{\mathbf{R}^3} tr(a \wedge *a) - \int_{\mathbf{R}^3} tr(\eta \wedge *\eta)
\end{aligned}
\tag{8.192}
$$

This metric descends to the configuration space $\mathcal{A}_{YMH}/\mathcal{G}$ if we project onto
those (a, η) orthogonal to the orbit through $m = (A, \phi)$; we shall regard
this projection as understood. The kinetic energy T of a slowly moving
monopole $m(t)$ with $m(t) = (A + ta, \phi + t\eta)$, say, is given by

$$
T = < \frac{dm}{dt}, \frac{dm}{dt} > = - \int_{\mathbf{R}^3} tr(a \wedge *a) - \int_{\mathbf{R}^3} tr(\eta \wedge *\eta)
\tag{8.193}
$$

Now the potential energy is

$$
E = \frac{1}{2}\{ < F, F > + < D_A\phi, d_A\phi > \}
\tag{8.194}
$$

and both E and T are functionals on the configuration space $\mathcal{A}_{YMH}/\mathcal{G}$. The
assumptions mentioned above are given substance by assuming that for low
velocities the subsequent motion does not change E much from its minimum
value. Thus the monopole moves approximately on the finite dimensional

subspace of $\mathcal{A}_{YMH}/\mathcal{G}$ on which E is constant: the moduli space M_k'. The equations of motion are then given by the critical points of the kinetic term T, restricted now to M_k'. But these equations are just those for the geodesics on M_k' with respect to the metric induced by g_{IJ} on its finite dimensional subspace M_k' (compare with the geodesics giving the critical points of the Morse energy functional on p. 206).

To calculate in this approximation of monopole dynamics by geodesic flow on the moduli space requires us to find the metric on M_k'. This metric is of a very special kind—it is hyperkähler. Recall from chapter 5 that a Kähler metric g_{ij} is a Hermitian metric on a complex manifold M for which the associated two form $\omega = (i/2)g_{ij}dz^i \wedge d\bar{z}^j$ is closed. For M to possess a hyperkähler metric there must be three complex structures on M specified by the matrices I, J, K, which then satisfy the quaternion algebra generating relations

$$I^2 = J^2 = K^2 = IJK = -1 \qquad (8.195)$$

and M must be Kähler with respect to the three possible symplectic ω that can be constructed using I, J and K. This means that the three forms

$$\omega_I(X,Y) = g(IX,Y), \quad \omega_J(X,Y) = g(JX,Y), \quad \omega_K(X,Y) = g(KX,Y)$$
$$(8.196)$$

are all closed. For a complex manifold M the existence of the matrix J gives the tangent space the structure of a complex vector space and ensures that M is even dimensional. If M is hyperkähler the three matrices I, J and K give the tangent space the structure of a quaternionic vector space and ensure that M has dimension $4n$.

Actually the reduced moduli space M_k^0, whose dimension is $4k - 4$, is also hyperkähler and the geodesic flow on M_k' induces geodesic flow on M_k^0; it is the flow on this latter space which contains the interesting scattering information. The hyperkählerity of the metric is a key factor in rendering it computable and, without the metric, the geodesics cannot be calculated, so that the scattering would be unknown.

For the $k = 2$ case a detailed picture of the scattering is possible (Atiyah and Hitchin [1,2]). The space M_2' has dimension 8 which is reduced to 4 after fixing the centre and the gauge. These 4 parameters arise in the following way. Suppose the two monopoles are far apart so that the particle interpretation applies. We locate the two monopoles a distance $2|x|$ apart by placing one at x and the other at $-x$; in addition to x we have a unit vector y perpendicular to x, y is the relative phase of the monopole pair and, for covering space reasons, y and $-y$ are identified. If $|x|$ is not large then we do not think of $\mp x$ as the location of the monopoles but the same parameters apply.

The two most interesting scatterings are where y is fixed and the monopoles follow a straight line path and have a head on collision, and where the scattering occurs in the x–y plane. In the first case the scattered monopoles emerge and recede from one another along a straight line which is at right angles to the incident line. In the second case, if the monopoles approach along parallel lines which are close enough together, the emergent monopoles have spin angular momentum; thus they are now electrically charged and have become dyons.

§ 6. Critical point theory and gauge theories

So far in this chapter we have considered the critical points of two functionals: the action functional $S(A)$ of the pure Yang–Mills theory in four dimensions and the energy functional $E(A, \phi)$ of the Yang–Mills–Higgs theory in three dimensions. In each case all the critical points that we have studied have been absolute minima. We would now like to consider the existence of non-minimal critical points and to examine in brief the mechanisms which determine the type of a critical point.

We know from chapter 7 that, in finite dimensions, functions $f : M \to \mathbf{R}$ with non-degenerate critical points have a close link between the number and type of their critical points and the topology of M. When this link exists, an M with a rich topological structure will guarantee that such Morse functions f have a rich critical point structure; in particular one does not expect the critical points to be limited to just minima. Our present interest is in knowing how much, if any, of this Morse theory carries over to gauge theories.

The gauge theory examples are all infinite dimensional and must be analysed with some generalisation of Morse theory to infinite dimensions cf. Palais [1,2,3] and Taubes [3]. It turns out that the four dimensional pure Yang–Mills theory has rather different properties from the Yang–Mills–Higgs system; in addition the two dimensional Yang–Mills theory has striking properties. In all of these theories the energy or action functional has a large gauge invariance and this will have to be factored out equivariantly, or otherwise, when investigating the topology of the critical points.

We begin with the $SU(2)$ Yang–Mills–Higgs system on $\mathbf{R}^3$ in the Prasad–Sommerfield limit. The energy functional

$$E(A, \phi) = \frac{1}{2}\{\|F(A)\|^2 + \|d_A\phi\|^2\} \tag{8.197}$$

is a real valued functional on $\mathcal{A}_{YMH}$ whose critical points are given, in general, by the solutions to

$$d_A^* F = [d_A \phi, \phi], \qquad d_A^* d_A \phi = 0 \qquad (8.198)$$

$E(A, \phi)$ descends to a functional on the quotient $\mathcal{A}_{YMH}/\mathcal{G}$ where $\mathcal{G}$ are the base point preserving gauge transformations. The Higgs field ϕ in the space $\mathcal{A}_{YMH}$ is subject to the usual boundary condition at infinity. This allows us to decompose $\mathcal{A}_{YMH}$ into components $\mathcal{A}_{YMH}^k$ according to the value of magnetic charge k. We write

$$\mathcal{A}_{YMH} = \bigcup_k \mathcal{A}_{YMH}^k \qquad (8.199)$$

and each $\mathcal{A}_{YMH}^k$ contains those (A, ϕ) which satisfy

$$\frac{1}{4\pi} < F, *d_A \phi > = k \qquad (8.200)$$

There is also a corresponding decomposition of the orbits into components $\mathcal{A}_{YMH}^k/\mathcal{G}^k$ which we denote by C_k. The C_k are topologically non-trivial and, on each C_k, there are an infinite number of critical points of $E(A, \phi)$; an infinite number of these are non-minimal and they are distributed so that there are an infinite number of them with energy larger than any given value. These results (Taubes [2,4]) establish that $E(A, \phi)$ has the properties of a good Morse function. The situation for the other gauge theories is different.

Next we take the action functional $S(A)$ for the pure Yang–Mills theory in two dimensions. The connection A is defined on a Riemann surface Σ and the group G is $U(n)$. For $S(A)$ we write

$$S(A) = \|F\|^2 = - \int_\Sigma tr\,(F \wedge *F) \qquad (8.201)$$

and the critical points are given, as usual, by

$$\begin{aligned} d_A^* F(A) &= 0 \\ \Rightarrow d_A * F(A) &= 0 \end{aligned} \qquad (8.202)$$

These equations are much simpler than in higher dimensions. To see this, remember that, in two dimensions, the dual of a 2-form is a 0-form and so $*F(A)$ is just an $ad\,P$-valued function on Σ. More precisely, the Yang–Mills equations 8.202 simply state that $*F(A)$ is a covariantly constant section of $ad\,P$ over Σ and holomorphic methods can be applied to translate this into a holomorphic problem. $S(A)$ is defined on the contractible space $\mathcal{A}$ but it descends to an orbit space whose topology is far from trivial. If the Riemann surface Σ has genus zero then Σ is just S^2 and then the orbit space $\mathcal{A}/\mathcal{G} = B\mathcal{G}$ is ΩG and we know that this space has a rich topology. In general, whatever the genus of Σ, a very attractive Morse theory holds for the functional $S(A)$, cf. Atiyah and Bott [4]. This is that $S(A)$ is equivariantly perfect where the equivariance is with respect to the group of gauge transformations $\mathcal{G}$. Using the notation of chapter 7 we record this by the equality

$$M_t^{\mathcal{G}}(S) = P_t^{\mathcal{G}}(\mathcal{A}) = P_t(B\mathcal{G}) \tag{8.203}$$

The Hessian for S is obtained by expanding the action about a critical point. Just as in 8.66 we write $A_t = A + ta$ and we have

$$S(A_t) = S(A) + 2t < F(A), d_A a > + t^2 \{ < d_A a, d_A a > + < a \wedge a, F(A) > \\ + < F(A), a \wedge a > \} + \cdots \tag{8.204}$$

Since A is a critical point the coefficient of t vanishes while the coefficient of t^2 gives the Hessian. More precisely the Hessian is the operator H defined by

$$\begin{aligned} < a, Ha > &= < d_A a, d_A a > + < a \wedge a, F(A) > + < F(A), a \wedge a > \\ &= < d_A a, d_A a > + < a, *(*F(A)a) > \end{aligned} \tag{8.205}$$

Thus H is the differential operator

$$H = d_A^* d_A + *(*F(A)\ ) \tag{8.206}$$

To investigate the type of the critical points we must find its index i.e. the dimension of the maximal space on which the quadratic form $< a, Ha >$ is negative; we must also impose a gauge condition, which we take to be the standard one, $d_A^* a = 0$. Hence the index of the critical point is now the dimension of the maximal negative subspace of $ker\ d_A^*$; this is finite because ellipticity renders $ker\ d_A^*$ finite dimensional. In fact there are critical points of arbitrarily large index so that we do not just have minima.

Lastly we come to four dimensions and the $SU(2)$ Yang–Mills action on S^4. In this case no Morse-type results hold; this is just as well because the configuration space $\mathcal{A}/\mathcal{G}$ has plenty of cohomology but we only know of critical points which are absolute minima, i.e. the Hessian has index zero. A special property of the Yang–Mills theory in four dimensions is its conformal invariance; the two dimensional sigma model is also a conformally invariant theory and it, too, only has absolute minima.

CHAPTER IX

The Elliptic Geometry of Strings

§ 1. The Bosonic string

In the present chapter we wish to show how some of the techniques that we have already developed can be applied to string theory. Our discussion will be fairly self-contained and will not require an extensive background in string theory. However, reference can be made, where necessary, to Green, Schwarz and Witten [1,2], D'Hoker and Phong [1] and the extensive bibliographies therein.

We restrict the discussion, for simplicity, to the Polyakov formulation (Polyakov [1]) of the closed oriented Bosonic string in d-dimensions. A string moving in space-time sweeps out a surface Σ or world sheet in the same way that a moving particle traces out a world line. The classical Nambu-Goto action for a free string is a constant times the area of its world sheet. This means that the critical points of the action are the surfaces of extremal area. This is in analogy with the fact that the critical points of the action for a free particle are the geodesics: the curves of extremal length. Let σ and τ be coordinates on the surface, and let the position of the string in space-time M be specified by a function $\phi^\mu \equiv \phi^\mu(\sigma, \tau)$, with $\mu = 1, \ldots, d$, then the string action is the expression

$$T \int_\Sigma d\sigma d\tau \sqrt{\left(\frac{\partial \phi^\mu}{\partial \sigma} \frac{\partial \phi_\mu}{\partial \sigma}\right) \left(\frac{\partial \phi^\nu}{\partial \tau} \frac{\partial \phi_\nu}{\partial \tau}\right) - \left(\frac{\partial \phi^\mu}{\partial \sigma} \frac{\partial \phi_\mu}{\partial \tau}\right)^2} \qquad (9.1)$$

where T is a constant known as the string tension. We take our string to be oriented, which has the consequence that, as it sweeps out the surface, it imparts an orientation to it also; thus Σ is orientable. The square root in this formula makes is difficult to use and, following Polyakov, one can replace it by the action S where

$$S = -\frac{T}{2} \int_\Sigma d^2 x \sqrt{g} \, g^{ab} \frac{\partial \phi^\mu}{\partial x^a} \frac{\partial \phi_\mu}{\partial x^b} \qquad (9.2)$$

The surface Σ is now equipped with a metric g^{ab} and the coordinates are denoted by x^a, $a = 1, 2$ instead of σ and τ. Actually, at a critical point, the two expressions for the action coincide.

To quantise this string theory we want to use a functional integral approach. This requires us to sum over all the variables that appear in the action S. These variables are the metric g_{ab} on Σ, and the string's position functions $\phi^\mu \equiv \phi^\mu(x^1, x^2)$—the ϕ^μ should be thought of as specifying an embedding of Σ in the space-time M. For this quantisation a fundamental object of interest is the partition function, a contribution to which is given by the quantity Z_p where

$$Z_p = \int \mathcal{D}g\mathcal{D}\phi \, \exp\left[-\frac{T}{2}\int_\Sigma d^2x \sqrt{g} \, g^{ab}\frac{\partial\phi^\mu}{\partial x^a}\frac{\partial\phi_\mu}{\partial x^b}\right] \qquad (9.3)$$

and $\mathcal{D}g$ and $\mathcal{D}\phi$ denote functional integration over the spaces of metrics and embeddings respectively. We shall usually abbreviate ϕ^μ to ϕ and g_{ab} to g.

Now a partition function is mathematically a trace; in quantum theoretic language it represents a vacuum-to-vacuum amplitude where a quantum object is emitted by the vacuum only to disappear by combining with itself, thus leaving us with the vacuum again. The consequence of this for us is that the surface Σ has no boundary—one can interpret the trace in the partition function as being the process of taking a surface whose boundary consists of two circles, an initial and final string state, and then joining the two circles together so as to produce the closed surface Σ; finally one sums over all possible configurations by integrating over g and ϕ.

Since a closed compact orientable surface with a metric g is just a Riemann surface with a complex structure determined by g, it makes sense to use complex coordinates on Σ. As well as doing this we set the string tension T to unity and write the action in the more concise form

$$S = (-1/2)\int_\Sigma < \partial\phi, \partial\phi >_g = (1/2)\int_\Sigma < \phi, \Delta_g\phi > \qquad (9.4)$$

with $< \partial\phi, \partial\phi >_g = g^{ab}\partial_a\phi^\mu\partial_b\phi_\mu\sqrt{g} \, dzd\bar{z}$, and $\Delta_g = (\partial\partial^* + \partial^*\partial)$

where $*$ denotes adjoint with respect to the inner product $< \, , \, >_g$ and we note in passing that the term $\partial\partial^*$ does not actually contribute anything to the Laplacian since Δ_g is only acting on functions.

To obtain the complete partition function Z we must sum over all possible Riemann surfaces Σ as well as integrating over g and ϕ. Each Σ has a genus p and so we can write

$$Z = \sum_{genera} \int \mathcal{D}g\mathcal{D}\phi \, \exp\left[-\frac{1}{2}\int_\Sigma < \partial\phi, \partial\phi >_g\right] = \sum_{p=0}^\infty Z_p \qquad (9.5)$$

From now on we simplify matters by choosing space-time M to be the flat Euclidean space $\mathbf{R}^d$. The calculation of the partition function Z necessitates a proper understanding of the symmetries of the string action $S = (-1/2)\int_\Sigma < \partial\phi, \partial\phi >_g$. Now ϕ belongs to the space of maps $Map(\Sigma, \mathbf{R}^d)$ and g belongs to $Met(\Sigma)$; thus we can consider S as a functional $S(\phi, g)$ on the string configuration space $Map(\Sigma, \mathbf{R}^d) \times Met(\Sigma)$. It is easy to see that S is invariant under translations in $\mathbf{R}^d$; however there are two further groups that leave S invariant. These groups, both of which we met in chapter 5, are the orientation preserving diffeomorphisms $Diff^+(\Sigma) \subset Diff(\Sigma)$, and the group of positive smooth functions $C_+^\infty(\Sigma)$. An element $f \in C_+^\infty(\Sigma)$ just acts on a metric g to give fg; in contrast, a diffeomorphism $\alpha \in Diff^+(\Sigma)$ acts on the *pair* $(\phi(x), g(x))$ to give $(\phi(\alpha(x)), g(\alpha(x)))$. The combined action of the two groups is via the semi-direct product $Diff^+(\Sigma) \ltimes C_+^\infty(\Sigma)$. This is defined by the map A below where we have set

$$B = Map(\Sigma, \mathbf{R}^d) \times Met(\Sigma)$$

so that A is the map:

$$A : \left(Diff^+(\Sigma) \ltimes C_+^\infty(\Sigma)\right) \times \left(Map(\Sigma, \mathbf{R}^d) \times Met(\Sigma)\right) \longrightarrow B$$
$$((\alpha, f), (\phi, g)) \mapsto (\phi(\alpha(x)), f(x)g(\alpha(x)))$$
(9.6)

The partition function Z_p contains two integrations; the ϕ integration, being Gaussian, can be realised as a suitable infinite dimensional determinant, e.g. by a zeta function expression, since the operator Δ_g is elliptic. Thus we have a second formula for Z_p:

$$Z_p = \int_{Met(\Sigma)} \mathcal{D}g \left(\frac{det(\Delta_g/2)}{(1,1)_g}\right)^{-d/2}$$
(9.7)

The notation det in the above expression denotes the zeta function determinant [1] and requires the restriction of $\Delta_g/2$ to its strictly positive subspace. Alternatively, one can include the zero eigenvalue; this will produce a factor of the 'volume' of $\mathbf{R}^d$. The removal of this infinity has then to be regarded as implicit in our formulae for Z_p. This removal can be carried out by applying a compactness cutoff and dividing by the resulting finite volume so

---

[1] We introduced the zeta function $\zeta_P(s)$ of a positive elliptic operator P in chapter 4; $\zeta_P(s)$ is always regular at the origin and this means that a regularised determinant of P can be defined as $\exp[-\zeta_P'(0)]$.

that the cutoff may then be discarded. Further, the expression $(1,1)_g$ is the total volume of Σ with respect to the metric g and occurs because of the invariance of S under translations in $\mathbf{R}^d$.

§ 2. The space of metrics

The remaining integration in the expression 9.7 for Z_p is over the space of metrics $Met(\Sigma)$. The measure $\mathcal{D}g$ on $Met(\Sigma)$ is at present a formal object whose properties we wish to understand. The study of Z_p will involve dividing up the metrics on Σ into conformal equivalence classes, i.e. it will involve considering all the complex structures on Σ for a given genus p—the moduli space—and then summing over p. It will turn out that, in the end, Z_p can be expressed as a *finite dimensional integral* over the moduli space $\mathcal{M}_p$. Note that it is relevant here to be aware that the action S is a conformal invariant only when the dimension d of space-time is equal to 26—when $d \neq 26$ the action S has a conformal anomaly. This is something that we analyse in chapter 10—for the rest of this chapter we set $d = 26$.

Our next task is to describe in some detail the geometric properties of the measure on $Met(\Sigma)$ that we have so far denoted by $\mathcal{D}g$. If $\mathcal{D}Met(\Sigma)$ denotes a measure on $Met(\Sigma)$ then the invariance of the action under $C_+^\infty(\Sigma)$ and $Diff^+(\Sigma)$ suggests that a formal specification of $\mathcal{D}g$ is

$$\mathcal{D}g = \frac{\mathcal{D}Met(\Sigma)}{vol\left(C_+^\infty(\Sigma)\right) vol\left(Diff^+(\Sigma)\right)} \tag{9.8}$$

where the denominator contains the volumes of the orbits of the invariance groups. It is clear from the RHS of this formula that such a measure $\mathcal{D}g$ would be supported on an appropriately defined quotient of $Met(\Sigma)$ by $C_+^\infty(\Sigma)$ and $Diff^+(\Sigma)$. However, we considered precisely quotients of this type in chapter 5 when we constructed the Teichmüller and moduli spaces of a Riemann surface Σ. We can make use of those constructions here. In particular we recall that $Diff^+(\Sigma)$ does not act freely on $Met(\Sigma)$ and has to be replaced by its identity component $Diff_0(\Sigma)$; in addition, the existence of conformal Killing vectors for genus $p = 0, 1$ means that one has to replace $Met(\Sigma)$ by the space $Met_{const}(\Sigma)$.

The preceding paragraph can be taken as an outline of our present task: we must define a measure $\mathcal{D}Met(\Sigma)$ on the infinite dimensional space $Met(\Sigma)$, and then use the quotient construction of $\mathcal{M}_p$ to induce a measure on this finite dimensional space $\mathcal{M}_p$.

We can always construct a measure on a finite dimensional Riemannian manifold by using the metric g: one simply takes a frame or basis $\{e_i\}$, coordinates $\{x\}$, and the corresponding Riemannian-invariant measure is

$\sqrt{det < e_i, e_j >_g} d^n x = \sqrt{g} \, d^n x$. We would like to do the same here and so we immediately look for a metric on $Met(\Sigma)$. In fact $Met(\Sigma)$ possesses a natural metric which we denote by $< \, , \, >_{met}$. To define $< \, , \, >_{met}$, let $g \in Met(\Sigma)$, then turn to the tangent space $T_g Met(\Sigma)$ and consider $X, Y \in T_g Met(\Sigma)$. For X and Y we have the formal local coordinate expressions:

$$X = X^I \frac{\partial}{\partial g^I} = X^I_{a_1 b_1} \frac{\partial}{\partial g^I_{a_1 b_1}}$$
$$Y = Y^J \frac{\partial}{\partial g^J} = Y^J_{a_2 b_2} \frac{\partial}{\partial g^J_{a_2 b_2}}$$

(9.9)

where $(g^I) \equiv (g^1, g^2, \ldots)$ are the coordinates in a patch $U \subset Met(\Sigma)$. Our definition of $< \, , \, >_{met}$ is that

$$< X, Y >_{met} = \int_\Sigma < X^I, Y^I >_{g(x)} \sqrt{g} \, dx$$

(9.10)

and the inner product $< \, , \, >_{g(x)}$ on the RHS is given by

$$< X^I, Y^I >_{g(x)} = g^{a_1 a_2}(x) g^{b_1 b_2}(x) X^I_{a_1 b_1}(x) Y^I_{a_2 b_2}(x)$$

(9.11)

Notice that X and Y can be thought of both as tangents to $Met(\Sigma)$ and as tensor fields on Σ. $Met(\Sigma)$ may now to be thought of as an infinite dimensional Riemannian manifold.

While on the subject of metrics it is convenient to point out for later use that $vect(\Sigma)$ possesses, for each g, a natural metric. If $V, W \in vect(\Sigma)$ its definition is

$$< V, W >_{vect} = \int_\Sigma < V(x), W(x) >_{g(x)} \sqrt{g} \, dx$$

(9.12)

and $< \, , \, >_{g(x)}$ denotes the inner product in $T_x \Sigma$.

Having constructed a metric on $Met(\Sigma)$ we can straightaway put it to use to decompose $T_g Met(\Sigma)$ in a way which reflects the quotient structure of the moduli space. To begin this decomposition process we consider the action of the group $Diff_0(\Sigma)$ on $Met(\Sigma)$. If we work infinitesimally we can use the fact that the Lie algebra of $Diff_0(\Sigma)$ is the space $vect(\Sigma)$ of vector fields. Then as in chapter 5 we can express an infinitesimal displacement of a metric g as

$$L_V g$$

(9.13)

where V is a vector field generating the one parameter family of diffeomorphisms $\exp[tV]$.

First we restrict ourselves to conformal changes of metrics. Since a conformal change of metric leaves the action invariant then we would like to project out the subspace C_g, say, of $T_g Met(\Sigma)$ which is tangent to those orbits of $C_+^\infty(\Sigma)$ that pass through the point $g \in Met(\Sigma)$—infinitesimally speaking, displacements from g along such orbits only multiply g by a function. To accomplish this we use the metric $< , >_{met}$ to make an orthogonal decomposition of $T_g Met(\Sigma)$:

$$T_g Met(\Sigma) = C_g \oplus C_g^\perp \tag{9.14}$$

To identify these spaces C_g and $C_g^\perp$ we return to the quantity $L_V g$ and write, for some function α

$$L_V g = \alpha g + h; \qquad g \in C_g, \ h \in C_g^\perp$$

Then because $h \in C_g^\perp$ we have

$$< g, h >_{met} = 0 \tag{9.15}$$

so that

$$
\begin{aligned}
0 &= \int_\Sigma dx \sqrt{g} < g(x), h(x) >_{g(x)} \\
&= \int_\Sigma dx \sqrt{g} \, g^{a_1 a_2}(x) g^{b_1 b_2}(x) g_{a_1 b_1}(x) h_{a_2 b_2}(x) \\
&= \int_\Sigma dx \sqrt{g} \, \delta_{b_1}^{a_2} g^{b_1 b_2}(x) h_{a_2 b_2}(x) = \int_\Sigma dx \sqrt{g} \, g^{a_2 b_2}(x) h_{a_2 b_2}(x) \\
&= \int_\Sigma dx \sqrt{g} \, tr(h)
\end{aligned}
\tag{9.16}
$$

i.e. $tr(h) = 0$ implies 9.15. With this information we can implement 9.15 by writing

$$L_V g + \left\{ L_V g - \frac{tr(L_V g)}{2} g \right\} + \frac{tr(L_V g)}{2} g \tag{9.17}$$

Thus h and α have been determined[2] according to

$$
\begin{aligned}
h &= L_V g - \frac{tr(L_V g)}{2} g \\
\alpha &= \frac{tr(L_V g)}{2}
\end{aligned}
\tag{9.18}
$$

---

[2] These formulae are sometimes written without using Lie derivatives by virtue of introducing the Levi–Civita connection Γ for the metric g. If ∇ denotes covariant derivative it is a routine tensorial calculation to show that $L_V g_{ab} = (\partial V_b / \partial x^a) + (\partial V_a / \partial x^b) - 2\Gamma_{ab}^i V_i = \nabla_a V_b + \nabla_b V_a$; this in turn shows that h and α are given by $h_{ab} = \nabla_a V_b + \nabla_b V_a - (\nabla^c V_c) g_{ab}$ and $\alpha = \nabla^c V_c$.

Having learned how to project out C_g we move on to $C_g^{\perp}$. Suppose $h \in C_g^{\perp}$; we can characterise $C_g^{\perp}$ by regarding h as being obtained by the linear operator A_g defined below:

$$A_g : vect\,(\Sigma) \longrightarrow C_g^{\perp}$$
$$V \mapsto (L_V g - \frac{tr\,(L_V g)}{2} g) \tag{9.19}$$

Such h will not be sufficient to span $C_g^{\perp}$—instead we have the standard linear algebraic fact that

$$C_g^{\perp} = Range\,A_g \oplus ker\,A_g^* \tag{9.20}$$

where the $*$ appearing in A_g^* is defined with respect to the inner products on $< \,,\, >_{met}$ and $< \,,\, >_{vect}$. Thus we have now decomposed $T_g Met\,(\Sigma)$ into three orthogonal pieces

$$T_g Met\,(\Sigma) = C_g \oplus C_g^{\perp}$$
$$= C_g \oplus Range\,A_g \oplus ker\,A_g^* \tag{9.21}$$

If we pass to the quotient of $Met(\Sigma)$ by $C_+^{\infty}(\Sigma)$ and $Diff_0\,(\Sigma)$ then the space that remains is $ker\,A_g^*$; but A_g can be seen to be an elliptic operator thus $ker\,A_g^*$ is a finite dimensional space. In fact it is easy to see that $ker\,A_g^*$ can be identified with the space of quadratic differentials; in the physics literature A_g is often denoted by P_1.

Now we already know from our study of Teichmüller space that

$$\mathcal{T}_p = Met_{const}\,(\Sigma)/Diff_0\,(\Sigma) \tag{9.22}$$

and we are now in a position to describe the metric on T_p that is induced by the Teichmüller construction. This metric is known as the Weil–Petersson metric.

§ 3. The Weil–Petersson metric

Since $Met\,(\Sigma)$ is endowed with the natural metric $< \,,\, >_{met}$, then viewing $Met_{const}\,(\Sigma)$ as a Riemannian sub-manifold of $Met\,(\Sigma)$ induces a metric on $Met_{const}\,(\Sigma)$ just as it would in a finite dimensional example. This metric is obtained simply by restricting to $Met_{const}\,(\Sigma)$. We shall denote this metric by $< \,,\, >_{const}$.

The vital property of $< , >_{const}$ is that the elements of $Diff_0(\Sigma)$ are isometries[3] of $Met_{const}(\Sigma)$. Therefore a metric is also induced on the quotient $Met_{const}(\Sigma)/Diff_0(\Sigma)$, i.e. on $\mathcal{T}_p$ itself. This metric is the Weil–Petersson metric, which we write as $< , >_{wp}$.

We would like to examine $< , >_{wp}$ in a little more detail. In terms of $< , >_{const}$ the definition of the Weil–Petersson metric is as follows: Let π be the projection

$$\pi : Met_{const}(\Sigma) \longrightarrow Met_{const}(\Sigma)/Diff_0(\Sigma) = \mathcal{T}_p \qquad (9.23)$$

and let the $Diff_0(\Sigma)$-equivalence class $[g_{const}]$ denote an element of $\mathcal{T}_p$. We write the tangent space to $\mathcal{T}_p$ at this point $[g_{const}]$ as

$$T_{[g_{const}]}\mathcal{T}_p \qquad (9.24)$$

Then for $U, V \in T_{[g_{const}]}\mathcal{T}_p$ we define $< U, V >_{wp}$ by

$$< U, V >_{wp} = < X, Y >_{const} \qquad (9.25)$$

where $U = \pi_* X$ and $V = \pi_* Y$, and of course $X, Y \in T_{g_{const}}Met_{const}(\Sigma)$. The only thing to check is that, since neither g_{const} nor X and Y are unique in the above, this definition is indeed independent of the choice made for these objects. To prove this, suppose

$$\begin{aligned} [g_{const}] &= [g'_{const}] \\ \Rightarrow \quad g'_{const} &= \alpha_* g_{const}, \quad \text{with} \quad \alpha \in Diff_0(\Sigma) \end{aligned} \qquad (9.26)$$

Now if $X', Y' \in T_{g'_{const}}Met_{const}(\Sigma)$ and also satisfy

$$\pi_* X' = U, \quad \pi_* Y' = V \qquad (9.27)$$

then it is evident from the fibration 9.23 that

$$X' = \alpha_* X, \quad \pi_* Y' = V \qquad (9.28)$$

Thus

$$\begin{aligned} < X', Y' >_{const} &= < \alpha_* X, \alpha_* Y >_{const} \\ &= < X, Y >_{const}, \quad \text{since } \alpha \text{ is an isometry} \end{aligned} \qquad (9.29)$$

---

[3] Note that we specified isometries of $Met_{const}(\Sigma)$ not Σ; this means that if $X, Y \in T_{g_{const}}Met_{const}(\Sigma)$ and $\alpha \in Diff_0(\Sigma)$, then $< \alpha_* X, \alpha_* Y >_{const} = < X, Y >_{const}$, this equality being readily verifiable using the definition 9.10 of $< , >_{met}$ and $< , >_{const}$ given above.

and so $< , >_{wp}$ is properly defined.

Finally we can write the partition function as an integral over the moduli space. The thing to do is to decompose the measure $\mathcal{D}g$ on $Met(\Sigma)$ into pieces which reflect the decomposition

$$T_g Met(\Sigma) = C_g \oplus Range\, A_g \oplus ker\, A_g^* \qquad (9.30)$$

If the coordinates for C_g, $Range\, A_g$, and $ker\, A_g^*$ are denoted by $\{\sigma\}$ (for $f = \exp \sigma$, $f \in C_+^\infty(\Sigma)$), $\{A_g V\}$, and $\{\mathcal{T}_p\}$ respectively then we can write

$$\mathcal{D}Met(\Sigma) = N(f, A_g V, T_p)\, \mathcal{D}\sigma\, \mathcal{D}(A_g V)\, d\mathcal{T}_p \qquad (9.31)$$

where $N(f, A_g V, T_p)$ is an appropriate normalising factor depending on the frames used. Next we would *like* to write

$$\mathcal{D}(A_g V) = \sqrt{det\,(A_g^* A_g)}\, \mathcal{D}V \qquad (9.32)$$

But this requires the operator $A_g^* A_g$ to be invertible. However, $A_g^* A_g$ is only positive semi-definite and will only be positive definite, and hence invertible, if $ker\, A_g = \emptyset$. In fact, for $p = 0, 1$ $ker\, A_g \neq \emptyset$: the elements of $ker\, A_g$ are the conformal Killing vectors, for if $V \in ker\, A_g$ then

$$A_g V = 0$$

$$\Rightarrow \quad L_V g = fg, \quad \text{with} \quad f = \frac{1}{2} tr(L_V g)$$

Now since $ker\, A_g \subset vect\,(\Sigma)$ then we can use the inner product $< , >_{vect}$ to decompose $vect\,(\Sigma)$ according to

$$vect\,(\Sigma) = ker\, A_g \oplus ker^{\perp} A_g \qquad (9.33)$$

Remember that $vect\,(\Sigma) = T_e Diff_0\,(\Sigma)$ so that 9.33 is a decomposition of $T_e Diff_0\,(\Sigma)$. Hence, if $p = 0, 1$, we restrict quantities to $ker^{\perp} A_g$ and integrate separately over the conformal Killing vectors, then we can proceed. We also know that

$$\mathcal{D}g = \frac{\mathcal{D}Met(\Sigma)}{vol\,(C_+^\infty(\Sigma))\, vol\,(Diff^+(\Sigma))} \qquad (9.34)$$

and when all this is put together we obtain (cf. D'Hoker and Phong [1]) expressions for the partition function of the form

$$\begin{aligned}
Z_p &= \frac{1}{|\Gamma_\Sigma|} \int_{\mathcal{T}_p} d\mu(\mathcal{T}_p) \left(\frac{det\,(\Delta_g/2)}{< 1, 1 >_g}\right)^{-13} \sqrt{det\,(\tilde{A}_g^* \tilde{A}_g)} \\
&= \int_{\mathcal{M}_p} d\mu(\mathcal{M}_p) \left(\frac{det\,(\Delta_g/2)}{< 1, 1 >_g}\right)^{-13} \sqrt{det\,(\tilde{A}_g^* \tilde{A}_g)}
\end{aligned} \qquad (9.35)$$

with $d\mu(\mathfrak{T}_p)$ and $d\mu(\mathcal{M}_p)$ measures on $\mathfrak{T}_p$ and $\mathcal{M}_p$ respectively and $\tilde{A}_g$ denoting the restriction of A_g needed for $p = 0, 1$. $|\Gamma_\Sigma|$ is the cardinality of the mapping class group Γ_Σ and its presence reflects the difference between using the space $\mathfrak{T}_p$ and $\mathcal{M}_p$; the restriction of $Diff^+(\Sigma)$ to $Diff_0(\Sigma)$ when constructing $\mathfrak{T}_p$ must be abandoned dealing with the full symmetry group of the action. Thus we have achieved our objective of writing the partition function Z_p as an integral over the finite dimensional moduli space $\mathcal{M}_p$. The operator $\tilde{A}_g$ is a real operator; in terms of the closely related $\bar{\partial}$-operators used in chapter 5 it is easy to verify that $det\,(\tilde{A}_g^*\tilde{A}_g) = (det\,(\bar{\partial}_{K_\Sigma^*}^*\bar{\partial}_{K_\Sigma^*}))^2$. With this change we can write Z as

$$Z = \sum_p Z_p = \sum_p \int_{\mathcal{M}_p} d\mu(\mathcal{M}_p) \left(\frac{det\,(\Delta_g/2)}{<1,1>_g} \right)^{-13} det\,(\bar{\partial}_{K_\Sigma^*}^*\bar{\partial}_{K_\Sigma^*}) \qquad (9.36)$$

CHAPTER X

Anomalies

§ 1. Introduction

The term anomaly, though not a very precise one, usually has to do with strange or even pathological behaviour of chiral Fermions when coupled to gauge fields. The first anomaly of this kind was the Adler–Bell–Jackiw triangle anomaly (Adler [1,2], and Bell and Jackiw [1]) in which the group G is just the Abelian group $U(1)$ familiar from quantum electrodynamics. Subsequently many other kinds of anomalies have been discussed. In this chapter we shall concentrate on these later developments.

These developments include Yang–Mills theories, theories of gravity and string theory. The Yang–Mills case represents a natural widening of the discussion to allow for non-Abelian G. An investigation of gravitational anomalies involves applying the analysis to the group of general (orientation preserving) coordinate transformations $Diff^+(M)$ instead of just to the group of gauge transformations $\mathcal{G}$. In dimensions other than the critical dimension, the action S of the Bosonic string theory has a conformal anomaly. In § 6 we shall also examine anomalies from a Hamiltonian standpoint. In this wider context it is more appropriate to consider an anomaly as some obstruction to the defining of a quantum theory with the same invariance groups as its underlying classical theory.

A common feature of many of these anomalies is that this obstruction has a topological origin. The topology involved is the cohomology of the invariance groups of the theory: for example, in the Yang–Mills case one has to consider certain elements of $H^*(\mathcal{G})$. In constructing these elements it is frequently useful to consider families of Dirac operators $\partial\!\!\!/_A$ parametrised by $A \in \mathcal{A}$ and look for an obstruction to the existence of a gauge invariant determinant for $\partial\!\!\!/_A$. The determinant of an operator O vanishes if $ker\, O$ is non-empty and so index theory enters; since we have a family of such objects, the index theorem for families becomes relevant. In the next section we see how this comes about.

§ 2. Anomalies and Yang–Mills theories

We shall actually distinguish two sorts of anomalies: local anomalies and global anomalies. The term global anomaly will be explained in § 5—for the remainder of this section the word anomaly will mean a local anomaly. We work with Riemannian manifolds throughout, so that space-time M is viewed as Euclidean—a Hamiltonian treatment is also possible, cf. § 6.

For the moment we shall assume that the dimension n of M is even and set $n = 2m$; we shall come to the properties of odd dimensional n in § 6. In even dimensions, where the notion of chirality exists, we can couple the gauge field to chiral Fermions. It is then natural to study a quantum field theory containing these fields. Let the gauge group be G and the action S for the theory be given by

$$S \equiv S(A, \psi) = \|F\|^2 + \langle \psi, \not{D}_A \psi \rangle$$
$$= -tr \int_M F \wedge *F + \frac{1}{2} \int_M \bar{\psi}(x)\gamma_\mu(\partial^\mu + \Gamma^\mu + A^\mu)(1 + \gamma_5)\psi(x) \tag{10.1}$$

where Γ_μ and A_μ are the Levi–Civita and Yang–Mills connections respectively. Since we are discussing Fermions M must be a spin-manifold so that we will require the vanishing of the first two Stiefel–Whitney classes $w_1(TM)$ and $w_2(TM)$ of M. This is not too severe a requirement and is true for S^n for example.

The anomaly has to do with chiral Fermions. A compelling way of seeing that these Fermions are not being treated even handedly is to compute the index of the Dirac operator $\not{D}_A$. Since the Fermions are coupled to the connection A, via a bundle F say, the Dirac operator is of the form

$$\not{D}_A : \Gamma(M, E^+ \otimes F) \longrightarrow \Gamma(M, E^- \otimes F) \tag{10.2}$$

where $E^\mp$ are the chiral spin bundles defined in chapter 4. This means that we must use the index formula 4.122 for the twisted Dirac operator. For simplicity we shall take M to be the sphere S^{2m}. If n_+ and n_- are the numbers of positive and negative chirality massless Fermions respectively, we have

$$index \not{D}_A = n_+ - n_-$$
$$= \int_{S^{2m}} \hat{A} \wedge ch(F)$$
$$= \int_{S^{2m}} ch(F), \quad \text{for spheres} \tag{10.3}$$
$$= k$$

where we have used the fact that $\hat{A}$ is expressible as an expansion in Pontrjagin classes cf. 4.110; this fact, coupled with the vanishing of all Pontrjagin classes $p_i(S^n)$ of spheres (p. 85), implies that $\hat{A} = 1$ on S^{2m}. Also k is the integer $c_m(F)$ we had in chapter 8 which classifies bundles over S^{2m}—clearly in four dimensions $-k$ is the familiar instanton number. Thus a non-zero k, which is the generic situation, produces an asymmetry in the massless chiral Fermion sector.

To make progress in uncovering the anomaly we examine the partition function Z which is given by the functional integral

$$\int \mathcal{D}A\mathcal{D}\psi\mathcal{D}\bar{\psi} \exp[-\|F\|^2 - \langle\psi, \partial\!\!\!/_A\psi\rangle] \tag{10.4}$$

We can carry out the Fermionic integration using the expression

$$\int \mathcal{D}\psi\mathcal{D}\bar{\psi} \exp[-\langle\psi, \partial\!\!\!/_A\psi\rangle] = \sqrt{\det(\partial\!\!\!/_A^*\partial\!\!\!/_A)} \tag{10.5}$$

This allows us to write

$$Z = \int \mathcal{D}A\sqrt{\det(\partial\!\!\!/_A^*\partial\!\!\!/_A)} \exp[-\|F\|^2] \tag{10.6}$$

The next step in the computation of the functional integral for Z would be to assume that the integrand is gauge invariant and then Z is naturally expressed as an integral over the space of gauge orbits $\mathcal{A}/\mathcal{G} = B\mathcal{G}$. An anomaly is said to have arisen if this is not the case. Unfortunately, our assumption requires justification and is generally false: although the expression $\|F\|^2$ is manifestly gauge invariant, the same cannot be said of the Fermionic determinant $\det(\partial\!\!\!/_A^*\partial\!\!\!/_A)$. Indeed, if g is a gauge transformation under which $A \mapsto A_g$ then, in general,

$$\det(\partial\!\!\!/_{A_g}^*\partial\!\!\!/_{A_g}) \neq \det(\partial\!\!\!/_A^*\partial\!\!\!/_A) \tag{10.7}$$

This is so despite the fact that, if we change the gauge in an expression such as $\langle\psi, \partial\!\!\!/_A\psi\rangle$, we can verify that the Dirac operator transforms according to

$$\partial\!\!\!/_{A_g} = g^{-1}\partial\!\!\!/_A g \tag{10.8}$$

The point is that we are in infinite dimensions and the Dirac determinant requires regularisation,[1] and, for such determinants, we cannot assume that $\det AB = \det A \det B$.

---

[1] The regularisation is most conveniently done using a zeta-function (cf. p. 261). If an infinite dimensional operator A has the property that $A - I$ is of trace class (T is of trace class if $tr(\sqrt{TT^*}) < \infty$) then $\det A$ exists without the need for regularisation, and operators of this kind do obey $\det AB = \det A \det B$.

We shall now verify this by direct calculation. Having done this we will be able to show that the variation of the Fermionic determinant under a gauge transformation has a natural topological interpretation as a cohomology class in $H^1_{de\,Rham}(\mathcal{G})$.

Choose $g = \exp(tf)$ where t is a real parameter and $f \in \mathfrak{LG}$. We can represent f locally by an expression of the form $f = t^a f^a(x)$ where $\{t^a\}$ is a basis for the spinor representation the Lie algebra of G. Now take the Fermionic integral 10.5 with $A = A_g$ and calculate the variation by evaluating

$$\frac{d}{dt}RHS\,|_{t=0}$$

The result is a routine calculation and one obtains

$$\frac{d}{dt}\sqrt{det\,(\slashed{\partial}^*_{A_{g(t)}}\slashed{\partial}_{A_{g(t)}})}\,|_{t=0} = \int \mathcal{D}\psi\mathcal{D}\bar{\psi}\exp[-\langle\psi,\slashed{\partial}_A\psi\rangle]\int_{S^n} f^a(x)\nabla^{ab}_\mu j_5^{b\mu}(x)$$

where ∇^{ab}_μ denotes covariant derivative and $j_5^{b\mu}(x)$ is the axial current

$$j_5^{b\mu}(x) = \frac{1}{2}\bar{\psi}\gamma^\mu(1 + \gamma_5)t^b\psi \qquad (10.9)$$

and we have used Stokes' theorem. In the case where G is the group $U(n)$ we can restrict ourselves to its $U(1)$ subgroup; this means that t^a is a one dimensional matrix and the covariant derivative expression $f^a(x)\nabla^{ab}_\mu j_5^{b\mu}(x)$ reduces to just $f(x)\partial_\mu j_5^\mu(x)$. The presence of an anomaly then forces the non-conservation of the well known $U(1)$, or singlet, axial current $j_5^\mu(x) = (1/2)\bar{\psi}\gamma^\mu(1 + \gamma_5)\psi$.

If we divide both sides of 10.9 by the standard normalisation factor $\sqrt{det\,(\slashed{\partial}^*_A\slashed{\partial}_A)}$ used to eliminate vacuum Feynman graphs (Nash [2]) then we obtain

$$\frac{1}{\sqrt{det\,(\slashed{\partial}^*_A\slashed{\partial}_A)}}\frac{d}{dt}\sqrt{det\,(\slashed{\partial}^*_{A_{g(t)}}\slashed{\partial}_{A_{g(t)}})}\Big|_{t=0} =$$
$$\frac{\int \mathcal{D}\psi\mathcal{D}\bar{\psi}\exp[-\langle\psi,\slashed{\partial}_A\psi\rangle]\int dx\,\nabla^{a\mu}j_\mu^{a5}(x)f^a(x)}{\sqrt{det\,(\slashed{\partial}^*_A\slashed{\partial}_A)}} \qquad (10.10)$$

The LHS now has the preferred structure and is the infinitesimal variation of $\ln\sqrt{det\,(\slashed{\partial}^*_A\slashed{\partial}_A)}$, while the RHS is just the vacuum expectation value of $\nabla^{ab}_\mu j_5^{b\mu}(x)$ smeared with an arbitrary f. Hence if $\ln\sqrt{det\,(\slashed{\partial}^*_A\slashed{\partial}_A)}$ really were gauge invariant we could conclude that

$$\nabla^{ab}_\mu j_5^{b\mu}(x) = 0 \qquad (10.11)$$

The catch is that when an anomaly is present the above equation is false.

A direct perturbative way of seeing this is to expand 10.10 in terms of the coupling constant and calculate the resultant Feynman graphs. If we do this we find at one loop the celebrated divergent triangle graph whose non-vanishing implies the existence of the anomaly. Alternatively, from the point of view of functional integration, since the integrand $\exp[-\langle\psi,\partial\!\!\!/_A\psi\rangle]$ of 10.5 is gauge invariant, the lack of gauge invariance of $det\,(\partial\!\!\!/_A^*\partial\!\!\!/_A)$ must occur somewhere in the integration process over ψ and $\bar{\psi}$. This is demonstrated by Fujikawa [1], who shows that the Fermionic 'measure' $\mathcal{D}\psi\mathcal{D}\bar{\psi}$ is not gauge invariant.

The rest of this section is devoted to a topological interpretation of the anomaly. This is made up of two parts. The first part consists of showing that the infinitesimal variation 10.10 of $\ln\sqrt{det\,(\partial\!\!\!/_A^*\partial\!\!\!/_A)}$ can be identified with an element of the cohomology group $H^1_{de\,Rham}(\mathcal{G})$; the second part is a demonstration that, if one uses the families index theorem, then this cohomology formulation ties in very naturally with the idea of an anomaly as an obstacle to the gauge invariance of the determinant $det\,(\partial\!\!\!/_A^*\partial\!\!\!/_A)$.

Next we define the operator T_g (Atiyah and Singer [7]) by

$$T_g = \partial\!\!\!/_A^*\partial\!\!\!/_{A_g} \tag{10.12}$$

The idea now is to define a regularised determinant $det\,T_g$; then, as g varies, this is viewed as a function on $\mathcal{G}$. The anomaly discussion above suggests a natural cohomology class associated with $det\,T_g$, namely that determined by the 1-form μ_1 where

$$\mu_1 = \frac{d\,det\,T_g}{det\,T_g} \tag{10.13}$$

and d denotes the exterior derivative acting on the infinite dimensional space $\mathcal{G}$.

To this end we must first explain how to define $det\,T_g$; this requires a slightly extended zeta function technique. It is clear that T_g is an elliptic operator since, if we use $\partial\!\!\!/_{A_g} = g^{-1}\partial\!\!\!/_A g$, we can verify that its leading symbol is just the product of the leading symbols of $\partial\!\!\!/_A^*$ and $\partial\!\!\!/_A$, that is the matrix

$$g_{\mu\nu}p^\mu p^\nu I \tag{10.14}$$

Ellipticity is immediate since this expression is invertible for $g_{\mu\nu}p^\mu p^\nu \neq 0$, which is equivalent to $p_\mu \neq 0$ since the metric is Riemannian.

Now, though elliptic, T_g need not be positive despite the fact that its leading term is positive semi-definite. This is because T_g has lower order terms containing derivatives of g. These lower order terms can perturb the spectrum off the non-negative part of the real axis: It is evident that the spectrum of T_g may have a finite number of negative and zero eigenvalues; but, apart from this finite set, the spectrum of this type of elliptic operator must lie on or near the real axis, and in any case will be contained in a wedge shaped region which contains the real axis (cf. Seeley [1]). We want to use a zeta function to define $det\,T_g$ but the lack of positivity of the spectrum requires some attention: for those λ_i in the wedge the zeta function power $\lambda_i^{-s} = \exp[-s\ln\lambda_i]$ can be defined by taking the branch of the logarithm outside this wedge. We write the restriction of T_g to this subspace as $\hat{T}_g$. Let us temporarily assume that there is no zero eigenvalue. Then if we denote the set of negative eigenvalues by $\lambda_1^-, \ldots, \lambda_k^-$, and the remainder of the eigenvalues by $\{\lambda_i\}$, we can define

$$det\,T_g = (\lambda_1^- \lambda_2^- \cdots \lambda_k^-)\exp[-\zeta_{\hat{T}_g}'(0)] \qquad (10.15)$$

When $T_g = \partial_A^* \partial_{A_g}$ *does* have a zero eigenvalue our strategy depends on whether $index\,\partial_A$ is zero or not: If $index\,\partial_A = 0$ then, since the index is a perturbation invariant, we can perturb A to a new connection for which $ker\,\partial_A = \phi$. Because g is invertible this automatically implies that $ker\,\partial_{A_g} = \phi$; also $ker\,\partial_A^*$ is now empty because the index is zero. Thus T_g now has no zero eigenvalue. On the other hand, if $index\,\partial_A \neq 0$, say $index\,\partial_A > 0$, then we can perturb A so as to make $ker\,\partial_A^*$ empty, but $ker\,\partial_A$ cannot be simultaneously empty, and this in turn implies that $ker\,\partial_{A_g}$ is non-trivial. In this event we restrict T_g to the orthogonal complement of $ker\,\partial_{A_g}$ and on this space T_g has no zero eigenvalue (clearly a similar argument applies if $index\,\partial_A < 0$). Finally, because we want to consider $det\,T_g$ as a function of g, we require our arguments to remain valid when g varies; this is not difficult to check. Note that, as g varies, $ker\,\partial_{A_g}$ will be a vector bundle over the orbit A_g of A. In any case our task of defining of $det\,T_g$ has been accomplished.

We return to the use of $det\,T_g$ to study the topology of $\mathcal{G}$. Since $det\,T_g$ is a non-vanishing function of $g \in \mathcal{G}$, we can certainly now make proper sense of the 1-form quoted above in 10.13. We had

$$\mu_1 = \frac{d\,det\,T_g}{det\,T_g} \qquad (10.16)$$

This 1-form would be exact and hence cohomologically trivial provided we could write

$$\mu_1 = d\ln det\,T_g$$

But we cannot do this unless $\ln det\,T_g$ exists. Recall now that a necessary condition for $\ln f$ to exist, for a map $f : W \to \mathbf{C} - \{0\}$, is that there is no loop α in W whose image $f(\alpha)$ circles the origin in $\mathbf{C}$. If W is simply connected this is avoided. But in our case W is $\mathcal{G}$ and

$$\pi_1(\mathcal{G}) = \pi_{2n+1}(G) \neq 0 \tag{10.17}$$

in general. For example if $M = S^4$ and $G = U(N)$ then, from chapter 8, we know that

$$\pi_1(\mathcal{G}) = \pi_5(G) = \begin{cases} 0, & N = 1 \\ \mathbf{Z}_2, & N = 2 \\ \mathbf{Z}, & N \geq 3 \end{cases} \tag{10.18}$$

Thus, for $N \geq 3$, $[\mu_1]$ is a non-trivial cohomology class in $H^1_{de\,Rham}(\mathcal{G})$. The final part of the topological argument is to show how the non-triviality of this cohomology class is related to the anomaly. This is where the families index theorem comes in.

Let us return to the functional integral for the partition function Z. We have

$$Z = \int_{\mathcal{A}} \mathcal{D}\mathcal{A} \sqrt{det\,(\partial^*_A \partial_A)} \exp[-\|F\|^2] \tag{10.19}$$

This expression contains both the Dirac operator ∂_A and a sum over all $A \in \mathcal{A}$. Thus it is natural to consider the family of elliptic operators given by ∂_A as A varies throughout $\mathcal{A}$. Now for the anomaly we have already seen in 10.3, and in the construction of $det\,T_g$, that $index\,\partial_A$ (for a fixed A) is important; thus, when A varies, we ought to consider the index of the whole Dirac family.

In chapter 4 we showed that the (analytic) index of a family of elliptic operators parametrised by a space Y is given by an element of $K(Y)$. In the present case $Y = \mathcal{A}$ and, denoting the index of the Dirac family by $Index\,\partial$, we obtain

$$\begin{aligned} Index\,\partial &= \{ker\,\partial_A : A \in \mathcal{A}\} - \{ker\,\partial^*_A : A \in \mathcal{A}\} \\ &= [ker\,\partial] - [ker\,\partial^*] \end{aligned} \tag{10.20}$$

Such a formal difference defines an element of the K-theory $K(\mathcal{A})$ of $\mathcal{A}$ which immediately projects to an element of $K(\mathcal{A}/\mathcal{G})$ because of the gauge invariance 10.8 of ∂_A.

Now because the anomaly centres round determinants constructed from the Dirac operator it is natural for us to consider a certain determinant line bundle $det\,Index\,\partial$ associated with $Index\,\partial$. To understand how $det\,Index\,\partial$ arises it is helpful to have some preliminary remarks on determinants and their generalisations.

Let V and W be finite dimensional vector spaces connected by a linear map $O : V \rightarrow W$ and suppose that $\dim V = \dim W = k$. Then if we form the maximum exterior powers $\wedge^k V$ and $\wedge^k W$, we induce a map between the one dimensional spaces $\wedge^k V$ and $\wedge^k W$ which it is natural to call the determinant of O: that is, we obtain $det\, O : \wedge^k V \rightarrow \wedge^k W$. If we write $O \in Hom\,(V, W)$ and use the standard linear algebraic isomorphism $Hom\,(V, W) \simeq V^* \otimes W$ then we have

$$O \in Hom\,(V, W) \quad \text{and} \quad det\, O \in (\wedge^k V)^* \otimes (\wedge^k W) \qquad (10.21)$$

Now we allow V and W to have differing dimensions and use O and O^* to decompose V and W according to

$$V = Range\, O^* \oplus ker\, O, \qquad W = Range\, O \oplus ker\, O^* \qquad (10.22)$$

Using the slightly better notation $\wedge^{max} V$ to denote the maximum exterior power of a vector space V we find that

$$
\begin{aligned}
det\, O &\in (\wedge^{max} V)^* \otimes (\wedge^{max} W) \\
&= (\wedge^{max}(Range\, O^* \oplus ker\, O))^* \otimes (\wedge^{max}(Range\, O \oplus ker\, O^*)) \\
&= (\wedge^{max} Range\, O^* \otimes \wedge^{max} ker\, O)^* \otimes (\wedge^{max} Range\, O \otimes \wedge^{max} ker\, O^*)
\end{aligned}
$$
$$(10.23)$$

However, in this expression for $(\wedge^{max} V)^* \otimes (\wedge^{max} W)$ the inclusion of both the range and the kernel of O and O^* is somewhat redundant. If we project out the factors containing the range we still obtain a one dimensional vector space which is a natural measure of the determinant of O. So now we write [2]

$$det\, O \in (\wedge^{max} ker\, O)^* \otimes (\wedge^{max} ker\, O^*) \qquad (10.24)$$

Next imagine that V and W are promoted to be vector bundles over another space Y and that O is promoted to be a family of operators O_y, $y \in Y$ connecting the fibres of V and W. In this case we can replace $\wedge^{max} V$ and $\wedge^{max} W$ by the determinant bundles [3] $det\, V$ and $det\, W$. If we use the notation $ker\, O$ and $ker\, O^*$ to denote the corresponding vector bundles over Y whose fibres are $ker\, O_y$ and $ker\, O_y^*$ then $det\, O$ is replaced by the line bundle

$$(det\, ker\, O)^* \otimes (det\, ker\, O^*) \qquad (10.25)$$

---

[2] A little calculation will show that, in the case where $det\, O$ exists so that $\wedge^{max} ker\, O$ and $\wedge^{max} ker\, O^*$ are both isomorphic to $\mathbf{C}$, this projection amounts to normalising $det\, O$ to unity.

[3] Recall that if E is a vector bundle of rank k over a space Y, then the determinant bundle $det\, E$ is the line bundle over Y whose fibre at y is just the one dimensional space $\wedge^k E_y$ where E_y is the fibre of E at y.

Proceeding still further we move on from vector bundles over Y to an element of the K-theory over Y. Suppose that an element of $K(Y)$ is given by the formal difference of vector bundles

$$[V] - [W] \tag{10.26}$$

then the natural line bundle associated with this K-theory element is still

$$(det\,V)^* \otimes (det\,W) \tag{10.27}$$

The reader can easily check that this line bundle is independent of the way the K-theory element is written: that is, if $[V] - [W] = [V'] - [W']$ then the same line bundle is obtained whether one uses the pair (V, W) or the pair (V', W').

Applying the preceding discussion to the index of the Dirac family $Index\,\partial\!\!\!/ = [ker\,\partial\!\!\!/] - [ker\,\partial\!\!\!/^*]$ gives us the determinant line bundle $det\,Index\,\partial\!\!\!/$ over $\mathcal{A}/\mathcal{G}$, whose definition is

$$det\,Index\,\partial\!\!\!/ = (det\,ker\,\partial\!\!\!/)^* \otimes (det\,ker\,\partial\!\!\!/^*) \tag{10.28}$$

One should be aware that, as pointed out in chapter 4, the dimensions of $ker\,\partial\!\!\!/_A$ and $ker\,\partial\!\!\!/_A^*$ can jump at certain A in $\mathcal{A}$; however, the deformation invariance of the index means that the difference in their dimensions remains constant and one still obtains an element of $K(\mathcal{A})$ which we then project onto $K(\mathcal{A}/\mathcal{G})$.

Since $det\,Index\,\partial\!\!\!/$ is a line bundle it is completely characterised by its Chern class $c_1(det\,Index\,\partial\!\!\!/) \in H^2(\mathcal{A}/\mathcal{G})$; however, we can calculate $c_1(det\,Index\,\partial\!\!\!/)$ from the standard cohomology formula for the index of families given in 4.126. This formula says that

$$ch\,(Index\,\partial\!\!\!/) = (-1)^n \int_M \{ch\,([\sigma(\partial\!\!\!/)]) \cdot td\,(T_F Z_{\mathbf{c}})\} \tag{10.29}$$

If we note that n is even, and specialise to the spin complex, we obtain

$$\begin{aligned} ch\,(Index\,\partial\!\!\!/) &= \int_M ch\,(\mathcal{E})\hat{A} \\ &= \int_M ch\,(\mathcal{E}), \qquad \text{when } M = S^n \end{aligned} \tag{10.30}$$

where $\mathcal{E}$ is the appropriate vector bundle over $Z = M \times \mathcal{A}/\mathcal{G}$. We do not want $ch\,(Index\,\partial\!\!\!/)$ but rather $c_1(det\,Index\,\partial\!\!\!/)$; however, if E is a vector bundle, it is straightforward to verify that $c_1(det\,E) = c_1(E)$. Thus

$c_1(det\,Index\,\partial\!\!\!/) = c_1(Index\,\partial\!\!\!/)$; and we can obtain our desired characteristic class by just selecting the dimension two cohomology class in the expansion of the RHS of 10.30.

Summing up our cohomology calculations we see that we have constructed two cohomology group elements, namely

$$\mu_1 = \frac{d\,det\,T_g}{det\,T_g} \in H^1_{de\,Rham}(\mathcal{G}) \qquad \text{and} \qquad c_1(det\,Index\,\partial\!\!\!/) \in H^2_{de\,Rham}(\mathcal{A}/\mathcal{G})$$

(10.31)

Actually, drawing on our work on $H^1_{de\,Rham}(\mathcal{G})$ and $H^2_{de\,Rham}(\mathcal{A}/\mathcal{G})$ in chapter 8 allows us to conclude that

$$\mu_1 = 0 \iff c_1(det\,Index\,\partial\!\!\!/) = 0 \qquad (10.32)$$

This same work also enables us to give formulae for μ_1 and $c_1(det\,Index\,\partial\!\!\!/)$. For example, if $M = S^4$ and $G = SU(N)$, then, referring to 8.58, we find that a formula for μ_1 is

$$\mu_1 = -\frac{i}{(2\pi)^3} \int_0^1 dt \int_{S^4} tr(A \wedge F_t \wedge F_t), \qquad \text{with} \qquad F_t = tdA + t^2 A \wedge A$$

(10.33)

where the notation is as used in chapter 8; this is the expression for the anomaly found in Bardeen [1].

The cohomology class $c_1(det\,Index\,\partial\!\!\!/)$ is the one we require for the anomaly. To see this suppose that there is no anomaly, then $det\,(\partial\!\!\!/_A^* \partial\!\!\!/_A)$ is gauge invariant and the partition function descends to an integral over the orbit space, that is

$$\int_{\mathcal{A}} \mathcal{D}A \sqrt{det\,(\partial\!\!\!/_A^* \partial\!\!\!/_A)} \exp[-\|F\|^2] = \int_{\mathcal{A}/\mathcal{G}} \mathcal{D}A_{\mathcal{G}}\,Jac \sqrt{det\,(\partial\!\!\!/)} \exp[-\|F\|^2]$$

(10.34)

where $\mathcal{D}A_{\mathcal{G}}$ is the measure on the orbits $\mathcal{A}/\mathcal{G}$, $det\,(\partial\!\!\!/)$ denotes $det\,(\partial\!\!\!/_A^* \partial\!\!\!/_A)$ projected onto $\mathcal{A}/\mathcal{G}$ and Jac stands for any Jacobian Fadeev–Popov factors. In this case $det\,(\partial\!\!\!/)$ is nothing but a global section of the determinant bundle $det\,Index\,\partial\!\!\!/$ over $\mathcal{A}/\mathcal{G}$. But a nowhere zero global section of a line bundle trivialises it and causes $c_1(det\,Index\,\partial\!\!\!/)$ to vanish. Thus the presence of the anomaly is detected by the non-triviality of the determinant line bundle $det\,Index\,\partial\!\!\!/$. We see that the index theorem for the Dirac family brings out very succinctly the anomaly as an obstacle to the definition of a gauge invariant determinant for $\partial\!\!\!/_A$.

It is convenient here to introduce the distinction between what is called *local cohomology* and conventional cohomology. According to 10.18, if $G = U(1)$ then $\pi_1(\mathcal{G})$ vanishes; hence $H^1_{de\ Rham}(\mathcal{G})$ also vanishes. But we know from the triangle graph that there is an anomaly in this theory. We know too that there is no local counterterm (local meaning a polynomial in fields and their derivatives) which can cancel the triangle graph and remove the anomaly. Thus the anomaly is, physically speaking, unremovable. Mathematically speaking, the vanishing of $H^1_{de\ Rham}(\mathcal{G})$ means that there is a counterterm, but it is non-local. This suggest that, using local in the specialised sense we just had above, one should define a *local cohomology theory* for $\mathcal{G}$ and $\mathcal{A}/\mathcal{G}$ as well as the conventional cohomology theory, and that one should distinguish cases where the two cohomologies differ. A significant start in constructing and studying such local cohomology structures has already been made in Bonora and Cotta-Ramusino [1].

§ 3. Gravitational anomalies

In the previous section we considered anomalies which arise when examining the properties of $det\,(\partial\!\!\!/_A^* \partial\!\!\!/_A)$ when A undergoes the change of gauge $A \mapsto A_g$ with $g \in \mathcal{G}$. We discovered a topological basis for these properties which rests on an element of $H^1_{de\ Rham}(\mathcal{G})$ or, equivalently, of $H^2_{de\ Rham}(\mathcal{A}/\mathcal{G})$. In this section we want to show that these properties have a gravitational counterpart (Alvarez-Gaumé and Witten [1]).

Because we wish to concentrate on purely metrical matters we temporarily omit the Yang–Mills connection A. Thus our interest is in the properties of the Dirac operator $(1/2)\gamma_\mu(\partial^\mu + \Gamma^\mu)(1 + \gamma_5)$ under a coordinate transformation $h \in Diff^+(M)$ rather than an element of $g \in \mathcal{G}$. Recall that the γ-matrices depend on a choice of metric; the same is therefore true of the Dirac operator. In order to make obvious our interest in this metric dependence we denote the present Dirac operator by $\partial\!\!\!/_\rho$ where ρ denotes a metric, that is $\rho \in Met(M)$, $Met(M)$ being the space of metrics on M. Now, as we know from chapters 5 and 9, any $h \in Diff^+(M)$ will act on this metric and thus on $\partial\!\!\!/_\rho$. If $h \cdot \rho$ denotes the new metric, then h produces the usual change in the Dirac operator, that is

$$\partial\!\!\!/_{h\cdot\rho} = h^{-1}\partial\!\!\!/_\rho h \qquad (10.35)$$

The next stage is to consider the Dirac determinant $det\,(\partial\!\!\!/_\rho^* \partial\!\!\!/_\rho)$ as a function of ρ and ask whether it is coordinate invariant so that it can descend to a function on the quotient $Met(M)/Diff^+(M)$.

Actually, just as in chapter 5, the quotient $Met(M)/Diff^+(M)$ has to be modified slightly as the existence of isometries or Killing vectors for some metrics in $Met(M)$ means that, in general, $Diff^+(M)$ does not act freely on $Met(M)$. What we do is to restrict to the identity component of

$Diff^+(M)$ and then restrict further to the subgroup which leaves fixed a basis at one point of M. We shall denote this subgroup by $Diff_0(M)$; the resulting quotient $Met(M)/Diff_0(M)$ is well behaved.

Having done all this, the analysis is analogous to that of the preceding section with $\mathcal{A}$ and $\mathcal{A}/\mathcal{G}$ replaced by $Met(M)$ and $Met(M)/Diff_0(M)$ respectively. Hence, apart from the proviso above about local cohomology, the anomaly corresponds to an appropriate element of $H^1_{de\ Rham}(Diff_0(M))$ or $H^2_{de\ Rham}(Met(M)/Diff_0(M))$. Applying the index theorem to the family $\partial\!\!\!/_\rho$ with parameter space $Y = Met(M)/Diff_0(M)$ instead of $\mathcal{A}/\mathcal{G}$, and denoting the index of the gravitational Dirac family by $Index\ \partial\!\!\!/_{grav}$, we have

$$Index\ \partial\!\!\!/_{grav} = \{ker\ \partial\!\!\!/_\rho : \rho \in Met(M)\} - \{ker\ \partial\!\!\!/_\rho^* : \rho \in Met(M)\}$$
$$= [ker\ \partial\!\!\!/_{grav}] - [ker\ \partial\!\!\!/_{grav}^*] \tag{10.36}$$

considered as an element of $K(Met(M)/Diff_0(M))$. The corresponding determinant bundle is

$$det\ Index\ \partial\!\!\!/_{grav} = (ker\ \partial\!\!\!/_{grav})^* \otimes (ker\ \partial\!\!\!/_{grav}^*) \tag{10.37}$$

and, following 10.29, its Chern character is given by

$$ch\ (det\ Index\ \partial\!\!\!/_{grav}) = \int_M \hat{A} \tag{10.38}$$

Now because the expansion of $\hat{A}$ only contains Pontrjagin classes, the only non-zero characteristic classes on the RHS of 10.38 occur in dimension $4k - n$, but the non-zero class on the RHS has dimension 2. Thus we only have an anomaly if $2 = 4k - n$ or

$$n = 4k - 2 \tag{10.39}$$

Let us adopt the alternative view where the anomaly corresponds to an element of the group $H^1_{de\ Rham}(Diff_0(M))$; setting $M = S^n$ and imposing the restriction on n above, means that we wish to compute an element of $H^1_{de\ Rham}(Diff_0(S^{4k-2}))$. It is known (Alvarez, Singer and Zumino [1]) that this de Rham group is zero for $k = 1$, and for $k > 6$, and is conjectured to be zero for all k. However, as we know, the vanishing of the de Rham cohomology is not the whole story as to be sure of no anomaly we would still have to calculate $H^1_{local}(Diff_0(S^{4k-2}))$, which is not expected to vanish.

There is also the possibility of enlarging the gravitational anomaly to include what may be called a Lorentz anomaly. To understand what this means we refer back to chapter 6 and our introduction of gauge transformations via the automorphisms $Aut(P)$ of a principal bundle P. In that discussion we obtained $Diff^+(M)$ by projecting the $Aut(P)$ onto the base M of P; the kernel of this projection was the group of gauge transformations $\tilde{\mathcal{G}}$, which we then restricted to its subgroup $\mathcal{G}$. Although P is usually used to discuss Yang–Mills theories, because we are dealing with gravity here we can take P to be an $SO(4k-2)$-bundle. The gauge transformations of this $SO(4k-2)$ gauge theory will be written as $\mathcal{G}_L$ with L standing for Lorentz. This extension of the invariance group of the problem amounts to replacing $Diff_0(M)$ by $Aut(P)$, which is the semi-direct product $\mathcal{G}_L \ltimes Diff_0(M)$. The full anomaly is now a Lorentz anomaly combined with the first gravitational anomaly and requires us to construct an element of $H^1(Aut(P))$. Interplay can now occur between the two kinds of anomaly so that one can be traded for the other and vice-versa (Alvarez-Gaumé and Ginsparg [1]). Finally one can now bring back the Yang–Mills field and vary the Yang–Mills connection, the metric, or both.

§ 4. The critical dimension for strings

For the Bosonic string the critical dimension may be characterised as the only dimension for which the conformal anomaly vanishes. Instead of having a family of Dirac operators one has a family of Cauchy–Riemann operators. Recall that a conformal anomaly arises in string theory if one cannot project from the space $Met(\Sigma)$ to the space of conformal equivalence classes $Met(\Sigma)/C^\infty_+(\Sigma)$.

Our final expression in chapter 9 for the string partition function is

$$Z = \sum_p Z_p = \sum_p \int_{\mathcal{M}_p} d\mu(\mathcal{M}_p) \left(\frac{det\,(\Delta_g/2)}{<1,1>_g} \right)^{-13} det\,(\bar\partial^*_{K_\Sigma} \bar\partial_{K_\Sigma}) \quad (10.40)$$

The integral over $\mathcal{M}_p$ contains the two determinant factors on which we will focus our interest. In constructing this expression we first projected[4] from $Met(\Sigma)$ onto $Met(\Sigma)/C^\infty_+(\Sigma)$ and then onto $\mathcal{M}_p$. For the first projection to be allowed it is necessary that there is no anomaly in the product of the two determinants considered as a 'function' on $Met(\Sigma)/C^\infty_+(\Sigma)$: in other words, the determinants are invariant under conformal changes of the metric. We can show that such an anomaly is absent using the families index theorem

---

[4] We know from chapter 5 that the quotient $Met(\Sigma)/C^\infty_+(\Sigma)$ is singular when Σ has genus zero or one. Since we have already explained there how to construct an improved quotient, we simplify the present discussion by assuming that Σ has genus greater than one.

(Alvarez [1] and Freed [1]). The parameter space Y for the family is the space $Met(\Sigma)/C_+^\infty(\Sigma)$. To identify the specific family of operators involved, we proceed as follows.

First we change the dimension of space-time from 26 to an arbitrary value d; this just changes the first determinant $det\,(\Delta_g/2)^{-13}$ to $det\,(\Delta_g/2)^{-(d/2)}$. The scalar Laplacian $\Delta_g/2$ is equal to $\bar\partial^*\bar\partial$ (cf. 5.32). Then the quantity $det\,(\bar\partial^*\bar\partial)$ can be thought of as a section of the determinant bundle $det\,Index\,\bar\partial$ over $Met(\Sigma)/C_+^\infty(\Sigma)$. To raise the section to the power $-(d/2)$ means just to take a section of the bundle obtained by tensoring the bundle on which $\bar\partial$ operates (the holomorphic tangent bundle K_Σ^*) with $-(d/2)I$ where I is the trivial bundle. The bundle $-(d/2)I$ has negative virtual dimension and so must be regarded as an element of $K(Met(\Sigma)/C_+^\infty(\Sigma))$. We write the $\bar\partial$ operator with coefficients in $-(d/2)I$ as $\bar\partial_{-(d/2)}$.

Now the second determinant contains the $\bar\partial$ operator coupled to K_Σ^*; hence if we replace the bundle $-(d/2)I$ by the formal difference $K_\Sigma^* - (d/2)I$ then it is natural to contemplate the operator $\bar\partial_{(K_\Sigma^*-(d/2)I)}$. The index bundle for this latter family of $\bar\partial$ operators is the one we need. Denoting this bundle by $Index\,\bar\partial_Y$, where $Y = Met(\Sigma)/C_+^\infty(\Sigma)$, we pass straightaway to its determinant bundle $det\,Index\,\bar\partial_Y$ and compute its Chern character. Applying the formulae of the previous section to the $\bar\partial$ complex instead of the spin complex we have

$$det\,Index\,\bar\partial_Y = \int_\Sigma ch\,(K_\Sigma^* - (d/2)I)\,td\,(\Sigma) \qquad (10.41)$$

But because K_Σ^* is a line bundle, if $x = c_1(K_\Sigma^*) \equiv tr\,(iF/2\pi)$, the splitting principle expansions for $td\,(\Sigma)$ and $ch\,(K_\Sigma^*)$ give us

$$td\,(\Sigma) \equiv td\,(K_\Sigma^*) = 1 + \frac{x}{2} + \frac{x^2}{12} + \cdots$$

$$ch\,(K_\Sigma^* - (d/2)I) = ch\,(K_\Sigma^*) - ch\,((d/2)I) = (1 + x + \frac{x}{2} + \cdots) - \frac{d}{2}$$
$$(10.42)$$

The Chern character of $det\,Index\,\bar\partial_Y$ is given by the dimension two element obtained from the expansion of 10.41. Hence we find that

$$ch\,(det\,Index\,\bar\partial_Y) = \int_\Sigma (1 - \frac{d}{2} + x + \frac{x}{2} + \cdots)(1 + \frac{x}{2} + \frac{x^2}{12} + \cdots)$$

$$\Rightarrow c_1(det\,Index\,\bar\partial_Y) = \int_\Sigma \frac{26 - d}{24}x^2$$

$$(10.43)$$

We notice at once that the determinant bundle vanishes precisely when $d = 26$ so that we have indeed identified the critical string dimension as being the one for which the conformal anomaly vanishes.

Another important property of the determinant bundle associated to the family a $\bar{\partial}$ operators is that Quillen [1,2] has used zeta functions so as to construct a smooth metric and connection on the determinant bundle. Belavin and Knizhnik [1] have further proved that, in the critical dimension, the partition function Z_p is a holomorphic square on $\mathcal{T}_p$, i.e. $Z_p = f\bar{f}$ with f holomorphic; also Manin [1] proves that f can be written in terms of elliptic Θ-functions.

§ 5. Global anomalies

Global anomalies are a further headache in the quest for well behaved quantum theories. Global anomalies enter when $\mathcal{G}$ or $Diff^+(M)$ have more than one connected component. This means that they contain elements not continuously connected to the identity. Such discreteness is not detectable using methods of curvature and de Rham cohomology—these methods are only sensitive to objects in the tangent space. The situation is closely analogous to the calculation of torsion in homology and cohomology. Unfortunately, torsion calculations are typically more difficult than free cohomology calculations for which the de Rham method may well apply. We shall deal first with examples where $\mathcal{G}$ is disconnected and then go on to examples where $Diff^+(M)$ is disconnected. In each case we will be able to interpret any anomaly as an index of an appropriate elliptic family of Dirac operators.

We start with an example described by Witten [3]. Let us take a Yang–Mills theory with action S given by 10.1 and group G. We can count the connected components with $\pi_0(\mathcal{G})$. In this way we have, with $M = S^{2m}$,

$$\begin{aligned} \pi_0(\mathcal{G}) &= \pi_0(\Omega^{2m}G) \\ &= \pi_{2m}(G) \end{aligned} \tag{10.44}$$

Hence if we take $G = SU(2)$, $2m = 4$ we find

$$\begin{aligned} \pi_0(\mathcal{G}) &= \pi_4(SU(2)) \\ &= \mathbf{Z}_2 \end{aligned} \tag{10.45}$$

i.e. $\pi_0(\mathcal{G}) \neq 0$, and this was the example taken in Witten [3]. This means that there are global gauge transformations under which $det\,\partial\!\!\!/_A$ is not invariant—in fact they change the sign of $\sqrt{det\,(\partial\!\!\!/_A^* \partial\!\!\!/_A)}$. Actually, if there are p left-handed Fermion doublets then $\sqrt{det\,(\partial\!\!\!/_A^* \partial\!\!\!/_A)}$ enters raised to the power p. Thus if p is even, the theory is free of global anomalies.

At the moment we still only have one Dirac operator $\not{D}_A$ rather than a family of operators. We cannot use the same family as we used in § 2 because the index of that family only detects the local anomaly in $H^1_{de\,Rham}(\mathcal{G})$, which we know is zero for $SU(2)$. A clue to the correct family is present in chapter 4, § 5 where we discussed families of real elliptic operators. The point is that, in four dimensions, the full Dirac operator $\not{D}_A = \gamma_\mu(\partial^\mu + A^\mu)$ when coupled to $SU(2)$ doublet can be taken to be a *real* skew-adjoint operator. Drawing on chapter 4, § 5 we know that, though the usual index of a skew-adjoint operator D is necessarily zero, nevertheless dim *ker D* mod 2 is an invariant. Also this quantity dim *ker D* mod 2 is given as the index of a *real family* of operators $\widetilde{D}$ parametrised by S^1. Witten [3] applies this analysis to $\not{D}_A$ and regards the S^1 *family* of four dimensional Dirac operators as a *single* five dimensional Dirac operator. The mod 2 index is then calculated and found to be non-zero. This non-zero value of the index is shown to be due to the existence of elements of $\mathcal{G}$, not continuously connected to the identity, which change the sign of $\sqrt{det\,(\not{D}_A^*\not{D}_A)}$. Two key techniques employed in the calculation are those of spectral flow and adiabatic approximation. We shall introduce adiabatic approximation later in this section; spectral flow will be introduced in the next section when we examine anomalies from a Hamiltonian standpoint.

Let us see how to obtain this S^1 family. Consider the usual Dirac family parametrised by $Y = \mathcal{A}/\mathcal{G}$. Select a map

$$\alpha : S^1 \to Y$$

This map pulls back the Dirac family $\not{D}_Y$ on Y to another family $\not{D}_{S^1}$ on S^1; it also pulls back any bundles over Y to corresponding bundles over S^1. To make it easier to distinguish one family from another we shall now denote the determinant bundle *det Index* $\not{D}$ by *det Index* $\not{D}_Y$. This pulls back to the determinant bundle for the S^1 family, which we write as *det Index* $\not{D}_{S^1}$: we have

$$\alpha^* det\,Index\,\not{D}_Y = det\,Index\,\not{D}_{S^1} = (ker\,\not{D}_{S^1})^* \otimes (ker\,\not{D}_{S^1}^*) \qquad (10.46)$$

However, pullbacks are classified by their homotopy class in $[S^1, Y]$, that is by an element of $\pi_1(Y)$. We know, too, that

$$\pi_1(Y) = \pi_1(\mathcal{A}/\mathcal{G}) = \pi_0(\mathcal{G}) = \mathbf{Z}_2, \quad \text{from } 10.45$$
$$\Rightarrow H_1(\mathcal{A}/\mathcal{G}) = \mathbf{Z}_2 \qquad (10.47)$$

But the universal coefficient theorem (Bott and Tu [1]) says that, when torsion is present, the (finitely generated) homology and cohomology of a general space M are related by

$$H^i(M; \mathbf{Z}) = F_i(M) \oplus T_{i-1}(M) \qquad (10.48)$$

where the $F_i(M)$ and $T_i(M)$ denote the free part and the torsion part respectively of the homology of $H_i(M; \mathbf{Z})$. Applying this to $\mathcal{A}/\mathcal{G}$ we see that

$$F_1(\mathcal{A}/\mathcal{G}) = 0, \; F_2(\mathcal{A}/\mathcal{G}) = 0 \; (\text{since } H^2_{de\,Rham}(\mathcal{A}/\mathcal{G}) = 0), \; T_1(\mathcal{A}/\mathcal{G}) = \mathbf{Z}_2$$
$$\Rightarrow T^2(\mathcal{A}/\mathcal{G}) = \mathbf{Z}_2$$
$$\Rightarrow H^2(\mathcal{A}/\mathcal{G}) = \mathbf{Z}_2$$

$$(10.49)$$

Thus $H^2(\mathcal{A}/\mathcal{G})$ is pure torsion and the same is therefore true for the Chern class $c_1(L)$ of any line bundle L over $\mathcal{A}/\mathcal{G}$. Returning to the line bundles $det\,Index\,\partial\!\!\!/_{\mathcal{A}/\mathcal{G}}$ and $det\,Index\,\partial\!\!\!/_{S^1}$ we see that the non-triviality of the latter is measured by an element of $\mathbf{Z}_2$.

But this time $det\,Index\,\partial\!\!\!/_{S^1}$ is a real line bundle instead of a complex one. Hence its group is $O(1)$, which is just the group $\mathbf{Z}_2$. The only invariant of a real line bundle over S^1 is $\pi_0(O(1)) = \mathbf{Z}_2$; thus when the bundle $det\,Index\,\partial\!\!\!/_{S^1}$ is non-trivial it has precisely the required properties. Further, suppose that an anomaly is present but, nevertheless, we still attempt to construct a global trivialising section '$det\,\partial\!\!\!/$' of $det\,Index\,\partial\!\!\!/_{S^1}$. We shall of course fail. However, the unsuccessful section '$det\,\partial\!\!\!/$' will be characterised by the property that if we go round S^1 exactly once then the section will not be periodic but will be multiplied by the non-trivial element of the structure group, i.e. by minus one. Also when this happens it can be traced back to the non-triviality of the corresponding line bundle over $\mathcal{A}/\mathcal{G}$, which is in turn due to the determinant $det\,(\partial\!\!\!/_A^* \partial\!\!\!/_A)$ changing sign under a (necessarily global) gauge transformation. This is just what we argued above. In addition, if we consider p doublets of Fermions then it is easy to verify that the effect is to raise the bundle $det\,Index\,\partial\!\!\!/_{S^1}$ to the power p, rendering it trivial precisely when p is even.

We now move on to gravity theory; again there is an important result due to Witten [4]. We have already seen that global gravitational anomalies arise when $\pi_0(Diff^+(M)) \neq 0$. We shall examine the case where $M = S^n$. The non-triviality of $\pi_0(Diff^+(M))$ when $M = S^n$ is extremely closely tied to the existence of exotic spheres—recall from chapter 1 that exotic spheres are spheres whose differentiable structure is distinct from the standard one.

The connection between $\pi_0(Diff^+(S^n))$ and exotic spheres is the following. An exotic sphere S^n can be constructed by an appropriate joining of two hemispheres: what one does is to cut S^n into two hemispheres H^+ and H^- both of which have boundary S^{n-1}; then act on the boundary of one of them with an element $f \in Diff^+(S^{n-1})$ and rejoin the transformed hemisphere to its untransformed partner. The resulting n-sphere S_f^n, say, is known to be exotic when f does *not* belong to the identity component of $Diff^+(S^{n-1})$. Further, two such spheres S_f^n and S_g^n are diffeomorphic precisely when f and g belong to the same component of $Diff^+(S^{n-1})$ (Cerf [1]). In short, the components of $Diff^+(S^{n-1})$ correspond exactly to the exotic n-spheres.

In chapter 1 we found that exotic spheres exist for the case $n = 7$; thus an example for which $Diff^+(S^n)$ becomes disconnected is $n = 6$. However, Green and Schwarz [1] have discovered a remarkable cancellation of local anomalies for $N = 1$ supersymmetric string theories provided the gauge group G is one or other of the two 496-dimensional groups $E_8 \times E_8$ or $O(32)$, and the dimension of space-time is 10. This seminal result makes it important to check for the presence or absence of global anomalies in these theories. This was done in Witten [4] and global anomalies were also found to be absent.

We would like to describe the method used to verify the absence of global gauge and gravitational anomalies in 10 dimensions. The vanishing of any global *gauge* anomalies is immediate because, if the gauge group G is $E_8 \times E_8$ or $O(32)$, we have $\pi_0(\mathcal{G}) = \pi_{10}(G) = 0$. We are thus left with the gravitational case.

It is clear that there is something to check because in chapter 1 we found that there are 991 exotic spheres in 11 dimensions. If we state this in terms of $Diff^+(S^n)$ it is just the assertion that

$$\pi_0(Diff^+(S^{10})) = \mathbf{Z}_{992} \tag{10.50}$$

Moreover, we can use this assertion to deduce that the parameter space $Y = Met(S^{10})/Diff^+(S^{10})$ satisfies

$$H^2(Y; \mathbf{Z}) = \mathbf{Z}_{992} \tag{10.51}$$

The above result follows at once from the fibre homotopy exact sequence

$$\cdots \to \pi_i(Diff^+(S^{10})) \to \pi_i(Met(S^{10})) \to \pi_i(Met(S^{10})/Diff^+(S^{10}))$$
$$\to \pi_{i-1}(Diff^+(S^{10})) \to \cdots \tag{10.52}$$

One uses the fact that $Met(S^{10})$, being a vector space, is contractible; this implies that $\pi_i(Y) = \pi_{i-1}(Diff^+(S^{10}))$ and, setting $i = 1$, we obtain the desired result by an exactly similar calculation to 10.49 above.

Now we can proceed to the calculation of the global anomaly. Consider the determinant bundle $det\,Index\,\partial\!\!\!/_{grav}$ for the gravitational family. This now has a Chern class which is pure torsion, that is

$$c_1(det\,Index\,\partial\!\!\!/_{grav}) \in H^2(Met(S^{10})/Diff^+(S^{10}); \mathbf{Z}) = \mathbf{Z}_{992} \tag{10.53}$$

In [4] Witten studied the variation of the Dirac determinant $det\,(\slashed{\partial}_\rho^* \slashed{\partial}_\rho)$ as the metric ρ changes to $f \cdot \rho$ where f is a global diffeomorphism belonging to some component of $Diff^+(S^{10})$. The change of ρ to $f \cdot \rho$ is accomplished by using a one parameter family of metrics defined by

$$\rho^t = (1 - t)\rho + tf \cdot \rho \qquad (10.54)$$

Clearly as t varies from 0 to 1 we have the desired state of affairs.

The next step is to realise that we can produce a one parameter family of metrics ρ^t on S^{10} by taking a *single metric* $\bar{\rho}$ on some 11-dimensional manifold N. Let us do this and also require the line element on N to have the simple form

$$ds^2 = dt^2 + \rho^t dx^2 = (1 - t)\rho + tf \cdot \rho \qquad (10.55)$$

where we are using (t, x) as local coordinates on N. We need to commit ourselves to a specific manifold N. A desirable feature for N to possess is that it has encoded into it some data about the diffeomorphism f; in fact we can construct N entirely from f alone. To do this, first construct the product $S^{10} \times [0, 1]$. This manifold has a boundary consisting of the two pieces $S^{10} \times \{0\}$ and $S^{10} \times \{1\}$; we shall get rid of this boundary by joining the two together but, before doing so, we act on $S^{10} \times \{1\}$ with f so that it becomes $S_f^{10} \times \{1\}$. Having done this the manifold N is the closed 11-dimensional manifold obtained by identifying $S^{10} \times \{0\}$ and $S_f^{10} \times \{1\}$. Notice that N is actually a fibre bundle over S^1 with fibre S^{10} and structure group $Diff^+(S^{10})$. This means that it is classified by an element of $\pi_0(Diff^+(S^{10}))$ and so there are precisely 992 such bundles or manifolds N. Clearly we have elegantly achieved our objective of incorporating the global diffeomorphisms of $Diff^+(S^{10})$ into the fabric of N.

A final point about the construction of N is that we can also obtain N by use of a pullback. Recall that in the setup for the families index theorem we have the bundle Z over Y with fibre M (note that we called the fibre X rather than M in the original treatment of chapter 4); also, to maintain invariance of any local calculations, we required Z to have structure group $Diff(M)$ or $Diff^+(M)$. The significance of Z in the present context is that if we have a map $\alpha : S^1 \to Y$ then $\alpha^* Z$ is a bundle over S^1 with fibre M and structure group $Diff^+(M)$; in other words, setting $M = Diff^+(S^{10})$, $\alpha^* Z$ is the same as N.

Let us move in on the global anomaly itself. The presence of an anomaly is detected by a change in $det\,(\slashed{\partial}_\rho^* \slashed{\partial}_\rho)$ as ρ varies from ρ^0 to ρ^1. However, it is far from obvious how to calculate such a change. The successful method introduced by Witten [4] is to replace the data on S^{10} by the corresponding data on N: this means replacing the family ρ^t by the single metric $\bar{\rho}$ on the manifold N, replacing the Dirac operator $\partial\!\!\!/_\rho$ on S^{10} by the Dirac operator[5] $D\!\!\!\!/$ on N, and calculating the change in $det\,(\partial\!\!\!/_\rho^*\partial\!\!\!/_\rho)$ by using an adiabatic approximation. In this context the term adiabatic approximation means scaling the family ρ^t to $\epsilon\rho^t$, where ϵ is small, and then calculating the spectrum of $D\!\!\!\!/$ by expansion in ϵ.

When this programme is carried through one finds the beautiful result that the change in $det\,(\partial\!\!\!/_\rho^*\partial\!\!\!/_\rho)$ can be expressed solely in terms of the η-function of the 11-dimensional Dirac operator $D\!\!\!\!/$. What Witten shows is that, for a diffeomorphism f under which $\rho \mapsto f \cdot \rho$, we have

$$det\,(\partial\!\!\!/_\rho^*\partial\!\!\!/_\rho) \longmapsto det\,(\partial\!\!\!/_{f\cdot\rho}^*\partial\!\!\!/_{f\cdot\rho}) = C\,det\,(\partial\!\!\!/_\rho^*\partial\!\!\!/_\rho)$$
$$\text{where} \quad C = \exp[\pi i\eta_{D\!\!\!/}(0)] \tag{10.56}$$

Now the functional integral contains $det\,(\partial\!\!\!/_\rho^*\partial\!\!\!/_\rho)$ in the square root form $\sqrt{det\,(\partial\!\!\!/_\rho^*\partial\!\!\!/_\rho)}$. Thus the anomaly is absent if

$$\sqrt{C} = \exp\left[\frac{\pi i}{2}\eta_{D\!\!\!/}(0)\right] = 1 \tag{10.57}$$
$$\Rightarrow \eta_{D\!\!\!/}(0) = 4k$$

with k an integer. The effective action S_ψ of the Fermions is $\ln\sqrt{det\,(\partial\!\!\!/_\rho^*\partial\!\!\!/_\rho)}$ and so we can also state the above condition as

$$\Delta S_\psi = 2\pi i k \tag{10.58}$$

where ΔS_ψ, stands for the change in S_ψ under the diffeomorphism f.

However, we know that η-functions appear in the index formula for some elliptic boundary value problems. Referring to the relevant theorem 4.164 we have that, if X is a *twelve* dimensional (spin) manifold whose boundary is N, then

$$index\,\partial\!\!\!/_X = \int_X \hat{A} - \frac{(h + \eta_{D\!\!\!/}(0))}{2} \tag{10.59}$$

---

[5] Remember N is an odd dimensional manifold and so, because chirality does not exist, the chiral Dirac operator is replaced by the full, self-adjoint, Dirac operator $D\!\!\!\!/_{\bar{\rho}}$ which for simplicity we denote by $D\!\!\!\!/$; moreover, $D\!\!\!\!/$ has a spectrum.

where $\partial\!\!\!/_X$ is the chiral Dirac operator on X with respect to a metric on X which is a product with $\bar\rho$ near its boundary N. This immediately provides us with a formula for $\eta_{\partial\!\!\!/}(0)$—this is somewhat of a vital necessity since calculating $\eta_{\partial\!\!\!/}(0)$ directly from the spectrum of $\partial\!\!\!/$ is not really practical. Using this formula we find that the change ΔS_ψ in the effective action is given by

$$\Delta S_\psi = \pi i \{ \int_X \hat A - index\, \partial\!\!\!/_X - \frac{h}{2} \} \tag{10.60}$$

Actually in 12 dimensions it is known that the index of the Dirac operator is always even, thus the term $index\, \partial\!\!\!/_X$ contributes a term of the form $2\pi i k$ to ΔS_ψ and so may be omitted. However, the $N = 1$ supersymmetric string theory, which is anomaly free, contains more fields that we have not yet mentioned. It is necessary to include these fields both to guarantee the absence of local anomalies (the result of Green and Schwarz) and to cancel the global anomaly. There are three contributions to the global anomaly and they come from Dirac spin $(1/2)$ fields, Rarita–Schwinger spin $(3/2)$ fields and a self-dual tensor field. In general, if the number of each type of field is n_d, n_{rs} and n_{sd} respectively then the total contribution (Witten [4]) to the global anomaly is

$$\frac{\pi i}{2} \{ n_d \eta_{\partial\!\!\!/}(0) + n_{rs}(\eta_{rs}(0) - \eta_{\partial\!\!\!/}(0)) - \frac{n_{sd}}{2}\eta_{sd}(0) \} \tag{10.61}$$

where $\eta_{rs}(0)$ and $\eta_{sd}(0)$ are the η-functions for the Rarita–Schwinger and self-dual tensor fields respectively.

For the $E_8 \times E_8$ or $O(32)$ theories we do not have to include the self-dual tensor fields and we have $n_d = 495$ and $n_{rs} = 1$. With these values the formula 10.61 gives the change in the Fermionic determinants under a global diffeomorphism. A last task in the calculation is to add the corresponding change in the other terms in the action. When this is done the complete contribution ΔS to the change in the action does indeed satisfy $\Delta S = 2\pi i k$ and so the theory is free of global anomalies.

There is a geometrical property underlying the formula 10.56 for the Fermionic determinants. It is that this formula can be interpreted (Bismut and Freed [1,2]) as giving the holonomy round a loop for a connection A on the determinant line bundle over Y. If we return to the map $\alpha : S^1 \to Y$ introduced above then the holonomy of A round S^1 is given by

$$\exp[\int_{S^1} A] \tag{10.62}$$

To calculate this holonomy Bismut and Freed [2] use the fact that $\alpha^* Z$ is N and introduce a limiting procedure similar to the adiabatic limit described above: If ds^2 is the line element on N then one writes

$$ds^2 = \frac{dt^2}{\epsilon} + \rho^t dx^2 \tag{10.63}$$

This line element only differs from that used in the adiabatic limit by the conformal factor ϵ. In this case the metric on S^1 has been scaled by $(1/\epsilon)$ and so becomes singular as $\epsilon \to 0$; for this reason this mathematical counterpart of the adiabatic limit is called blowing up the metric. The result of this limiting procedure is the formula

$$\exp[\int_{S^1} A] = (-1)^{index\,\not{\partial}_\rho} \lim_{\epsilon \to 0} \exp[\pi i(h_\epsilon + \eta_{\not{D}^\epsilon}(0)] \tag{10.64}$$

where $index\,\not{\partial}_\rho$ is the index of the ordinary chiral Dirac operator on S^{10}, and the ϵ on h and $\not{D}$ are reminders that these quantities depend on the metric on N and thus on ϵ. Thus the holonomy is determined by the global change in the Fermionic determinant, that is by the global anomaly.

The methods used in this section on S^{10} are not restricted to this particular manifold. The method works equally well for many other manifolds M of arbitrary (even) dimension n and first order elliptic operators L, say; in each case the main task is to construct the analogues of the manifolds N and X. Then one uses an adiabatic limit, or blow up of the metric on S^1, to show that the change in the appropriate determinants under a global diffeomorphism in $Diff^+(M)$ is given by $(\pi i/2)\eta_{L_N}(0)$ where L_N is the operator on N induced by L.

Example *Complex anomalies*

We have seen that in considering anomalies of the global type it is necessary to distinguish manifolds (exotic spheres) which are homeomorphic but not diffeomorphic. A separate point is that in addition to considering distinct differentiable structures on manifolds one can go one step further and consider distinct complex structures as we do in string theory. Invariants in terms of characteristic classes which would distinguish such structures seem difficult to find. The first examples seem to occur in dimension 10. In 10 dimensions there is a complex manifold

$$M = \frac{U(4)}{U(2) \times U(1) \times U(1)} = \frac{G}{H} \tag{10.65}$$

Following the technique we introduced when discussing the complex manifold G/T in chapter 6 we can use the roots of G and H, and a decomposition into positive and negative parts, to assign a complex structure to M. But the decomposition into positive and negative parts is done relative to some Weyl chamber or lexicographic ordering, and is not unique. If we exploit this fact we can endow M with two distinct complex structures and thus define two complex manifolds M_1 and M_2 both homeomorphic to M. Further, the Chern classes distinguish M_1 from M_2 for one can show that, Nash [3],

$$c_1^5(M_1) = 4500$$
$$c_1^5(M_2) = 4860 \tag{10.66}$$

Thus, although M_1 is homeomorphic to M_2, their Chern numbers differ.

To see how this can affect physics is quite straightforward; we introduce a simple model with action S given by

$$S = \frac{i}{(2\pi)^5} \int_M (tr F)^5$$

Actually M is in fact a Kähler manifold and F is its curvature form, and because of this it possesses a metric g with the property that g is invariant under the complex structure selected. The anomaly comes from the following: The action S is invariant under differentiable coordinate transformations. Now if g and $\tilde{g}$ are the two Kähler metrics associated to the two complex structures, then these two metrics are differentiably but not complex analytically equivalent. Thus if we consider a family of metrics g^t parametrised by t, with $g^0 = g$ and $g^1 = \tilde{g}$, then the action is not in fact invariant under this coordinate transformation but changes by an amount ΔS where

$$\Delta S = 360$$

This example is only chosen to illustrate the potential destructiveness of a complex anomaly and the action S need not be taken seriously as being that of a realistic physical model.

§ 6. Anomalies from a Hamiltonian perspective

We end this chapter by giving some idea of how one sees anomalies in a Hamiltonian formulation of the quantum field theory. Anomalies should not disappear in an alternative formulation, though they may manifest themselves in slightly different ways.

The central idea is still to use families of Dirac operators. Suppose that we have a family of Dirac operators on an n-dimensional spin manifold. We shall further suppose that this family depends on p parameters and a typical member of this family is denoted by

$$\mathcal{D}^{(n)}_{(t_1,\ldots,t_p)} \tag{10.67}$$

where $(t_1,\ldots,t_p) = t$ denotes dependence on the parameters. Using this family we can always construct a *single* Dirac operator $\mathcal{D}^{(n+p)}$ in $(n+p)$ dimensions by writing

$$\mathcal{D}^{(n+p)} = \mathcal{D}^{(n)}_{(t_1,\ldots,t_p)} + \sum_{i=1}^{p} \gamma_i \frac{\partial}{\partial t_i} \tag{10.68}$$

The $\{\gamma_1,\ldots,\gamma_p\}$ are just the extra γ-matrices needed in $(n+p)$ dimensions. Alternatively, if we are interested in a Hamiltonian approach, we can construct a $(p+1)$-parameter family of Dirac operators $\mathcal{D}^{(n-1)}_t$ in $(n-1)$ dimensions, the extra parameter being the time. The main point is that, though the topology of the anomaly displays itself differently as n and p vary, the anomaly really only depends on $(n+p)$ rather than on n and p separately. As we shall now show, the way to prove this last assertion is to use Fredholm operators.

In chapter 2, § 5 we showed that elliptic operators can be realised as Fredholm operators on a suitable Hilbert space. Applying this result here allows us to regard the various Dirac operators as Fredholm operators. We begin by supposing that H is a (complex) Hilbert space with an associated space of Fredholm operators $\mathcal{F}(H)$. It is now opportune to expand on an allusion that we made earlier (chapter 3, § 6) to a relation between $\mathcal{F}(H)$ and $K(X)$. The result that we need is the following isomorphism

$$[X, \mathcal{F}(H)] \simeq K(X) \tag{10.69}$$

We can easily give an explicit description of this isomorphism: Let us consider the continuous map

$$\alpha : X \to \mathcal{F}(H) \tag{10.70}$$

Notice that α provides us at once with a *family* of Fredholm operators parametrised by X; we shall denote this family by α_X. Now, for each $x \in X$, the image $\alpha(x)$ is a Fredholm operator with kernel and cokernel $ker\,\alpha(x)$ and $ker\,\alpha^*(x)$ respectively. As x varies, these (finite dimensional) spaces can be considered to be the fibres of vector bundles over the parameter space X. More precisely, we define the vector bundle $ker\,\alpha$ over X as being that bundle whose fibre over x is the space $ker\,\alpha(x)$, with a similar definition for the bundle $ker\,\alpha^*$. The difference of these vector bundles is the K-theory element

$$[ker\,\alpha] - [ker\,\alpha^*] \tag{10.71}$$

In the terminology introduced in § 2 this element is clearly the index bundle $Index\,\alpha_X$ for the family of Fredholm operators α_X. Now we have all that we need for the isomorphism 10.69—which it is natural to denote by $Index$ — proceeding to the definition we have

$$
\begin{aligned}
Index\,&: [X, \mathcal{F}(H)] \longrightarrow K(X) \\
&[\alpha] \longmapsto [ker\,\alpha] - [ker\,\alpha^*] \equiv Index\,\alpha_X
\end{aligned}
\tag{10.72}
$$

This isomorphism can be interpreted as saying that $\mathcal{F}(H)$ is a *classifying space* [6] for the functor K. Because of this the homotopy type of $\mathcal{F}(H)$ is determined and, referring to chapter 3, it is easy to verify the existence of the homotopy equivalence

$$\mathcal{F}(H) \simeq \Omega U(\infty) \tag{10.73}$$

So far the Dirac–Fredholm operators we have discussed are fairly unrestricted. However, because the Dirac operator in *odd dimensions* is always self-adjoint, we need a little of the theory of *self-adjoint* Fredholm operators. Let $\mathcal{F}^1(H)$ denote the set of self-adjoint Fredholm operators, then we would like to know the homotopy type of $\mathcal{F}^1(H)$. It turns out that $\mathcal{F}^1(H)$ contains two contractible components usually denoted by $\mathcal{F}^1_+(H)$ and $\mathcal{F}^1_-(H)$. The space $\mathcal{F}^1_+(H)$ is the space of self-adjoint Fredholm operators whose (necessarily real) spectrum has only a finite number of negative eigenvalues, while $\mathcal{F}^1_-(H)$ is the corresponding subspace of $\mathcal{F}^1(H)$ whose spectrum has only a finite number of positive eigenvalues. The remainder of $\mathcal{F}^1(H)$, that is the complement of $\mathcal{F}^1_+(H)$ and $\mathcal{F}^1_-(H)$, is *not* contractible and is denoted by $\mathcal{F}^1_*(H)$. Hence the homotopy type of $\mathcal{F}^1(H)$ is carried by $\mathcal{F}^1_*(H)$. Fortunately the homotopy type of $\mathcal{F}^1_*(H)$ is known (Atiyah and Singer [6]) to be that of $\Omega \mathcal{F}(H)$. Summarising, we have the homotopy equivalences

$$\mathcal{F}^1(H) \simeq \mathcal{F}^1_*(H) \simeq \Omega \mathcal{F}(H) \tag{10.74}$$

Also, since we know that $\mathcal{F}(H) \simeq \Omega U(\infty)$ and that Bott periodicity says that $\Omega^2 U(\infty) \simeq U(\infty)$, then we see that

$$\mathcal{F}^1(H) \simeq U(\infty) \tag{10.75}$$

[6] We have already met classifying spaces for bundles: for example, we know that the isomorphism $[X, Gr(k, \infty, \mathbf{C})] \simeq Vect_k(X, \mathbf{C})$ of 3.67 means that $Gr(k, \infty, \mathbf{C})$ is a classifying space for the functor $Vect_k$.

We now have in place all the required Fredholm theory which we observe is all determined by the space $U(\infty)$. Using the information on the periodicity of its homotopy groups given in chapter 3 we deduce that

$$\pi_k(\mathcal{F}(H)) = \begin{cases} \mathbf{Z} & \text{if } k \text{ is even} \\ 0 & \text{if } k \text{ is odd} \end{cases} \quad \text{and} \quad \pi_k(\mathcal{F}^1(H)) = \begin{cases} 0 & \text{if } k \text{ is even} \\ \mathbf{Z} & \text{if } k \text{ is odd} \end{cases}$$
$$(10.76)$$

Now we return to anomalies. For simplicity we restrict ourselves to the case where the parameter space X is a p-sphere S^p. If $\partial^{(n)}_{(t_1,\ldots,t_p)}$ is a typical family of Dirac–Fredholm operators then

$$\begin{aligned} \partial^{(n)}_{(t_1,\ldots,t_p)} &\in \mathcal{F}(H) \qquad \text{when } n \text{ is even} \\ \partial^{(n)}_{(t_1,\ldots,t_p)} &\in \mathcal{F}^1(H) \qquad \text{when } n \text{ is odd} \end{aligned} \qquad (10.77)$$

The topological properties of the anomaly reside in the index bundle of this family. But using 10.72 with $X = S^p$ we see that this index bundle is just an appropriate element of the homotopy groups

$$\begin{aligned} \pi_p(\mathcal{F}(H)) & \quad \text{if } n \text{ is even} \\ \pi_p(\mathcal{F}^1(H)) & \quad \text{if } n \text{ is odd} \end{aligned} \qquad (10.78)$$

Now suppose that $(n + p)$ is even, then n and p must be either both odd or both even. In either case we can combine the homotopy results 10.76 and 10.78 to conclude that the relevant homotopy group is $\mathbf{Z}$. On the other hand, if $(n + p)$ is odd then n and p form an odd-even or an even-odd pair. Again we find that the homotopy group for the anomaly is always zero. In fact, whatever the values of n and p the anomaly is measured by the same homotopy group, namely

$$\pi_{n+p}(\Omega U(\infty)) \qquad (10.79)$$

So the only thing that matters is whether $(n + p)$ is even or odd.

All this applied to the case where $\mathcal{F}(H)$ is the space of Fredholm operators for a complex Hilbert space. However, we know that global anomalies can involve real Dirac operators and thus real Fredholm operators. The thing to do here is to investigate the homotopy type of $\mathcal{F}(H_r)$ where H_r denotes a real Hilbert space. The replacement of H by H_r causes the infinite unitary group $U(\infty)$ to be replaced by the infinite orthogonal group $O(\infty)$. We can use the real version of the index map 10.72 to show that $\mathcal{F}(H_r)$ is a classifying space for the functor KO. The global anomaly for $SU(2)$ that we studied required us to use a single skew-adjoint real Dirac operator. But the space of real skew-adjoint Fredholm operators $\mathcal{F}^1_{skew}(H_r)$ has the homotopy type of $O(\infty)$ (Atiyah and Singer [6]). The topology of this anomaly is captured by $\pi_0(\mathcal{F}^1_{skew}(H_r)) = \pi_0(O(\infty))$; however $\pi_0(O(\infty)) = \mathbf{Z}_2$ and so we are in agreement with our previous calculation.

We can now look at some examples.

Example *Spectral flow and an anomaly in two dimensions*

Let us take an Abelian $U(1)$ gauge theory on a space-time which is the two dimensional torus $S^1 \times S^1$. First we consider it from a non-Hamiltonian standpoint. Since the space-time is even dimensional we have the chiral Dirac operator $\partial\!\!\!/_A$. But the local anomaly calculated in § 2 is of course zero; what is left is the fact that there are non-trivial $U(1)$-bundles over the torus corresponding to the connection A being an 'Abelian' instanton. Our previous work tells us that these bundles are classified by a single integer, and this integer is equal to the ordinary integer index of the single Dirac operator $\partial\!\!\!/_A$. In terms of the Fredholm treatment we are dealing with the case where the parameter space X is a point. This means that the index isomorphism of 10.72 reduces to the ordinary assignment of a Fredholm operator to its index.

Now we switch to the Hamiltonian view. This time we take space to be the circle S^1 on which is defined a self-adjoint Dirac operator. On S^1 this Dirac operator becomes simply $(-id/d\theta)$. The other S^1 of the torus is no longer a Euclidean time but a parameter which labels a family of Dirac operators. Our family of Dirac operators is given by

$$\partial\!\!\!/^1_t = -i\frac{d}{d\theta} + kt \tag{10.80}$$

where k is some integer and the other S^1 coordinate t varies from 0 to 1. According to our Fredholm analysis the anomaly is measured by an element $\pi_1(\mathcal{F}^1(H)) = \mathbf{Z}$. We shall calculate this integer by using the spectral flow technique of Atiyah and Lusztig.

Spectral flow is an integer associated to a 'periodic' family of operators. In the present case the family of operators $\partial\!\!\!/^1_t$ is not periodic in t but it is true that the *spectrum* of the family is periodic; i.e. the spectra of the operators $\partial\!\!\!/^1_0$ and $\partial\!\!\!/^1_1$ coincide. The power of the concept of spectral flow lies in utilising the easily ignored fact that, though these two spectra coincide as sets of eigenvalues, the evolution, or flow, of the former into the latter as t varies may entail some rich rearrangement of the eigenvalues. It is much easier to understand this by studying the details of this simple example— this example being one in which we can calculate the relevant eigenvalues explicitly.

The eigenvalues of the operator $\partial\!\!\!/^1_t$ are just $\lambda_n(t)$ where

$$\lambda_n(t) = n + kt, \quad n = \ldots, -1, 0, 1, \ldots \tag{10.81}$$

Thus the spectra of the two operators ∂_0^1 and ∂_1^1 are the pair of sets

$$\{n : n \in \mathbf{Z}\} \quad \text{and} \quad \{n + k : n \in \mathbf{Z}\} \tag{10.82}$$

which we observe do coincide. However, when displayed in this explicit fashion we also notice the relationship

$$\lambda_n(1) = \lambda_n(0) + k \tag{10.83}$$

between the eigenvalues. What has happened is that k of the negative eigenvalues of ∂_0^1 have flowed up and become positive eigenvalues of ∂_1^1. Notwithstanding this flow, the two spectra still coincide since they are un-bounded above and below. The spectral flow of the family of operators ∂_t^1 is defined to be the number of negative eigenvalues which flow over to positive counterparts as t varies from 0 to 1. Note that the spectral flow is just an integer and in this example it is equal to k.

We remarked above that the family of operators ∂_t^1 are not periodic in t. Nevertheless, we can check that they are related by

$$\partial_1^1 = e^{-ik\theta} \partial_0^1 e^{ik\theta} \tag{10.84}$$

This relation is just the usual gauge covariance of the Dirac operator. Hence e^{ikt} is just a $U(1)$ gauge transformation and if we quotient by this gauge transformation the family ∂_t^1 descends to a periodic family. Having done this we have a loop on $\mathcal{F}^1(H)$ or, equivalently, an element of $\pi_1(\mathcal{F}^1(H))$. The homotopy type of this loop is then given by the spectral flow k. This integer k is also the index of the chiral Dirac operator ∂_A on the torus. To check this explicitly one would have to calculate the null spaces of ∂_A and ∂_A^* on the torus using

$$\partial_A = \partial_t^1 + \frac{\partial}{\partial t} \tag{10.85}$$

The rearrangement of eigenvalues under the flow may be more complex than the simple shift discovered here. In more general cases there may be both positive eigenvalues which flow into negative eigenvalues as well as negative eigenvalues which flow into positive ones. Thus the general definition of the spectral flow of a family is that it is the integer which is equal to the number of negative eigenvalues which become positive *minus* the number of positives which become negative. It is also useful to note that, for the spectral flow to be non-vanishing, the spectra of the family must be unbounded: in the present example, if the Dirac operator had a lowest bound no flow could take place.

Example *Fock space and Gauss's law*

In this example we take space-time to be S^4 on which we have a 2-parameter family of chiral Dirac operators with parameter space S^2. The anomaly then corresponds to an element of $\pi_2(\mathcal{F}(H)) = \mathbf{Z}$. On the other hand, the Hamiltonian viewpoint is to take space to be S^3 and consider a 3-parameter family of self-adjoint Dirac operators. Then the anomaly is an element of $\pi_3(\mathcal{F}^1(H))$ which is also $\mathbf{Z}$.

Now the Euclidean S^4 formulation of this anomaly is essentially the one that we described in § 2, where the anomaly was evaluated as a Chern class obtained by restricting to a 2-sphere in $\mathcal{A}/\mathcal{G}$. In the Hamiltonian picture on S^3 the cohomology generators shift from even dimensions to odd dimensions and the 2-sphere in $\mathcal{A}/\mathcal{G}$ becomes a 3-sphere. However, we now turn to the Hamiltonian picture. The interesting feature here is that a non-trivial kernel for the Dirac operator is an obstacle to the gauge invariant definition of the Fermionic Fock space.

First we must describe this Fock space (cf. Segal [**3**] and Pressley and Segal [**1**]). Let our $\not{D}_A$ be the self-adjoint Dirac operator in 3 dimensions coupled to a connection A. Let H be the Hilbert space of eigenfunctions of $\not{D}_A$. In general the spectrum of $\not{D}_A$ has a non-empty kernel, but suppose for a moment that A is such that $ker\not{D}_A = \phi$. In this event we can decompose H into its strictly positive and negative subspaces H^+ and H^- respectively, giving

$$H = H^+ \oplus H^- \tag{10.86}$$

Anti-particles are represented by the complex conjugate space $\bar{H}^-$ and to second quantise the theory we must construct the Fock space. The Fock space is the anti-symmetrised exterior algebra space defined by

$$\mathcal{F}(A) = \wedge^*(H^+ \oplus \bar{H}^-) = \bigoplus_{p,q} \wedge^p(H^+) \otimes \wedge^q(\bar{H}^-) \tag{10.87}$$

where the sum is the usual one over states containing p particles and q anti-particles. Note that the Fock space depends on A and, as A varies, the collection $\{\mathcal{F}(A) : A \in \mathcal{A}\}$ form a bundle $\mathcal{F}$ over the parameter space $\mathcal{A}$. But since gauge transformations act on connections, and thus on $\mathcal{F}$, it is natural to project $\mathcal{F}$ to obtain the bundle $\mathcal{F}/\mathcal{G}$ over $\mathcal{A}/\mathcal{G}$. The two bundles are thus

$$\begin{array}{ccc} \mathcal{F} & & \mathcal{F}/\mathcal{G} \\ \downarrow & \text{and} & \downarrow \\ \mathcal{A} & & \mathcal{A}/\mathcal{G} \end{array} \tag{10.88}$$

The physical states of this theory should then be the space of square integrable sections of the bundle $\mathcal{F}/\mathcal{G}$ over the gauge orbit space $\mathcal{A}/\mathcal{G}$.

Unfortunately this picture has a flaw: note first that as A varies we cannot always avoid those A for which $\rlap{/}{D}_A$ does have a non-empty kernel. The flaw arises because when $\rlap{/}{D}_A$ has zero eigenvalues the decomposition into H^+ and H^- is no longer defined. Also we cannot expect the Fock spaces $\mathcal{F}(A)$ to vary smoothly when A passes through one of these troublesome A's. We shall discover that the most serious consequence of this flaw is that its existence leads to the fact that the group $\mathcal{G}$ does not actually act on $\mathcal{F}$: to get a group which does act on $\mathcal{F}$ we must pass to a non-central extension $\bar{\mathcal{G}}$ of $\mathcal{G}$. This is the same extension we met in chapter 6 and requires the specification of a cocycle in $H^2(\mathcal{G})$ which we know corresponds under transgression to an element of $H^3(\mathcal{A}/\mathcal{G}) = \mathbf{Z}$. This obstruction cocycle is the anomaly.

We finish by describing the emergence of the group extension in the Fock space picture. It is necessary to return to the matter of the kernel of $\rlap{/}{D}_A$. Suppose that $\rlap{/}{D}_A$ has zero eigenvalues, then we shall decompose the Hilbert space H of eigenstates slightly differently: We choose a real number $\epsilon > 0$ which is *not* an eigenvalue of $\rlap{/}{D}_A$; then we define H_ϵ^+ to be the spaces of eigenstates whose eigenvalues are greater than ϵ, similarly H_ϵ^- consists of the states whose eigenvalues are less than ϵ. Let us implement these definitions a second time using another real number $\epsilon' > 0$ which is also not an eigenvalue. Then we have available two decompositions of H, namely

$$H = H_\epsilon^+ \oplus H_\epsilon^- = H_{\epsilon'}^+ \oplus H_{\epsilon'}^- \qquad (10.89)$$

We now have two possible Fock spaces given by

$$\mathcal{F}_\epsilon(A) = \wedge^*(H_\epsilon^+ \oplus \bar{H}_\epsilon^-) \quad \text{and} \quad \mathcal{F}_{\epsilon'}(A) = \wedge^*(H_{\epsilon'}^+ \oplus \bar{H}_{\epsilon'}^-) \qquad (10.90)$$

Now suppose for definiteness that $\epsilon' > \epsilon$. Then $H_\epsilon^+ \supset H_{\epsilon'}^+$ and the two spaces can only differ by a finite dimensional space $J_{\epsilon,\epsilon'}$, say, so that

$$
\begin{aligned}
H_\epsilon^+ &= H_{\epsilon'}^+ \oplus J_{\epsilon,\epsilon'} \\
\Rightarrow H &= H_{\epsilon'}^+ \oplus J_{\epsilon,\epsilon'} \oplus H_\epsilon^-
\end{aligned}
\qquad (10.91)
$$

Applying this to our Fock spaces gives

$$
\begin{aligned}
\mathcal{F}_\epsilon(A) &= \wedge^*(H_{\epsilon'}^+ \oplus J_{\epsilon,\epsilon'} \oplus \bar{H}_\epsilon^-) \\
&= \wedge^*(J_{\epsilon,\epsilon'}) \otimes \wedge^*(H_{\epsilon'}^+ \oplus \bar{H}_\epsilon^-)
\end{aligned}
\qquad (10.92)
$$

Similarly

$$\mathcal{F}_{\epsilon'}(A) = \wedge^*(\bar{J}_{\epsilon,\epsilon'}) \otimes \wedge^*(H_{\epsilon'}^+ \oplus \bar{H}_\epsilon^-) \qquad (10.93)$$

We see that the difference between the two Fock spaces is measured by the difference between the two spaces $\wedge^*(J_{\epsilon,\epsilon'})$ and $\wedge^*(\bar{J}_{\epsilon,\epsilon'})$. However, these latter two spaces are almost isomorphic (Pressley and Segal [1]): To see this let $n = \dim J_{\epsilon,\epsilon'}$ and let us choose any non-zero element λ, say, of the one dimensional space $\wedge^n(J_{\epsilon,\epsilon'})$, then we have a λ-dependent isomorphism I^λ defined by its action on p-forms, that is

$$
\begin{aligned}
I_p^\lambda : \wedge^p (J_{\epsilon,\epsilon'}) &\longrightarrow \wedge^{n-p}(\bar{J}_{\epsilon,\epsilon'}) \\
\omega &\longmapsto I_p^\lambda(\omega) \qquad \text{where } \omega \text{ satisfies } I_p^\lambda(\omega) \wedge \omega = \lambda
\end{aligned}
\tag{10.94}
$$

But if we were to replace the Fock spaces by their *projective counterparts* then all such $\lambda \in \wedge^n(J_{\epsilon,\epsilon'})$ would be equivalent and trivial; hence the isomorphism induced between the *projective* Fock spaces is independent of λ—i.e. it is canonical. Therefore the projective Fock spaces are identical. Of course in quantum theory the physical states *are* identifiable with a space of rays, and thus we now replace the Fock space $\mathcal{F}_\epsilon(A)$ by its projective version, which we write as $\mathcal{F}P(A)$; note that we drop the ϵ label from the projective version since it is no longer necessary.

The projective Fock spaces fit together to form the projective bundle

$$
\mathcal{F}P = \{\mathcal{F}P(A) : A \in \mathcal{A}\}
\tag{10.95}
$$

This time the group $\mathcal{G}$ does act smoothly on $\mathcal{F}P$ and we have no trouble in constructing the quotient bundle

$$
\begin{aligned}
\mathcal{F}P/\mathcal{G} \\
\downarrow \\
\mathcal{A}/\mathcal{G}
\end{aligned}
\tag{10.96}
$$

The group extension arises in the following way. Choose a Fock space $\mathcal{F}$ whose projective version is $\mathcal{F}P$ and consider the action of an element $g \in \mathcal{G}$ on $\mathcal{F}P$. We can lift this action from $\mathcal{F}P$ to a unitary action on $\mathcal{F}$, but on $\mathcal{F}$ it will only be a projective action: i.e. we must include a projective multiplier λ with $|\lambda| = 1$. This means that if $f_A \in \mathcal{F}$ then, under $g, h \in \mathcal{G}$, we have

$$
g \cdot (h \cdot f_A) = \lambda(A)(gh) \cdot f_A
\tag{10.97}
$$

Hence we have an extension of the group $\mathcal{G}$ by the group of projective multipliers $\lambda(A)$ which clearly belong to $Map(\mathcal{A}, S^1)$. We recognise this as being the non-central extension $\bar{\mathcal{G}}$ that we described in chapter 6. When the current algebra commutation relations are calculated, this extension is responsible for the Schwinger term.

Gauss's law is violated for theories with this anomaly. The reason for this is that the non-triviality of the extension cocycle means that the identity of $\mathcal{G}$ is covered by a non-trivial element of $Map\,(\mathcal{A}, S^1)$. Gauss's law says that the generators of the Lie algebra of $\mathcal{G}$ must annihilate the vacuum and this implies that the physical states of the theory must be invariant under $\mathcal{G}$. However, this is now impossible because the identity element acts non-trivially and will change the vacuum.

CHAPTER XI

Conformal Quantum Field Theories

§ 1. Conformal invariance and quantum field theory

Conformal invariance is a recurrent topic in theoretical physics. Since a conformally invariant universe is one with no masses, we do not expect conformal invariance to be a symmetry of a complete physical theory. Nevertheless, the conformal invariance present in certain physical systems has proved to be of great value in understanding their structure.

In statistical mechanics as a system approaches a second order phase transition its correlation length diverges. At the critical point the theory possesses no dimensional parameter and is scale or dilation invariant; in two dimensions the field theory describing the critical point turns out to be, not just dilation invariant, but conformally invariant.

The operator product $\phi(x)\phi(0)$ of some quantum field theories becomes independent of mass in the limit of small x. This has led to the suggestion that the elementary constituents of matter, which are the relevant degrees of freedom at very small distance, may be described by theories with conformally invariant small distance limits. The short distance behaviour of field theories is intimately related to their renormalisation properties. The renormalisation properties of correlation functions are constrained by the need to obey the Callan–Symanzik renormalisation group equations, cf. Nash [1] for a discussion. A necessary, but insufficient, condition for a theory to possess conformal invariance is that the renormalisation group flow has a fixed point; this means that the Callan–Symanzik function $\beta(g)$ has a zero. Wilson [1] introduced the operator product expansion as a tool to study the scale invariant properties of such theories. This is an expansion of the product $A(x)B(0)$ of two local quantum fields. The expansion is valid for small x and takes the form

$$A(x)B(0) \longrightarrow \sum_n C_n(x)O_n(0)$$

$$x \longmapsto 0$$

(11.1)

where $C_n(x)$ are a set of functions, singular at $x = 0$, and O_n are a set of local operators. If the field theory has a scale invariant small distance limit then both sides of the expansion have the same scale dimension;[1] in that case, if S_{C_n} denotes the strength of the singularity of the function C_n so that

$$C_n(\lambda x) = \frac{C_n(x)}{\lambda^{S_n}} \qquad (11.2)$$

then

$$d_A + d_B = d_{O_n} - S_{C_n}, \qquad \text{for each } n \qquad (11.3)$$

We shall see below that this operator product expansion plays a prominent part in the theory of conformally invariant fields in two dimensions.

In general, non-trivial theories with conformal invariant correlation functions are very difficult to find. However, in two dimensions, where the conformal group is infinite dimensional, the situation is somewhat better. The central example here is string theory but the other conformally invariant theories in two dimensions are of independent interest; as we have said above they occur in the statistical mechanics of critical phenomena, they also turn up in the quantum field theoretic approach to knots of chapter 12. In this chapter we shall limit our discussions to conformally invariant theories in two dimensions; for additional background, see Belavin, Polyakov and Zamolodchikov [1] and Zamolodchikov [1,2].

§ 2. Conformal field theories in two dimensions

To begin with we consider field theories on the complex plane, or perhaps on the Riemann sphere; later, in § 5 we shall be more general and consider theories defined on a Riemann surface Σ of genus p. Let $\phi(z, \bar{z})$ be a smooth quantum field on $\mathbf{C}$ with the tensorial transformation property

$$\phi(z, \bar{z}) \longmapsto \phi'(z', \bar{z}')$$
$$z \longmapsto z'$$

where $\qquad \phi(z, \bar{z}) = \left(\frac{dz'}{dz} \right)^h \left(\frac{d\bar{z}'}{d\bar{z}} \right)^{\bar{h}} \phi'(z', \bar{z}'), \quad \text{and} \quad h, \bar{h} \in \mathbf{R}$

$$(11.4)$$

The field $\phi(z, \bar{z})$ is called a *primary field* of conformal weight $(h, \bar{h})$. The sum of the weights $h + \bar{h}$ is the scale dimension of the primary field while

---

[1] The scale dimension d_A of a field A is defined by $U(\lambda)A(x)U^{-1}(\lambda) = \lambda^{d_A} A(\lambda x)$ where $U(\lambda)$ represents the dilation operator $\exp[\ln \lambda(x\partial/\partial x)]$; when the scale dimension differs from the canonical dimension d_A^c the difference $d_A^c - d_A$ is usually called the anomalous dimension of A and is written as γ_A.

the difference $h - \bar{h}$ is the spin; the difference $h - \bar{h}$ is therefore restricted to integer or half-integer values.

Example *A free Fermion*

An example of a primary field is provided by taking the field theory of a free Majorana (self-conjugate) Fermion ψ on the complex plane. The action for this theory is

$$S = \int dz d\bar{z}\, \psi(z)\partial\psi(z) \tag{11.5}$$

For the product of two Fermion fields we have

$$< \psi(z)\psi(0) >= \frac{C}{z} \tag{11.6}$$

where C is a numerical constant. This shows that $\psi(z)$ is a primary field of weight $(1/2, 0)$.

It is not sufficient to take any massless theory in order to get a primary field. The Bosonic counterpart of the previous example is *not* a primary field.

Example *A free Boson*

This time S is the action for a free Boson $\varphi(z, \bar{z})$ so that

$$S = \frac{1}{4\pi} \int dz d\bar{z}\, \partial\varphi(z, \bar{z})\bar{\partial}\varphi(z, \bar{z}) \tag{11.7}$$

The two point correlation function is now

$$< \varphi(z, \bar{z})\varphi(0) >= C \ln(z\bar{z}) \tag{11.8}$$

with C another constant; since the correlation function has no scaling properties, φ is not a primary field. Actually, in two dimensions, a scalar field is dimensionless so this result might have been anticipated.

On the other hand, if $\phi(z, \bar{z})$ *is* a primary field of weight $(h, \bar{h})$, then scale invariance alone constrains the two point function to be of the form

$$< \phi(z, \bar{z})\phi(0, 0) >= \frac{C}{z^h \bar{z}^{\bar{h}}} \tag{11.9}$$

We now want to make contact with the operator product expansion for a conformal theory. An instructive product to consider is

$$T_{ab}\phi \tag{11.10}$$

where ϕ is a primary field of weight $(h, \bar{h})$ and T_{ab} is the conserved energy momentum tensor. Before calculating the product we need to introduce more suitable coordinates to describe T_{ab} and, in addition, we must obtain the Ward identities associated with a conformal transformation. We deal first with the energy momentum tensor.

The scale invariance of the field theory means that T_{ab} is traceless so

$$T_{11} + T_{22} = 0 \qquad (11.11)$$

and, because T_{ab} is symmetric, it possesses only two independent components. Using the complex coordinates $z = x_1 + ix_2$ and $\bar{z} = x_1 - ix_2$ we find that

$$
\begin{aligned}
T_{z\bar{z}} &= T_{\bar{z}z} = T_{11} + T_{22} = 0 \\
T_{zz} &= T_{11} - T_{22} - 2iT_{12} \\
T_{\bar{z}\bar{z}} &= T_{11} - T_{22} + 2iT_{12}
\end{aligned}
\qquad (11.12)
$$

But the energy momentum conservation condition $\partial^a T_{ab} = 0$ implies that

$$\partial_{\bar{z}} T_{zz} = \partial_z T_{\bar{z}\bar{z}} = 0 \qquad (11.13)$$

Thus it is natural to define

$$T(z) = T_{zz}, \qquad \bar{T}(\bar{z}) = T_{\bar{z}\bar{z}} \qquad (11.14)$$

and observe that $T(z)$ and $\bar{T}(\bar{z})$ are holomorphic and anti-holomorphic respectively. Henceforth we shall use $T(z)$ and $\bar{T}(\bar{z})$ to represent the two independent components of T_{ab}. Now we turn to the matter of the Ward identity.

Let us temporarily use the real coordinates $x_\mu = (x_1, x_2)$ and make the infinitesimal coordinate transformation

$$x_\mu \longmapsto x'_\mu = x_\mu + \epsilon f_\mu(x) \qquad (11.15)$$

where ϵ is small. The change $\delta\phi(x)$ in the primary field $\phi(x)$ is $\phi'(x') - \phi(x)$. It is standard in quantum field theory that the change in the correlation functions of ϕ is related to the energy momentum tensor by the expression

$$
\begin{aligned}
\sum_i &< \phi(x_1) \cdots \delta\phi(x_i) \cdots \phi(x_n) > \\
&= \int_R d^2x < T_{\mu\nu}(x)\phi(x_1) \cdots \phi(x_i) \cdots \phi(x_n) > \partial^\mu f^\nu(x) \\
&\quad + \int_{\partial R} dl_\mu < T_{\nu\sigma}(x)\phi(x_1) \cdots \phi(x_n) > \epsilon^{\mu\nu} f^\sigma(x)
\end{aligned}
\qquad (11.16)
$$

where R is a compact region containing the x_i and $\epsilon^{\mu\nu}$ is the permutation tensor. Now we change to complex coordinates and specialise to a conformal coordinate transformation of the form

$$z \longmapsto z' = z + \epsilon f(z) \tag{11.17}$$

With this choice of coordinate transformation only the boundary term in 11.16 survives and we thereby obtain

$$\sum_i < \phi(z_1, \bar{z}_1) \cdots \delta\phi(z_i, \bar{z}_i) \cdots \phi(z_n, \bar{z}_n) >$$

$$= \epsilon \int_C \frac{dw}{2\pi i} < T(w)\phi(z_1, \bar{z}_1) \cdots \phi(z_i, \bar{z}_i) \cdots \phi(z_n, \bar{z}_n) > f(w)$$

$$+ \epsilon \int_C \frac{d\bar{w}}{2\pi i} < \bar{T}(\bar{w})\phi(z_1, \bar{z}_1) \cdots \phi(z_i, \bar{z}_i) \cdots \phi(z_n, \bar{z}_n) > \bar{f}(\bar{w}) \tag{11.18}$$

where $C \simeq \partial R$ is a contour containing the z_i. It is natural to separate the equation into its holomorphic and anti-holomorphic pieces; quoting only the equation for the holomorphic part $\delta^h \phi$ of $\delta\phi$, we deduce the operator equation

$$\delta^h \phi(z, \bar{z}) = \frac{\epsilon}{2\pi i} \int_C T(w)\phi(z, \bar{z})f(w) \tag{11.19}$$

This completes our use of the Ward identity.

To obtain the operator product expansion we shall have to compute explicitly the holomorphic properties of both sides of 11.19. First let us observe that if $z \mapsto z + \epsilon f(z)$ in the primary field transformation law 11.4, we obtain

$$\delta\phi(z, \bar{z}) = \epsilon(h(\partial_z f(z))\phi(z, \bar{z}) + f(z)\partial_z\phi(z, \bar{z})) \tag{11.20}$$

Hence we have the equality

$$h(\partial_z f(z))\phi(z, \bar{z}) + f(z)\partial_z\phi(z, \bar{z}) = \frac{1}{2\pi i} \int_C T(w)\phi(z, \bar{z})f(w) \tag{11.21}$$

Now we require the following expansions

$$f(z) = f_0 + f_1 z + f_2 z^2 + \cdots$$
$$f(w) = \hat{f}_0(z) + \hat{f}_1(z)(w - z) + \hat{f}_2(z)(w - z)^2 + \cdots$$
$$T(w)\phi(z, \bar{z}) = \cdots + \frac{T_{-2}(\phi)}{(w - z)^2} + \frac{T_{-1}(\phi)}{(w - z)} + T_0(\phi) + T_1(\phi)(w - z) + \cdots$$
$$\tag{11.22}$$

Inserting these into 11.21 gives

$$(f_0 + f_1 z + \cdots)\partial_z \phi + h(f_1 + 2f_2 z + \cdots)\phi = T_{-1}(\phi)\hat{f}_0(z) + T_{-2}(\phi)\hat{f}_1(z)$$
$$+ T_{-3}(\phi)\hat{f}_2(z) + \cdots$$
$$(11.23)$$

If we work at $z = 0$, which allows us to use the equality $\hat{f}_n(0) = f_n$, we discover that

$$T_{-1}(\phi) = \partial_z \phi, \quad T_{-2}(\phi) = h\phi, \quad T_{-3}(\phi) = T_{-4}(\phi) = 0 = \cdots \quad (11.24)$$

and this establishes the short distance operator product expansion of $T(w)$ with the primary field ϕ as

$$T(w)\phi(0,0) = \frac{h}{w^2}\phi(0,0) + \frac{1}{w}\partial_z\phi(0,0) + R(w) \quad (11.25)$$

with $R(w)$ a regular function in some neighbourhood of $w = 0$. Equivalently we can write

$$T(w)\phi(z,\bar{z}) = \frac{h}{(w-z)^2}\phi(z,\bar{z}) + \frac{1}{(w-z)}\partial_z\phi(z,\bar{z}) + R(w-z) \quad (11.26)$$

and the corresponding product of $\bar{T}$ with ϕ is

$$\bar{T}(\bar{w})\phi(z,\bar{z}) = \frac{\bar{h}}{(\bar{w}-\bar{z})^2}\phi(z,\bar{z}) + \frac{1}{(\bar{w}-\bar{z})}\partial_{\bar{z}}\phi(z,\bar{z}) + \cdots \quad (11.27)$$

Next we consider the product of $T(z)$ with itself. Taking the example of the free Boson field above, for which $T(z) = -(\partial_z\phi)^2$, it is easy to calculate that

$$T(w)T(z) = \frac{I}{2(w-z)^4} + \frac{2}{(w-z)^2}T(z) + \frac{1}{(w-z)}\partial_z T(z) + \cdots \quad (11.28)$$

where I is the identity operator. Thus $T(z)$ is *not* a primary field because of the presence of the term proportional to the identity. For an arbitrary conformal field theory this expansion is replaced by the more general form

$$T(w)T(z) = \frac{cI}{2(w-z)^4} + \frac{2}{(w-z)^2}T(z) + \frac{1}{(w-z)}\partial_z T(z) + R'(w-z) \quad (11.29)$$

with $R'(w - z)$ regular. We shall see below that the number c is actually the central charge[2] for a representation of the Virasoro algebra. This operator product expansion of $T(w)T(z)$ is the consequence of the following behaviour of $T(z)$ under the infinitesimal change $z \mapsto z + \epsilon f(z)$

$$\delta T(z) = \epsilon(2(\partial_z f(z))T(z) + f(z)\partial_z T(z) + \frac{c}{12}\partial_z^{(3)} f(z)) \tag{11.30}$$

If it were not for the number c then $T(z)$ would be a primary field of weight $(2,0)$ thus having a non-anomalous scaling dimension of 2. The finite form of the transformation law for $T(z)$ is not quite tensorial being

$$T(z) = \left(\frac{dz'}{dz}\right)^2 T'(z') + \frac{c}{12}\{z, z'\}$$

where $\qquad \{z', z\} = \frac{(d^3 z'/dz^3)}{(dz'/dz)} - \frac{3}{2}\frac{(d^2 z'/dz^2)^2}{(dz'/dz)^2}$ \tag{11.31}

and $\{z', z\}$ is known as the Schwarzian derivative of z'. The Schwarzian derivative vanishes if $z \mapsto z'(z)$ is a Möbius transformation. This can be proved by direct calculation: using the more convenient notation $\{f, z\}$ we see that

$$\{f, z\} = 0$$

$$\Rightarrow \frac{\ddot{g}}{g} - \frac{3}{2}\frac{\dot{g}^2}{g^2} = 0, \qquad \text{where } g = \dot{f}$$

$$\Rightarrow \frac{\ddot{g}}{\dot{g}} - \frac{3}{2}\frac{\dot{g}}{g} = 0$$

$$\Rightarrow \ln \dot{g} - \frac{3}{2}\ln g = A \tag{11.32}$$

$$\Rightarrow g = \frac{4}{(e^A z + B)^2}$$

$$\Rightarrow f = C\frac{az + b}{cz + d}$$

with $ad - bc = 1$ and C an undetermined constant.

§ 3. Relation to the Virasoro algebra

We know from chapter 6 that the conformal group in two dimensions consists of the group of holomorphic and anti-holomorphic transformations.

---

[2] It follows that for a single free Boson $c = 1$; for a free Fermion the reader can check that $c = (1/2)$; thus the value of c can be used to detect whether or not a theory is free.

More precisely we can realise the two dimensional conformal group as the product $Diff(S^1) \times Diff(S^1)$; one S^1 corresponds to the boundary values of the holomorphic transformations and the other to the boundary values of the anti-holomorphic transformations. Using the holomorphic and anti-holomorphic energy momentum tensors $T(z)$ and $\overline{T}(\bar{z})$ we can construct the generators of the corresponding algebra. We write

$$T(z) = \sum_{-\infty}^{\infty} L_n z^{-(n+2)}, \qquad L_n = \int_C \frac{dw}{2\pi i} T(w) w^{n+1} \qquad (11.33)$$

If we combine this with the operator product expansion 11.29 we can verify that these generators have the commutation relations

$$[L_m, L_n] = (m-n)L_{m+n} + \frac{c}{12} m(m^2 - 1)\delta_{m,-n} \qquad (11.34)$$

and thus the $\{L_n\}$ generate a Virasoro algebra, Vir, say, showing that the number c appearing in 11.29 is indeed a central charge. Clearly if we write

$$\overline{T}(\bar{z}) = \sum_{-\infty}^{\infty} \overline{L}_n \bar{z}^{-(n+2)} \qquad (11.35)$$

then the $\{\overline{L}_n\}$ generate another Virasoro algebra $\overline{Vir}$ commuting with the first.

Each primary field $\phi(z)$ of weight $(h, 0)$ can be used to construct an irreducible representation of the Virasoro algebra; moreover, c and h can be used to label this representation and, in so doing, they retain precisely the same meaning as they had in the description of the representations given in chapter 6.

A natural Hamiltonian $\mathcal{H}$ in a conformally invariant theory is the generator of dilations $L_0 + \overline{L}_0$. This change in $\mathcal{H}$ from the generator of time translations to the generator of dilations causes the usual time ordering to be replaced by radial ordering: the points $z_1, \ldots, z_n$ in a correlation function $< \phi(z_1, \bar{z}_1) \cdots \phi(z_n, \bar{z}_n) >$ are ordered by their differences $|z_i - z_j|$. Let the vacuum state for $\mathcal{H}$ be $|0\rangle$. We define

$$|h\rangle = \phi(0)|0\rangle \qquad (11.36)$$

and we can apply the L_n's to $|h\rangle$ to verify that

$$L_0|h\rangle = h|h\rangle, \qquad L_n|h\rangle = 0, \quad n \geq 1 \qquad (11.37)$$

Hence $|h\rangle$ is a highest weight vector and therefore the infinite tower of states of the form $L_{-1}^{\alpha_1} L_{-2}^{\alpha_2} \cdots L_{-s}^{\alpha_s} |h\rangle$ comprise an irreducible representation of the Virasoro algebra Vir. If we want a representation of both Virasoro algebras at once—i.e. a representation of $Vir \oplus \overline{Vir}$—we use a single valued primary field $\phi(z, \bar{z})$ of weight $(h, \bar{h})$ and define

$$|h, \bar{h}\rangle = \phi(0, 0) |0\rangle \qquad (11.38)$$

and this gives us an irreducible representation of $Vir \oplus \overline{Vir}$.

Apart from $|h\rangle$ itself the elements $L_{-1}^{\alpha_1} L_{-2}^{\alpha_2} \cdots L_{-s}^{\alpha_s} |h\rangle$ of the tower of states are referred to as secondary or descendant states. Just as the highest weight state $|h\rangle = \phi(0) |0\rangle$ is associated with the primary field $\phi(z)$, a secondary *state* is associated with a secondary *field*. For example, a simple secondary state is

$$L_n |h\rangle, \qquad n < 0 \qquad (11.39)$$

where we note that the condition $n < 0$ follows from 11.37; the associated secondary field is

$$L_n(z)\phi(z) \qquad (11.40)$$

where $L_n(z)$ is defined by

$$L_n(z) = \int_C \frac{dw}{2\pi i} T(w)(w - z)^{n+1} \qquad (11.41)$$

with C enclosing the point $w = z$. As might be expected these $L_n(z)$ generate a Virasoro algebra; we also have $L_n(0) = L_n$. Notice that $L_{-1}\phi(z) = \partial_z \phi(z)$ and so the derivative of a primary field is not primary. A more general secondary field together with its associated state are

$$L_{-1}^{\alpha_1}(z) \cdots L_{-s}^{\alpha_s}(z)\phi(z), \qquad L_{-1}^{\alpha_1} \cdots L_{-s}^{\alpha_s} |h\rangle \qquad (11.42)$$

and the whole collection is sometimes referred to as the conformal family $[\phi]$. Still more generally we can bring in the $\overline{L}_n$'s as well as the L_n's and, applying these to a primary field $\phi(z, \bar{z})$ of weight $(h, \bar{h})$, we obtain descendant data of the form

$$\begin{aligned} &\overline{L}^{\alpha_1}(\bar{z}) \cdots \overline{L}_{-t}^{\alpha_t}(\bar{z}) L_{-1}^{\alpha_1}(z) \cdots L_{-s}^{\alpha_s}(z)\phi(z, \bar{z}) \\ &\overline{L}_{-1}^{\alpha_1} \cdots \overline{L}_{-t}^{\alpha_t} L_{-1}^{\alpha_1} \cdots L_{-s}^{\alpha_s} |h, \bar{h}\rangle \end{aligned} \qquad (11.43)$$

A secondary state $L_{-1}^{\alpha_1} \cdots L_{-s}^{\alpha_s} |h\rangle$ is an eigenvector of L_0, for it is easy to use the Virasoro algebra to show that

$$L_0(L_{-1}^{\alpha_1} \cdots L_{-s}^{\alpha_s} |h\rangle) = (h + \sum_i i\alpha_i) L_{-1}^{\alpha_1} \cdots L_{-s}^{\alpha_s} |h\rangle \qquad (11.44)$$

Since the Virasoro algebra representation is unitary, the eigenvectors of L_0 are mutually orthogonal; hence secondary states with differing values of the integer $|\alpha| = \sum i\alpha_i$ are orthogonal. This suggests that $|\alpha|$ can be used to provide a grading [3] of the conformal family $[\phi]$; the integer $|\alpha|$ is called the *level* of the secondary state.

Descendant states are usually all linearly independent; when this is not the case—as happens for families $[\phi]$ corresponding to certain values of c and h—there exist non-trivial null states which have zero norm and are orthogonal to all other states. These null states must be projected out in order to obtain an irreducible representation of Vir. States of negative squared norm can also occur and are disallowed; the need to construct the representation on a Hilbert space of positive norm plays an important part in finding those values of c and h which give rise to unitary irreducible representations of Vir.

Using secondary fields we can identify formally all the terms in the operator product expansion of $T(w)\phi(z)$. Up to now we have just identified the singular terms. However, making use of the $L_n(w)$, we have

$$T(w)\phi(z) = \sum_{n=0}^{\infty} \frac{1}{(w-z)^{n+2}} L_n(w)\phi(z) \tag{11.45}$$

and it is not difficult to check that the singular first two terms of this expansion agree with those given in 11.26. If we take any two fields A and B in the same conformal family $[\phi]$ then the product $A(w)B(z)$ is expressible in a like manner over the fields in $[\phi]$. The form 11.29 of the product $T(w)T(z)$ shows that the energy momentum tensor $T(z)$ is a secondary field belonging to $[I]$, the conformal family of the identity.

A representation of the Möbius algebra $sl(2, \mathbf{C})$ is generated by the Virasoro sub-algebra whose elements are $\{L_{-1}, L_0, L_1\}$. The vacuum is annihilated by these elements for we know already that

$$L_0 |0\rangle = L_1 |0\rangle = 0 \tag{11.46}$$

and, for $L_{-1} |0\rangle$, we have

$$\begin{aligned} \|L_{-1} |0\rangle\|^2 &= \langle 0| L_1 L_{-1} |0\rangle \\ &= \langle 0| [L_1, L_{-1}] |0\rangle \\ &= 2 \langle 0| L_0 |0\rangle \\ &= 0 \end{aligned} \tag{11.47}$$

---

[3] By a grading of $[\phi]$ we mean a decomposition $[\phi] = \oplus_{i \in I} [\phi]^i$ where I is some indexing set, which in this case is the non-negative integers.

Hence the true invariance of the theory is that of the Möbius group $SL(2, \mathbf{C})/\{\mp I\}$; in fact if we consider the whole complex plane, i.e. the Riemann sphere, then the only invertible conformal transformations are the elements of the Möbius group. In this connection we note that the Schwarzian derivative term in the transformation law of $T(z)$ vanishes for Möbius transformations and, with respect to these transformations, $T(z)$ is a true tensor.

We have seen that representations of the Virasoro algebra are an essential ingredient of a conformal field theory. Since Virasoro algebra representations occur in a number of different ways, a conformal field theory may originate in a variety of ways. For example, we know that representations of Kac–Moody algebras may be used to construct representations of Virasoro algebras; the conformal field theories corresponding to the level k representations for various Kac–Moody algebras have been considered by Gepner and Witten [1]. In addition, supersymmetry can rather easily be combined with conformal symmetry to give two dimensional superconformal field theories and these provide another source of conformal field theories (Friedan, Martinec and Shenker [1]). In the next section we turn to statistical mechanics and obtain an example of a conformal field theory of genus one.

§ 4. Statistical mechanics

In statistical mechanics the imposition of periodic boundary conditions leads to partition functions Z defined on a torus; at a second order phase transition the conformal invariance of the theory means that Z should be invariant under the modular transformations

$$\tau \longmapsto \frac{a\tau + b}{c\tau + d} \tag{11.48}$$

where τ is the modular parameter for the torus and, of course, $(ad - bc) = 1$ and $a, b, c, d \in \mathbf{Z}$. The partition function is a trace over the Hilbert space $\oplus_i(H_i \otimes \overline{H}_i)$ of the theory. Hence it can be written as a sum over the characters of the representations of the two Virasoro groups

$$Z(\tau) = \sum_{h,\bar{h}} m_{h,\bar{h}} \chi_h(\tau) \chi_{\bar{h}}(\bar{\tau}) \tag{11.49}$$

where χ_h or $\chi_{\bar{h}}$ denotes a character and $m_{h,\bar{h}}$ is the multiplicity of the representation. For example, if we choose the two dimensional Ising model it turns out that the conformal field theory has $c = 1/2$ and contains only three primary fields whose conformal weights[4] are $(0,0)$, $(1/16, 1/16)$ and $(1/2, 1/2)$. The partition function is

$$Z = \chi_0 \bar{\chi}_0 + \chi_{1/2} \bar{\chi}_{1/2} + \chi_{1/16} \bar{\chi}_{1/16} \tag{11.50}$$

To verify that the partition function is modular invariant it is sufficient to check that Z is invariant under the generators S and T: i.e we must show that

$$Z(\tau) = Z(-1/\tau) = Z(\tau + 1) \tag{11.51}$$

For the Ising model the squared moduli $|\chi_h|^2$ can be written in terms of the Dedekind η-function, whose definition is

$$\eta(\tau) = \exp[\pi i \tau / 12] \prod_{n=1}^{\infty} (1 - \exp[2\pi n i \tau])$$
$$= q^{1/24} \prod_{n=1}^{\infty} (1 - q^n), \qquad \text{with } q = \exp[2\pi i \tau] \tag{11.52}$$

The relation between the characters and η is

$$|\chi_0|^2 + |\chi_{1/2}|^2 = \frac{1}{2} \left\{ \left| \frac{\eta^2(\tau/2)}{\eta^2(\tau)} \right| + \left| \frac{\eta^4(\tau)}{\eta^2(2\tau)\eta^2(\tau/2)} \right| \right\}$$
$$|\chi_{1/16}|^2 = 2 \left| \frac{\eta^2(2\tau)}{\eta^2(\tau)} \right| \tag{11.53}$$

Now the η-function is not actually invariant under modular transformations; in fact, $\eta^{24}(\tau)$ is a modular form of weight 6.

We digress briefly to obtain some facts about modular forms: To say that $f(\tau)$ is a modular form of weight k (or dimension $-2k$) means that

$$F(\tau) = f(\tau)d\tau^k \tag{11.54}$$

is invariant under elements of $SL(2, \mathbf{Z})/\{\mp I\}$; thus F is an *automorphic form* of weight k. Hence we have

$$F(m(\tau)) = F(\tau) \qquad \text{where } m(\tau) = \frac{a\tau + b}{c\tau + d}, \qquad ad - bc = 1$$
$$\Rightarrow f(m(\tau))(m'(\tau))^k = f(\tau)$$
$$\Rightarrow f(m(\tau)) = (c\tau + d)^{2k} f(\tau) \tag{11.55}$$

---

[4] Using the Virasoro algebra representation formula 6.58 one can check that, if $m = 2$, then $c = 1/2$ and h can take the values $0, 1/2$ and $1/16$.

Applying this to η^{24} gives

$$\eta^{24}(m(\tau)) = (c\tau + d)^{12}\eta^{24}(\tau) \qquad (11.56)$$

Choosing c and d appropriately we find that

$$\eta^{24}(\tau + 1) = \eta^{24}(\tau), \qquad \eta^{24}(-1/\tau) = (\tau)^{12}\eta^{24}(\tau) \qquad (11.57)$$

Finally it is easy to see directly from its definition that

$$\eta(\tau + 1) = \exp[\pi i/12]\eta(\tau) \qquad (11.58)$$

and some additional work shows that

$$\eta(-1/\tau) = (-i\tau)^{1/2}\eta(\tau) \qquad (11.59)$$

Returning to the formulae 11.53 it is now a routine matter to verify that, if we add the contributions together, their sum $Z(\tau)$ obeys $Z(\tau) = Z(-1/\tau) = Z(\tau + 1)$, thereby establishing the modular invariance of the Ising model partition function.

In general, modular invariance is a considerable restriction on Z and can be used to considerable effect, cf. Cardy [1]. Modular invariance has also been used to obtain a classification of a large class of theories; in addition, this classification has some intriguing correspondences with pairs of simply laced simple Lie algebras taken from the Cartan A, D and E series, cf. Cappelli, Itzykson and Zuber [1,2].

§ 5. Operator products, fusion rules and axiomatics

Let us consider a theory in which there are F primary fields $\phi_1, \ldots, \phi_F$ giving rise to the F conformal families $[\phi_1], \ldots, [\phi_F]$. A theory with a finite number of primary fields is sometimes referred to as a rational conformal field theory. The Hilbert space of the theory is the space associated to the fields in the factorised sum $\oplus_i (H_i \otimes \overline{H}_i)$ where the H_i and $\overline{H}_i$ are irreducible representation spaces for the Virasoro group and its complex conjugate. Let A be a local field belonging to the family $[\phi_i]$ for some i. If we introduce the more concise notation $L^I(z)$ and $\overline{L}^{\bar{J}}(\bar{z})$ where

$$L^I(z) = L^{i_1}_{-1} \cdots L^{i_s}_{-s}(z), \qquad \overline{L}^{\bar{J}}(\bar{z}) = \overline{L}^{j_1}_{-1} \cdots \overline{L}^{j_t}_{-t} \qquad (11.60)$$

then we can denote A by $\Phi^{I,\bar{J}}_i(z, \bar{z})$ where

$$\Phi^{I,\bar{J}}_i(z, \bar{z}) = L^I(z)\overline{L}^{\bar{J}}(\bar{z})\phi_i(z, \bar{z}) \qquad (11.61)$$

An arbitrary correlation function of the theory is

$$< \Phi_{i_1}^{I_1,J_1}(z_1,\bar{z}_1) \cdots \Phi_{i_n}^{I_n,J_n}(z_n,\bar{z}_n) > \qquad (11.62)$$

By using the operator product expansions deduced above from the conformal Ward identities this correlation function may be expressed in terms of the correlation functions $< \phi_{i_1}(z_1,\bar{z}_1) \cdots \phi_{i_n}(z_n,\bar{z}_n) >$ which contain only *primary fields*. Hence the $< \phi_{i_1}(z_1,\bar{z}_1) \cdots \phi_{i_n}(z_n,\bar{z}_n) >$ are the essential objects to calculate. If we impose crossing symmetry on the correlation functions this implies that the operator product algebra is associative (Belavin, Polyakov and Zamolodchikov [1]).

When $n = 2$ or 3 conformal invariance determines the primary field correlation functions up to a numerical constant: using $z_{ij} = (z_i - z_j)$ and $\bar{z}_{ij} = (\bar{z}_i - \bar{z}_j)$ one has

$$< \phi_i(z_1,\bar{z}_2)\phi_j(z_2,\bar{z}_2) > = \frac{C_{ij}}{z_{12}^{h_i+h_j}\,\bar{z}_{12}^{\bar{h}_i+\bar{h}_j}}$$

$$< \phi_i(z_1,\bar{z}_2)\phi_j(z_2,\bar{z}_2)\phi_k(z_3,\bar{z}_3) > = \qquad (11.63)$$

$$\frac{C_{ijk}}{z_{12}^{h_i+h_j-h_k} z_{13}^{h_k+h_i-h_j} z_{23}^{h_j+h_k-h_i} \bar{z}_{12}^{\bar{h}_i+\bar{h}_j-\bar{h}_k} \bar{z}_{13}^{\bar{h}_k+\bar{h}_i-\bar{h}_j} \bar{z}_{23}^{\bar{h}_j+\bar{h}_k-\bar{h}_i}}$$

By contrast, if $n = 4$, conformal invariance only determines the correlation function $< \phi_i(z_1,\bar{z}_2)\phi_j(z_2,\bar{z}_2)\phi_k(z_3,\bar{z}_3)\phi_l(z_4,\bar{z}_4) >$ up to an arbitrary function F of the Möbius invariant anharmonic ratios

$$z_{12,34} = \frac{(z_1-z_2)(z_3-z_4)}{(z_1-z_3)(z_2-z_4)}, \qquad \bar{z}_{12,34} = \frac{(\bar{z}_1-\bar{z}_2)(\bar{z}_3-\bar{z}_4)}{(\bar{z}_1-\bar{z}_3)(\bar{z}_2-\bar{z}_4)} \qquad (11.64)$$

Thus for $n \geq 4$ further information is required. Some of this information is provided if we can compute the operator product expansion of two, possibly distinct, primary fields. That is, we need to know the RHS of

$$\phi_i(z,\bar{z})\phi_j(0,0) = \sum_k \sum_{L,\bar{L}} \frac{C_{ij}^{k,L,\bar{L}}}{z^{h_i+h_j-h_k-|L|}\bar{z}^{\bar{h}_i+\bar{h}_j-\bar{h}_k-|\bar{L}|}}\Phi_k^{L,\bar{L}}(0,0) \qquad (11.65)$$

It is possible (Belavin, Polyakov and Zamolodchikov [1]) to factorise the constants $C_{ij}^{k,L,\bar{L}}$ into three factors, the first of which depends only on the primary fields while the remaining two are determined only by the conformal weights. This is usually written as

$$C_{ij}^{k,L,\bar{L}} = C_{ij}^k \beta_{ij}^{kL} \bar{\beta}_{ij}^{k\bar{L}} \qquad (11.66)$$

This fact allows us to define the field $\Psi^k_{ij}(z, \bar{z}, 0, 0)$ by

$$\Psi^k_{ij}(z, \bar{z}, 0, 0) = \sum_{L, \bar{L}} \frac{\beta^{kL}_{ij} \bar{\beta}^{k\bar{L}}_{ij}}{z^{h_i + h_j - h_k - |L|} \bar{z}^{\bar{h}_i + \bar{h}_j - \bar{h}_k - |\bar{L}|}} \Phi^{L, \bar{L}}_k(0, 0) \qquad (11.67)$$

Having done this we can write the operator product expansion 11.65 as

$$\phi_i(z, \bar{z})\phi_j(0, 0) = \sum_k C^k_{ij} \Psi^k_{ij}(z, \bar{z}, 0, 0) \qquad (11.68)$$

We see that $\Psi^k_{ij}(z, \bar{z}, 0, 0)$ is the contribution of the family $[\phi_k]$ to the product $\phi_i \phi_j$. Returning to the 4-point function we have

$$< \phi_i(z_1, \bar{z}_2)\phi_j(z_2, \bar{z}_2)\phi_k(z_3, \bar{z}_3)\phi_l(z_4, \bar{z}_4) >$$

$$= \sum_m \frac{C^m_{kl}}{z^{h_k + h_l - h_m} \bar{z}^{\bar{h}_k + \bar{h}_l - \bar{h}_m}} < \phi_i(z_1, \bar{z}_2)\phi_j(z_2, \bar{z}_2)\Phi^m_{kl}(z_3, \bar{z}_3, z_4, \bar{z}_4) >$$

$$= \sum_m C^m_{kl} C^m_{ij} \mathcal{F}^{ij}_{kl}(m, z) \bar{\mathcal{F}}^{ij}_{kl}(m, \bar{z})$$

where

$$\mathcal{F}^{ij}_{kl}(m, z) = \sum_L \frac{\beta^{mL}_{kl}}{z^{h_k + h_l - h_m - |L|} \sqrt{C^m_{ij}}} < \phi_1(\infty, \infty)\phi_j(1, 1)\Phi^L_m(0, 0) >$$

$$(11.69)$$

and we have adjusted the four points $(z_1, \ldots, z_4)$ to have the values $(\infty, 1, z, 0)$ respectively; we can do this because Möbius transformations enable us to fix three of the points. The functions $\mathcal{F}^{ij}_{kl}(m, z)$ used to construct the 4-point function are known as *conformal blocks*.

For $n \geq 4$ we have a similar result and Möbius invariance shows that the conformal block will now depend on $(n - 3)$ independent coordinates; we represent this as

$$< \phi_{i_1}(z_1, \bar{z}_1) \cdots \phi_{i_n}(z_n, \bar{z}_n) >= \sum_{I, \bar{J}} \mathcal{F}^I(z_1, \ldots, z_{n-3}) h_{I\bar{J}} \bar{\mathcal{F}}^{\bar{J}}(\bar{z}_1, \ldots, \bar{z}_{n-3})$$

$$(11.70)$$

with I and $\bar{J}$ appropriate generalised indices and $h_{I\bar{J}}$ a Hermitian metric constructed from the C^k_{ij}. A general correlation function can be thought of as defining a function which is finite on the Riemann sphere with n punctures. But the conformal blocks are conformally invariant and only depend on the underlying complex structure; thus they are holomorphic on the Riemann moduli space for this punctured surface. We shall return to these holomorphic properties below.

If ϕ_i and ϕ_j are two primary fields we have yet to discuss the matter of which primary fields can occur in the expansion of the product $\phi_i(z_1, \bar{z}_1)\phi(z_2, \bar{z}_2)$. This can be deduced from a calculation of the 3-point function $< \phi_i(z_1, \bar{z}_2)\phi_j(z_2, \bar{z}_2)\phi_k(z_3, \bar{z}_3) >$; a non-vanishing 3-point function can thought of as a fusing together of three representations of the conformal group. Data that describe which representations can be fused together are called *fusion rules*. A precise definition of the fusion rules is given in Verlinde [1] , who introduces the formal product $\phi_i \times \phi_j$ of two primary fields; this product is defined by the equation

$$\phi_i \times \phi_j = N_{ij}^k \phi_k \qquad (11.71)$$

where summation takes place over the repeated index k and the N_{ij}^k are integers which are equal to the multiplicity of ϕ_k in the operator product $\phi_i \phi_j$. The fusion rules are now regarded as a specification of the formal product $\phi_i \times \phi_j$.

A simple example of a fusion rule is available from the theory of a free scalar field. We have seen already in 11.8 that, in two dimensions, the free scalar field φ is not a primary field. However, if $p \in \mathbf{R}$, the vertex operator $V_p(z, \bar{z})$ defined by

$$V_p(z, \bar{z}) = {}^{\circ}_{\circ} \exp[ip\varphi(z, \bar{z})] {}^{\circ}_{\circ}$$

is a primary field of weight $(p^2/4, p^2/4)$. Since φ is free the operator product of two vertex operators is easily calculated and we find that

$$V_p(z, \bar{z})V_q(0, 0) = |z|^{(pq/2)} {}^{\circ}_{\circ} \exp[i(p + q)\varphi(0, 0)] {}^{\circ}_{\circ} + \cdots$$
$$= z^{(pq/4)} \bar{z}^{(pq/4)} V_{p+q}(0, 0) + \cdots \qquad (11.72)$$

Hence, if we write $\phi_p = V_p$, the fusion rules are simply

$$\phi_p \times \phi_q = \phi_{p+q} \qquad (11.73)$$

so that $N_{pq}^r = \delta_{r,p+q}$. This example has an obvious extension to d-dimensional string theory in which V_p is given by $V_p = \exp[ip_\mu \phi^\mu]$ and ϕ^μ is now a d-component scalar field. The same operator product applies and the fusion rule now has a physical interpretation: it is just the statement of momentum conservation.

In string theory we work on Riemann surfaces of arbitrary genus and we can do the same here. To come to grips with the non-trivial topological structure of Riemann surfaces of arbitrary genus will require a slightly more abstract approach.

Suppose that our F primary fields introduced above now constitute a conformal field theory on the surface Σ of genus p. The various expressions derived above must now be interpreted as being valid in some neighbourhood of the origin of a local coordinate on Σ. A general conformal block, which we write schematically as $\mathcal{F}^J(\mathbf{z})$, is now defined on $\Sigma_{p,n}$, by which we mean a Riemann surface of genus p with n distinct punctures. The collection of linearly independent conformal blocks form a vector space $V(\Sigma_{p,n})$. Now each primary field in the product $\phi_i \times \phi_j$ will give rise to a conformal block and so, applying this when $n = 3$, we have

$$N_{ij}^k = \dim V(\Sigma_{p,3}) \tag{11.74}$$

If we introduce $\mathcal{M}_{p,n}$—the moduli [5] space of Riemann surfaces of genus p with n punctures—then the vector spaces $V(\Sigma_{p,n})$ will vary holomorphically over $\mathcal{M}_{p,n}$ and the conformal invariance of the theory means that the conformal blocks can be considered as holomorphic sections of a vector bundle $V_{p,n}$ over the moduli space $\mathcal{M}_p$; the bundle $V_{p,n}$ being formed by fitting together the holomorphic family $\{V(\Sigma_{p,n})\}$. This is the generalisation of the remark we made on the conformal blocks for genus zero on p. 315.

Verlinde [1] showed that the N_{ij}^k form a representation of the fusion algebra 11.71; he then computed the characters of the representation of the Virasoro algebra and examined their behaviour under the action of elements of the modular group $SL(2, \mathbf{Z})/\{\mp I\}$. A distinguished rôle is played by the element $S \in SL(2, \mathbf{Z})/\{\mp I\}$, which we recall from chapter 5 is the map $z \mapsto -(1/z)$. The matrix representing the action of S on the characters can be used to diagonalise the fusion algebra and a simple formula results expressing N_{ij}^k in terms of matrix elements of S. Further, since any punctured Riemann surface $\Sigma_{p,n}$ can be assembled by joining together $\Sigma_{0,3}$'s—i.e. Riemann spheres with three punctures—then the fusion rules allow the dimension of a general $V(\Sigma_{p,n})$ to be determined in terms of the N_{ij}^k.

As we observed at the end of § 3, conformal field theories can be obtained from representations of $\overline{LG}$ or its corresponding Kac–Moody algebra. We know that a representation of $\overline{LG}$ requires us to specify a representation R of G and a level k. For use in the next chapter we shall require some of the values of $\dim V(\Sigma_{p,n})$ for these theories. We only need the case where $\Sigma = S^2$ and we modify the notation $V(S^2_{p,n})$ to $V(S^2_{R_1,\dots,R_n})$ so that we can

---

[5] We calculated *index* $\overline{\partial}$ for a Riemann surface with n holes in chapter 4; if we combine this with our result 5.99 for $\dim \mathcal{M}_p$ it is not difficult to show that $\dim \mathcal{M}_{p,n} = 3p - 3 + n$ provided this number is non-negative.

display the representation content of each puncture. The dimensions that we need are then

$$\dim V(S^2_{R_i}) = \begin{cases} 1 & \text{if } R_i \text{ is the trivial representation} \\ 0 & \text{otherwise} \end{cases}$$

$$\dim V(S^2_{R_i,R_j}) = g_{ij} \equiv N^0_{ij} \tag{11.75}$$

$$\text{with} \quad g_{ij} = \begin{cases} 1 & \text{if } R_i = R^*_j \\ 0 & \text{otherwise} \end{cases}$$

and R^*_j is the dual of the representation R_j.

The Weierstrass gap theorem is a constraint on the nature of the mero-morphic singularities when $p > 0$. This is also a constraint on the properties of some of the correlation functions of conformally invariant theories, cf. for example Vafa [1] and Alvarez-Gaumé, Gomez and Vafa [1].

Axiomatic approaches to conformal field theories have also been developed. We close this chapter with a few remarks about these approaches. For a proper account see Friedan and Shenker [1,2] and Segal [4].

Friedan and Shenker's approach is designed to formulate a conformal field theory as an analytic geometry over a suitable space $\mathcal{R}$ known as the *universal moduli space*. A conformal field theory constructed in this way is called a *gauge system*. A key application is to string theory, which we know to be a conformal field theory with $c = 26$. To describe the universal moduli space $\mathcal{R}$ we start with the conventional Riemann moduli space $\mathcal{M}_p$. The partition function for the closed Bosonic string is expressed as

$$Z = \sum_{p=0}^{\infty} \int_{\mathcal{M}_p} I_p \tag{11.76}$$

where I_p denotes the integrand. The expansion of Z contains only connected string vacuum diagrams. It is standard in quantum field theory that if we form $\exp[Z]$ we obtain the set of all string vacuum diagrams, including the disconnected ones. We therefore write

$$\exp[Z] = \exp\left[\sum_{p=0}^{\infty} \int_{\mathcal{M}_p} I_p\right]$$

$$= \prod_{p=0}^{\infty} \exp\left[\int_{\mathcal{M}_p} I_p\right] \tag{11.77}$$

$$= \prod_{p=0}^{\infty} \sum_{n=0}^{\infty} \frac{\left\{\int_{\mathcal{M}_p} I_p\right\}^n}{n!}$$

Now Friedan and Shenker [1] define the universal moduli space $\mathcal{R}$ by

$$\mathcal{R} = \prod_{p=0}^{\infty} \left\{ \bigcup_{n=0}^{\infty} Sym^n \left(\mathcal{M}_p \right) \right\} \tag{11.78}$$

where Sym means the symmetrised product of n factors and thus produces the $n!$ factors in the expansion of $\exp[Z]$. Having defined $\mathcal{R}$ the full string vacuum amplitude can be written as

$$\exp[Z] = \int_{\mathcal{R}} I \tag{11.79}$$

Note that $\mathcal{R}$ contains connected and disconnected Riemann surfaces; to deal satisfactorily with this feature of $\mathcal{R}$ it is desirable to allow the theory to have Riemann surfaces with nodes. This permits one to formulate a natural factorisation condition for the partition function: this condition asserts that the partition function for a surface with nodes is the product of the corresponding partition functions of the (possibly disconnected) surfaces that are created by pinching all the nodes so as to remove them. To include Riemann surfaces with nodes the moduli space $\mathcal{M}_p$ is replaced by a compactification[6] $\overline{\mathcal{M}}_p$ which is called the moduli space of *stable* Riemann surfaces. The set $\overline{\mathcal{M}}_p - \mathcal{M}_p$ is the moduli space of the surfaces with nodes and, since conformal blocks are generically singular on such surfaces, it can be thought of as a locus of singularities or a *divisor;* for this reason $\overline{\mathcal{M}}_p - \mathcal{M}_p$ is known as the *compactification divisor.* When nodes are included, the corresponding *stable* universal moduli space is $\overline{\mathcal{R}}$ and we replace 11.79 by

$$\exp[Z] = \int_{\overline{\mathcal{R}}} I \tag{11.80}$$

The energy momentum tensor $T(z)$ is used to define a connection on a line bundle over $\overline{\mathcal{M}}_p$. This is done by exploiting the fact that $T(z)$ has the affine transformation law

$$T'(z')(dz')^2 = T(z)dz^2 + \frac{c}{12}\{z, z'\}(dz')^2 \tag{11.81}$$

Thus the difference of two $T(z)dz^2$'s is a tensor like the difference of two gauge potentials. More precisely, after applying Serre duality, this difference

---

[6] The particular compactification used is quite natural and is described in Friedan and Shenker [1]

becomes a 1-form on $\overline{\mathcal{M}}_p$ and so $T(z)dz^2$ defines a connection 1-form on $\overline{\mathcal{M}}_p$.

Actually $T(z)$ is a projective connection and gives rise to a projective line bundle over $\overline{\mathcal{M}}_p$ and this in turn induces another projective bundle E_c on the space $\overline{\mathcal{R}}$; also, since the energy momentum tensor $T(z)$ is got by varying the moduli on which the partition function depends, then Z can viewed as being obtained by formally integrating this variation over $\overline{\mathcal{R}}$. It turns out that $E_c = (det\, T^* \overline{\mathcal{R}})^{c/2}$ and that the partition function can be realised as the norm squared of a holomorphic section of a bundle $V = E_c \otimes W$ over $\overline{\mathcal{R}}$ where W is a certain vector bundle over $\overline{\mathcal{R}}$ furnished with a projectively flat Hermitian metric h; for further details see Friedan and Shenker [1]. In principle, non-perturbative effects may be accessible by evaluating a $p \to \infty$ limit of the theory and then expanding in inverse powers of p. Thus if $\overline{\mathcal{R}}$ is extended to a space $\overline{\mathcal{R}}_\infty$ which includes surfaces of infinite genus it is possible that non-perturbative effects may be incorporated.

In Segal's approach one considers a Riemann surface Σ with boundary $\partial \Sigma$ which consists of n circles; homotopically this is the same as a surface with n punctures.

From the string point of view it is natural to deem q of the n circles to be incoming and r of them to be outgoing. The evolution of the q incoming strings into r outgoing ones is meant to be described by an operator

$$U_{\Sigma,\partial\Sigma} : \underbrace{H \otimes \cdots \otimes H}_{q \text{ factors}} \longrightarrow \underbrace{H \otimes \cdots \otimes H}_{r \text{ factors}} \tag{11.82}$$

where H is a Hilbert space of states which is defined as follows. Let $\Sigma_{p,2}$ be a Riemann surface of genus p whose boundary consists of precisely two circles, one incoming and one outgoing. We can think of the evolution from the in-state to the out-state as being given in terms of the loop space $Map\,(S^1, \Sigma_{p,2}) \equiv L\Sigma_{p,2}$: an element of $L\Sigma_{p,2}$ gives a string positioned somewhere on $\Sigma_{p,2}$ and at the boundary the string is either the in-state or the out-state. Thus we define

$$\alpha : [0, t] \longrightarrow L\Sigma_{p,2} \tag{11.83}$$

such that $\alpha(0) = \alpha_0$ and $\alpha(t) = \alpha_t$ give the boundary of $\Sigma_{p,2}$ and the period of time over which the evolution takes place is t. $L\Sigma_{p,2}$ is clearly the configuration space of the system and the Hilbert space H is the corresponding space of wave functions

$$H = L^2\,(L\Sigma_{p,2}) \tag{11.84}$$

Now if S is some conformally invariant action, such as the string action, then, following Segal, we form the integral

$$\int_{[\alpha_0, \alpha_t]} \exp[-S(\alpha)]\mathcal{D}\alpha = K(\alpha_0, \alpha_t) \qquad (11.85)$$

where $[\alpha_0, \alpha_t]$ denotes the set of α subject to the boundary condition that $\alpha(0)$ and $\alpha(t)$ comprise $\partial \Sigma_{p,2}$. Having done this the operator $U_{\Sigma_{p,2}}$ acting on a wave function $\psi(\alpha)$ is formally defined via a functional integral

$$U_{\Sigma_{p,2}}(\psi(\alpha)) = \int K(\alpha_0, \alpha_t)\psi(\alpha_t)\mathcal{D}\alpha_t \qquad (11.86)$$

If we join the two boundary circles of $\Sigma_{p,2}$ together we get a closed surface $\tilde{\Sigma}_{p,2}$; the operator $U_{\tilde{\Sigma}_{p,2}}$ associated to $\tilde{\Sigma}_{p,2}$ is then required to be a trace so that

$$tr\,(U_{\Sigma_{p,2}}) = U_{\tilde{\Sigma}_{p,2}} \qquad (11.87)$$

When $p = 0$, $\Sigma_{0,2}$ is a cylinder, $tr\,(U_{\Sigma_{0,2}})$ is a partition function and $\tilde{\Sigma}_{0,2}$ is a torus; the conformal invariance of S then implies modular invariance for the partition function. Alternatively, if two surfaces $\Sigma_{p,2}$ and $\Sigma'_{p,2}$, both of whose boundaries consist of a pair of circles, are sewed together, another such surface results and this gives rise to a condition on their associated operators

$$U_{\Sigma_{p,2}} \circ U_{\Sigma'_{p,2}} = \lambda U_{\Sigma_{p,2} \cup \Sigma'_{p,2}} \qquad (11.88)$$

where $\Sigma_{p,2} \cup \Sigma'_{p,2}$ denotes the surface obtained by the sewing operation and λ is a scalar multiplier.

The more general operator $U_{\Sigma, \partial\Sigma}$ of 11.82 can now be defined in an analogous manner. Segal uses functorial methods to define a conformal field theory. The collection of all finite disjoint unions of circles form the objects of a category $\mathcal{C}$, these are the boundary circles. A conformal field theory is a *modular functor* from $\mathcal{C}$ to a category $\mathcal{H}$ of topological vector spaces whose objects include the spaces $H \otimes \cdots \otimes H$. A morphism between a pair of objects in $\mathcal{C}$ is a Riemann surface which joins the two objects such that they form its boundary, morphisms in $\mathcal{H}$ are operators of the form $U_{\Sigma, \partial\Sigma}$. To compose morphisms in $\mathcal{C}$ we note that two composable morphisms are two surfaces with part of their boundaries in common and one simply sews the two surfaces together along this common part of their boundary; in $\mathcal{H}$ one simply composes the operators. The whole structure can be made quite concrete and fundamental notions such as conformal blocks and fusion rules can be identified, cf. Segal [4].

CHAPTER XII

Topological Quantum Field Theories

§ 1. Introduction

The initial data for the conformal field theories considered in the last chapter is a Riemann surface Σ and its complex structure; once these have been assigned, the Hilbert space of the particular theory can be constructed. One can conceive of starting with even less data and trying to construct a corresponding quantum field theory. An obvious piece of data to attempt to discard is the complex structure on Σ; a quantum field theory constructed from Σ alone, using neither complex structure nor metric, is a topological quantum field theory. Without a metric there are no distance measurements or forces and so no conventional dynamics. The Hamiltonian $\mathcal{H}$ of the theory has only zero eigenstates and the Hilbert space of the theory, unlike the conformal case, is usually finite dimensional. The non-triviality of the theory is reflected in the existence of tunnelling between vacua.

Topological field theories are not restricted to space-times of dimension 2; when the space-time is 3-dimensional we shall describe a striking application to knot theory. We shall see below that there is also an axiomatic approach which is analogous to that used for conformal field theories. It turns out that a rather natural topological field theory is encountered in trying to construct a quantum field theoretic generalisation of the quantum mechanical Morse theory that we discussed in chapter 7. This forms the subject of the next section.

§ 2. Floer theory and the Chern-Simons function

In this section we shall meet our first topological field theory. However, we shall postpone its introduction until the end of the section. In fact the bulk of the section is taken up with an extremely interesting generalisation of Witten's Morse theory complex of chapter 7.

In chapter 7 we used supersymmetric quantum mechanics both to derive the Morse inequalities and to construct the de Rham cohomology of a manifold M. The Hilbert space of the theory is

$$H = \bigoplus_{p \geq 0} \Omega^p(M) \qquad (12.1)$$

so that H is graded by p with even and odd p corresponding to Bosons and Fermions respectively. The de Rham complex is realised as the formal sum of the critical points with instantons being employed to construct the coboundary operator.

To generalise this example to quantum field theory we replace the finite dimensional manifold M by the infinite dimensional manifold of gauge orbits $\mathcal{A}/\mathcal{G}$. However, unlike the finite dimensional case, it is too difficult to work with an arbitrary function f on any $\mathcal{A}/\mathcal{G}$; in order to have a tractable problem it is necessary to make quite specific choices for f and $\mathcal{A}/\mathcal{G}$. Floer [1,2] chooses [1]

$$\mathcal{A} = \{\text{The } SU(2)\text{-connections on a closed orientable 3-manifold } M\} \qquad (12.2)$$

Before giving the function f considered by Floer we digress briefly to supply some useful background on principal $SU(2)$-bundles over 3-manifolds. Such bundles are classified by elements of $[M, BSU(2)]$; but $SU(2)$ are the unit quaternions and so, if $\mathbf{HP}^\infty$ is the infinite dimensional quaternionic projective space, we have $BSU(2) = \mathbf{HP}^\infty$ by exactly the same reasoning that we used in chapter 7 to show that $BU(1) = \mathbf{CP}^\infty$. We can also calculate $P_t(BSU(2))$ using the method employed there to calculate the Poincaré polynomial $P_t(BU(1))$. The result is that

$$P_t(BSU(2)) = 1 + t^4 + t^8 + \cdots = \frac{1}{1 - t^4} \qquad (12.3)$$

from which we deduce that a cell decomposition of $BSU(2)$ contains cells only in dimensions divisible by 4. Now we consider $[M, BSU(2)]$, bearing in mind that M is 3-dimensional. The cellular approximation theorem (Whitehead [1]) says that any map in $[M, BSU(2)]$ is homotopic to a map whose image is in the 3-skeleton [2] of $BSU(2)$; we have just seen that this

---

[1] We shall also assume that $H_1(M; \mathbf{Z}) = 0$ which, by Poincaré duality, means that $H_2(M; \mathbf{Z}) = 0$ and therefore M has the same homology as S^3; such manifolds are called (oriented) homology 3-spheres. This assumption is made to avoid difficulties with reducible connections, cf. p. 325.

[2] The n-skeleton of a CW complex X is the union of those cells whose dimension is at most n.

3-skeleton is trivial so that all elements of $[M, BSU(2)]$ are homotopically trivial. Hence all $SU(2)$ bundles over M are isomorphic to the trivial bundle $M \times SU(2)$. This fact has other useful consequences, among which we mention the following: the group of gauge transformations $\mathcal{G}$ is of the untwisted form $\mathcal{G} = Map\,(M, SU(2))$ and the tangent space to $\mathcal{A}$ is given by $T_A\mathcal{A} = \Omega^1(M) \times su(2)$.

Returning now to the Morse theory we must describe the function f studied by Floer; f is simply the Chern–Simons function obtained by integrating the Chern–Simons secondary characteristic class of chapter 8. We have

$$f : \mathcal{A} \longrightarrow \mathbf{R}$$
$$A \longmapsto f(A)$$
$$\text{with} \quad f(A) = -\frac{1}{8\pi^2} \int_M tr(A \wedge dA + \frac{2}{3} A \wedge A \wedge A) \tag{12.4}$$

This function f would descend to the quotient $\mathcal{A}/\mathcal{G}$ if it were gauge invariant; however, this is not so: instead f satisfies (Nash and Sen [1])

$$f(A_g) = f(A) + n, \quad n \in \mathbf{Z} \tag{12.5}$$

where A_g is the gauge transform of A under the gauge transformation g. However, if we modify the f of 12.4 by composing it with the natural projection from $\mathbf{R}$ to $\mathbf{R}/\mathbf{Z}$ then the resulting function takes values in $\mathbf{R}/\mathbf{Z}$; it is then gauge invariant and so does descend to $\mathcal{A}/\mathcal{G}$.

In the usual way the critical points of f are given by

$$df(A) = 0 \tag{12.6}$$

where the exterior derivative is now taken to be acting in the space $\mathcal{A}$. If A is such a critical point then we can write $A_t = A + ta$ and obtain

$$f(A_t) = -\frac{1}{8\pi^2} \int_M tr(A_t \wedge dA_t + \frac{2}{3} A_t \wedge A_t \wedge A_t)$$
$$= f(A) - \frac{t}{8\pi^2} \int_M tr\{(a \wedge dA + A \wedge da) + \frac{2}{3}(a \wedge A \wedge A)$$
$$+ \frac{2}{3}(A \wedge a \wedge A) + \frac{2}{3}(A \wedge A \wedge a)\} + \cdots$$
$$= f(A) - \frac{t}{4\pi^2} \int_M tr(F(A) \wedge a) + \cdots$$
$$\tag{12.7}$$

Hence we can conclude that [3]

$$df(A) = -\frac{F(A)}{4\pi^2} \tag{12.8}$$

and so the critical points of the Chern–Simons function are the *flat connections* on M.

If $\pi_1(M) \neq 0$ then flat connections on M are not trivial, since they can have non-zero holonomy round a non-trivial loop on M. The holonomy of each flat connection is an $SU(2)$ element parametrised by a loop on M; in this way it defines a representation of $\pi_1(M)$ in $SU(2)$. In fact this is precisely how these connections are characterised: Representations of $\pi_1(M)$ in $SU(2)$ are given by

$$Hom\,(\pi_1(M), SU(2)) \tag{12.9}$$

However, the group $Ad\,SU(2)$ acts on a representation by conjugation to give an equivalent one; thus the set of flat connections is the quotient

$$Hom\,(\pi_1(M), SU(2))/Ad\,SU(2) \tag{12.10}$$

These flat connections are also all irreducible because we required M to be a homology 3-sphere: A reducible flat connection would give rise to an Abelian representation of $\pi_1(M)$ in a $U(1) \subset SU(2)$; but, since $H_1(M; \mathbf{Z}) = 0$, the Abelian part of $\pi_1(M)$ is zero and so this representation is trivial.

Having found a critical point we would like to calculate its index and so we must also calculate the Hessian; expanding 12.7 to order t^2 shows that the coefficient of $(t^2/2)$ is

$$-\frac{1}{4\pi^2} \int_M \{tr(a \wedge da) + \frac{2}{3}tr(a \wedge a \wedge A + a \wedge A \wedge a + A \wedge a \wedge a)\}$$

$$= -\frac{1}{4\pi^2} \int_M tr\{a \wedge (da + 2a \wedge A)\}$$

$$= -\frac{1}{4\pi^2} \int_M tr\{a \wedge da + a \wedge A \wedge a + a \wedge a \wedge A\} \tag{12.11}$$

$$= -\frac{1}{4\pi^2} \int_M tr\{a \wedge d_A a\}$$

$$= \frac{1}{4\pi^2} < a, *d_A a > = < a, H(A)a >$$

---

[3] This means that we should be able to interpret $F(A)$ as a 1-form on the space $\mathcal{A}$. We can indeed do this because a 1-form is a linear functional on the tangent space $T_\mathbf{A}\mathcal{A}$; we denote the action of $F(A)$ on an arbitrary $a \in T_\mathbf{A}\mathcal{A}$ by $F_a(A)$ where $F_a(A) = \int_M tr(F(A) \wedge a)$.

where $H(A)$ is the Hessian and use of the usual inner product on forms identifies it as the differential operator $(*d_A/4\pi^2)$ acting on 1-forms. The trouble is that the index of a critical point is the number of negative eigenvalues of the Hessian but, in this case, $4\pi^2 H(A) = *d_A$ and a *first order* linear operator such as this has no upper or lower bounds; thus $H(A)$ has generically an infinite number of positive and negative eigenvalues, giving our critical point an *infinite* Morse index. By contrast the other infinite dimensional Morse functions f that we considered—the energy functional for geodesics and the various Yang–Mills examples—all have a Hessian which is a *second order* operator of Laplace type and this always has a lower bound and thus a *finite* Morse index.

To see how to get round this difficulty we recall how the Morse index enters into the de Rham cohomology construction of chapter 7: C_p is the set of critical points of index p and C^p is its dual. The de Rham cohomology of the relevant manifold is then constructed by defining a coboundary operator δ

$$\cdots \xrightarrow{\delta} C^p \xrightarrow{\delta} C^{p+1} \xrightarrow{\delta} \cdots \tag{12.12}$$

connecting critical points whose index differs by one. For the present example of the Chern–Simons function, Floer bypassed the difficulty of infinite Morse index by just defining the *difference* of the Morse indices between a pair of critical points; this *relative Morse index* is sufficient to define a cohomology complex. It is defined using spectral flow and is always finite. The resulting homology or cohomology theory is of an entirely new kind and we shall examine it below.

Before coming to the homology theory we must define the relative Morse index. In analogy with § 4 of chapter 7 we take two critical points A_P and A_Q and join them with a steepest descent path $A(t)$; that is, $A(t)$ is a path on $\mathcal{A}$ with end points A_P and A_Q which obeys the equation

$$\frac{dA(t)}{dt} = -grad\, f(A(t)) \tag{12.13}$$

with $grad$ denoting the gradient operator on the space $\mathcal{A}$. If we extend the definition of the Hessian $H(A)$ to all A on $A(t)$ then $4\pi^2 H(A(t)) = *d_{A(t)}$ is a one parameter family of self-adjoint operators on $\mathcal{A}$. But drawing on chapter 10, § 6, we consider the spectral flow of $H(A(t))$ from A_P to A_Q. This is the net number of negative eigenvalues of $H(A_P)$ which flow into positive eigenvalues of $H(A_Q)$. This number is finite and is a 'renormalised' measure of the difference between the Morse indices of A_P and A_Q. Denoting the spectral flow by $\sigma(A_P, A_Q)$ we define

$$\lambda_{A_P, A_Q} = \sigma(A_P, A_Q) \tag{12.14}$$

The integer λ_{A_P, A_Q} is the relative Morse index and, if the problem were finite dimensional, we would simply have $\lambda_{A_P, A_Q} = \lambda_{A_P} - \lambda_{A_Q}$.

There can be more than one steepest descent path $A(t)$ from A_P to A_Q and thus it is important to know how $\sigma(A_P, A_Q)$ depends on $A(t)$. The difference between two paths $A(t)$ and $A'(t)$ is a loop, and from chapter 10 we know that the spectral flow round a loop is given by the Atiyah–Singer index of an associated elliptic operator; since the index is a deformation invariant the spectral flow only depends on the homotopy class of the loop. Thus if $A(t)$ is homotopic to $A'(t)$ then the relative Morse index is unchanged. However, the Chern–Simons function f is really defined on the quotient $\mathcal{A}/\mathcal{G}$ and when the paths are taken on this space the question arises as to whether $\mathcal{A}/\mathcal{G}$ is simply connected or not. In fact[4] $\pi_1(\mathcal{A}/\mathcal{G}) = \mathbf{Z}$ and so there are infinitely many homotopically distinct loops passing through A_P and A_Q. Hence $\sigma(A_P, A_Q)$ is only defined modulo the flow round loops through A_P and A_Q. The fundamental loop is the one that generates $\pi_1(\mathcal{A}/\mathcal{G})$; the flow round any other loop is just a multiple of the flow round the generating loop.

Thus, to complete our definition of λ_{A_P, A_Q}, we must calculate the spectral flow round a generating loop and, as we have already observed, this is an index calculation: the spectral flow of $*d_{A(t)}$ on M is given by the index of $*d_{A(t)} + (\partial/\partial t)$ on $M \times S^1$ with t a local coordinate on S^1.

This index has an interpretation in terms of instantons which is the key to its calculation. We return to the gradient flow

$$\frac{dA(t)}{dt} = -grad\, f(A(t)) \qquad (12.15)$$

To define $grad$ we must supply a metric or inner product on the vector space $T_A\mathcal{A}$. Since the vectors in $T_A\mathcal{A}$ are the $su(2)$-valued 1-forms on M we have available the usual metric

$$< a, b > = -\int_M tr(a \wedge *b), \qquad a, b \in T_A\mathcal{A} \simeq \Omega^1(M \times su(2)) \qquad (12.16)$$

With this metric $grad\, f$ is the vector field on $T_A\mathcal{A}$ defined by

$$< grad\, f, a > = df_a, \qquad a \in T_A\mathcal{A} \qquad (12.17)$$

---

[4] This is because $\mathcal{G} = Map(M, SU(2))$ and, since both M and $SU(2)$ are 3-dimensional, this means that $\mathcal{G}$ has infinitely many components with each component consisting of elements of $Map(M, SU(2))$ with the same degree. Now if we take the homotopy sequence of the fibration $\mathcal{G} \to \mathcal{A} \to \mathcal{A}/\mathcal{G}$ we obtain $\pi_1(\mathcal{A}) \to \pi_1(\mathcal{A}/\mathcal{G}) \to \pi_0(\mathcal{G}) \to \pi_0(\mathcal{A})$; but $\mathcal{A}$ is contractible, yielding $\pi_{1,2}(\mathcal{A}) = 0$, and hence $\pi_1(\mathcal{A}/\mathcal{G}) = \pi_0(\mathcal{G}) = \mathbf{Z}$.

and on the RHS we have used the notation introduced in the footnote to p. 325 so that df_a denotes the action of the 1-form df on the vector a. But

$$df = -\frac{F(A)}{4\pi^2}$$

$$\Rightarrow df_a = -\frac{F_a(A)}{4\pi^2}$$

$$= -\frac{1}{4\pi^2}\int_M tr(F(A) \wedge a) \tag{12.18}$$

$$= <\frac{F(A)}{4\pi^2}, *a>$$

$$= <\frac{*F(A)}{4\pi^2}, a>$$

Thus combining the two preceding equations shows that

$$grad\, f(A) = \frac{*F(A)}{4\pi^2} \tag{12.19}$$

and so the gradient flow equation becomes

$$\frac{dA(t)}{dt} = - * F(A) \tag{12.20}$$

where we have absorbed the $4\pi^2$ in a redefinition of t.

Now we consider the 4-manifold $M \times \mathbf{R}$ and extend the connection $A(t)$ to a connection $\mathbf{A}$ on $M \times \mathbf{R}$ by letting its fourth component be zero, i.e. the covariant derivative in the $\mathbf{R}$-direction coincides with the partial derivative $(\partial/\partial t)$. Next we give $M \times \mathbf{R}$ a metric, namely the metric which is the product of the flat metric on $\mathbf{R}$ times the metric on M. This allows us to define a $*$ operation on the 4-manifold and so we can consider the instantons

$$\mathbf{F} = \mp * \mathbf{F} \tag{12.21}$$

where $\mathbf{F}$ is the curvature of $\mathbf{A}$ on $M \times \mathbf{R}$. Actually the gradient flow equation 12.20 is simply the anti-self-duality equation for $\mathbf{A}$: this can easily be verified by an elementary calculation since the curvature components $\mathbf{F}_{4i}$ reduce to $\mathbf{F}_{4i} \equiv dA/dt$.

Since $M \times \mathbf{R}$ is not closed, 12.21 needs boundary conditions; the natural choice is to set

$$A(\infty) = A_Q, \qquad A(-\infty) = A_P \tag{12.22}$$

These connections A_P and A_Q are flat; this is consistent with the fact that if, as usual, we require the four dimensional Yang–Mills action to be finite then the curvature $\mathbf{F}$ tends to zero at infinity. Having found the connection of the gradient flow with instantons we return once more to the matter of the index of $*d_{a(t)} + (\partial/\partial t)$ on $M \times S^1$. In fact we shall show now that the same operator $*d_{a(t)} + (\partial/\partial t)$ turns up in the instanton deformation problem, i.e. we perturb the instanton A to a new instanton $A + \epsilon a$. If we do this we get

$$\frac{d}{dt}(A(t) + \epsilon a(t)) = - * F(A(t) + \epsilon a(t))$$

$$\Rightarrow \frac{dA(t)}{dt} + \epsilon \frac{da(t)}{dt} = - * F(A) - \epsilon * d_{A(t)} a(t) + \cdots \qquad (12.23)$$

$$\Rightarrow \frac{da(t)}{dt} = - * d_{A(t)} a(t)$$

to first order in ϵ. Exploiting the instanton interpretation we write the above equation as

$$\left(\frac{d}{dt} + *d_{A(t)} \right) a(t) = 0 \qquad (12.24)$$

and observe that, after taking proper account of gauge equivalent a's, the index of the above operator is calculable from formulae 8.115 and 8.124 of chapter 8. Proceeding to do this we find that

$$index \left(\frac{d}{dt} + *d_{A(t)} \right) = \dim \mathcal{M}_k$$

$$= p_1(ad_{\mathbf{c}}P) - \frac{\dim G}{2}(\chi(M \times S^1) - \tau(M \times S^1)) \qquad (12.25)$$

But since $G = SU(2)$ and M is a homology 3-sphere for which $H_2(M; \mathbf{Z}) = 0$ we see that

$$p_1(ad_{\mathbf{c}}P) = 8k, \qquad \dim G = 3$$

$$H_2(M \times S^1) = H_2(M) = 0 \qquad (12.26)$$

$$\Rightarrow \tau(M \times S^1) = 0$$

Also it is an elementary observation that $\chi(M) = \chi(S^1) = 0$ so

$$\chi(M \times S^1) = \chi(M)\chi(S^1)$$

$$= 0 \qquad (12.27)$$

Thus the index result is

$$index \left(\frac{d}{dt} + *d_{A(t)} \right) = 8k \qquad (12.28)$$

Moving back to three dimensions we have therefore shown that the spectral flow round a loop is

$$8k \qquad (12.29)$$

If follows that $k = 1$ corresponds to a generating loop of $\pi_1(\mathcal{A}/\mathcal{G})$. The consequence of all this for the Morse theory construction is that the relative Morse index of the Chern–Simons function is only well defined $mod\,8$.

This property of the relative Morse index has immediate consequences for the homology theory constructed from the critical point set C_p. The point is that if C_p and C_q are two critical point sets, the difference $p - q$ is the relative Morse index and therefore p can only take 8 distinct values. The homology complex is then of length 8 (alternatively ∂_{-1} can map C_0 onto C_7 rendering the complex truly periodic)

$$0 \xrightarrow{\partial_7} C_7 \xrightarrow{\partial_6} C_6 \xrightarrow{\partial_5} \cdots \cdots \xrightarrow{\partial_1} C_1 \xrightarrow{\partial_0} C_0 \xrightarrow{\partial_{-1}} 0 \qquad (12.30)$$

and there are only 8 homology groups. These are known as the Floer homology groups of M and we denote them by $HF_p(M)$. We have

$$HF_p(M) = \frac{ker\,\partial_{p-1}}{Im\,\partial_p}, \qquad p = 0, \ldots, 7 \qquad (12.31)$$

Both the definition of the boundary operators ∂_p and the verification that $\partial_{p-1} \circ \partial_p = 0$ can be done (formally) with exactly the same instanton tunnelling method used in chapter 7. Considerable extra analysis needs to be used to deal rigorously with features present due to the infinite dimensionality of the space $\mathcal{A}/\mathcal{G}$; these are dealt with by Floer [1].

An important point to be clear on is the rôle played by the various metrics in the construction of the $HF_p(M)$. Although use is made of metrics on M, $M \times S^1$ and $\mathcal{A}$, the Floer groups are independent of these metrics and constitute a new topological invariant of homology 3-spheres. We have seen, too, that despite the fact that the homology complex is constructed on the *infinite dimensional* space $\mathcal{A}/\mathcal{G}$, the complex is of *finite* length and is graded by $\mathbf{Z}_8$, the integers modulo 8.

Homology 3-spheres also figure in an integer valued invariant $\lambda(M)$ of Casson. The definition of $\lambda(M)$ is complicated (Taubes [5]) but we wish merely to note that it involves a signed sum over the inequivalent irreducible representations of $\pi_1(M)$ in $SU(2)$, i.e. over the elements of $Hom\,(\pi_1(M), SU(2))/Ad\,SU(2)$. Taubes [5] has investigated Casson's invariant using gauge theory methods, and a close relation to Floer homology emerges. Taubes defines an Euler characteristic $\chi_F(M)$ for the Floer complex and finds that

$$\chi_F(M) = 2\lambda(M) \qquad (12.32)$$

The infinite dimensionality of the space $\mathcal{A}/\mathcal{G}$ provides problems for the definition of $\chi_F(M)$ but Taubes uses spectral flow to overcome these difficulties. In finite dimensions we know that the standard Euler characteristic of a manifold X can be found using a 1-form on X: the algebraic sum of its (non-degenerate) zeros is equal to $\chi(X)$. But in the present situation we know that the curvature F can be regarded as a 1-form on $\mathcal{A}$; further, the gauge properties of F allow us to project it to a 1-form on $\mathcal{A}/\mathcal{G}$. Now the zeros of F are the flat connections or the elements of

$$Hom\,(\pi_1(M), SU(2))/Ad\,SU(2) \tag{12.33}$$

a set which, when graded, gives the Floer homology complex. Hence their appropriately signed sum is the Euler characteristic $\chi_F(M)$ of the Floer complex. Taubes uses a Fredholm perturbation, if necessary, to obtain non-degenerate zeros and spectral flow to detect any change of sign in the orientation of the zeros. We notice that $\chi_F(M)$ is always even.

Finally we come to the topological field theory associated with the Chern–Simons function. The theory has action

$$S = -\frac{1}{8\pi^2} \int_M tr(A \wedge dA + \frac{2}{3} A \wedge A \wedge A) \tag{12.34}$$

Note that S is constructed out of M and the connection A—there is no metric. As we have seen already, the critical points of S are given by the flat connections

$$F(A) = 0 \tag{12.35}$$

and these need not be trivial if $\pi_1(M) \neq 0$. However, S is not single valued since, according to 12.5, it changes by an integer under a gauge transformation. Nevertheless, if we consider a Lorentzian space-time of $2+1$ dimensions then the action enters into functional integrals in the form

$$\exp[iS] \tag{12.36}$$

which is still multiply valued. But if we replace S by $2\pi kS$, with $k \in \mathbf{Z}$, then the exponential becomes

$$\exp\left[-\frac{ik}{4\pi} \int_M tr(A \wedge dA + \frac{2}{3} A \wedge A \wedge A)\right] \tag{12.37}$$

which 12.5 shows *is* single valued; variants of this action are also of some interest, cf. Nash [5].

We can go on to construct the field theoretic generalisation of the supersymmetric Hamiltonian of chapter 7. With d now denoting an exterior derivative acting on $\mathcal{A}$ we write

$$\mathcal{H}_t = d_t d_t^* + d_t^* d_t = e^{-ft}(dd^* + d^*d)e^{ft}$$

$$= dd^* + d^*d + t\frac{D^2 f}{Dx^i Dx^j}[a_i^*, a_j] + t^2 < grad\, f, grad\, f >$$

with $\qquad f = -\frac{1}{8\pi^2}\int_M tr(dA \wedge A + \frac{2}{3}A \wedge A \wedge A)$

$$\text{(12.38)}$$

To make the expression for $\mathcal{H}_t$ more explicit we need to use a functional derivative to represent the action of d on the space $\mathcal{A}$. If we refer to equation 12.7 it is clear that d is representable as the operator

$$d = \int_M tr(a \wedge \frac{\delta}{\delta A}) \qquad\qquad (12.39)$$

and $\mathcal{H}_t$ now becomes

$$\mathcal{H}_t = \int_M \left\{ -tr\left(\frac{\delta}{\delta A}\right)^2 - 2t\,tr(a \wedge d_A a^*) - t^2 tr(F \wedge *F) \right\} \qquad (12.40)$$

where a^* is a 1-form dual to a. Viewed as operators on $\mathcal{A}$ the 1-forms a and a^* act by wedge product, e.g. a acts on F (which is a 1-form on $\mathcal{A}$) to give $a \wedge F$; this is an anti-commutative action: if we decompose a using the $su(2)$ basis $\{\lambda^i\}$ by writing $a = a_i \lambda^i$ then the a_i anti-commute and this suggests that we identify a_i and a_i^* with Fermions. This is done in Witten [5] but we should realise that these Fermions have the unconventional property of possessing integral spin. Pursuing the analogy with chapter 7 we decompose $\mathcal{H}_t$ according to

$$\mathcal{H}_t = \bigoplus_{p=0}^{7} \Delta_p \qquad\qquad (12.41)$$

where the Laplacians act on $\mathcal{A}$. Hodge theory then suggests that the cohomology is given by the ground states of the Hamiltonian. Thus the Floer groups have a physical interpretation as the quantum ground states of the non-relativistic system whose Hamiltonian is $\mathcal{H}_t$.

There is a relativistic generalisation (Witten [5]) of this system to four dimensions which is closely related to some remarkable facts about instantons. We turn to these matters in the next section.

§ 3. Donaldson's polynomial invariants

In chapter 1 we described Donaldson's use of the first moduli space $\mathcal{M}_1$ to derive smoothability results about 4-manifolds. In addition to this, by using all the moduli spaces $\mathcal{M}_k$, $k = 1, 2, \ldots$, Donaldson [3,4] has constructed powerful differential topological invariants of simply connected 4-manifolds. These can also be approached from quantum field theory by using the relativistic generalisation we have just referred to at the end of the last section. However, we shall begin by describing Donaldson's invariants from the purely moduli space point of view.

To this end let M be a smooth, simply connected, orientable four manifold without boundary. Let P be a principal G-bundle over M on which is defined a connection A. M is endowed with a Riemannian metric g and this enables us to define a Hodge $*$ and to consider instantons via the anti-self-duality equation

$$F = - * F \tag{12.42}$$

Since M is simply connected the dimension of the moduli space $\mathcal{M}_k$ is the integer

$$\dim \mathcal{M}_k = p_1(ad_{\mathbf{c}}P) - \dim G(1 + b_2^+) \tag{12.43}$$

in the notation of chapter 8. Now we know already that $p_1(ad_{\mathbf{c}}P) = -2c_2(ad_{\mathbf{c}}P)$ so

$$\dim \mathcal{M}_k = -2c_2(ad_{\mathbf{c}}P) - \dim G(1 + b_2^+) \tag{12.44}$$

and we observe that, no matter what G we choose, $\dim \mathcal{M}_k$ is always *even* when b_2^+ is *odd*. The reason that we draw attention to this fact is that we require $\dim \mathcal{M}_k$ to be even to obtain Donaldson's invariants. This will emerge shortly below; in any case, from now on, we shall ensure that $\dim \mathcal{M}_k$ is even by requiring b_2^+ to be odd.

We also commit ourselves to the choice $G = SU(2)$, for which we know that $c_2(ad_{\mathbf{c}}P) = -4k$ and thus we have

$$\dim \mathcal{M}_k = 8k - 3(1 + b_2^+) \tag{12.45}$$

Donaldson's invariants are certain polynomials of degree d in $H_2(M; \mathbf{Z})$; their construction is easy to describe provided we can simplify matters slightly by using de Rham cohomology.

A Donaldson invariant $q_d(M)$ is a symmetric integer polynomial of degree d in the 2-homology $H_2(M; \mathbf{Z})$ of M

$$q_d(M) : H_2(M) \times \cdots \times H_2(M) \longrightarrow \mathbf{Z} \tag{12.46}$$

It is defined using a certain map

$$m : H_2(M) \to H^2(\mathcal{M}_k) \tag{12.47}$$

Let $E_\mathcal{G}$ be the universal $\mathcal{G}$-bundle over $\mathcal{BG}$ with Chern class $c_2(E_\mathcal{G}) \in H^4(\mathcal{BG})$, then for an element $a \in H_2(M)$, whose Poincaré dual[5] is represented by a 2-form $a^* \in H^2(M)$, we can construct the product

$$c_2(E_\mathcal{G}) \wedge a^* \in H^6(M \times \mathcal{BG}) \tag{12.48}$$

where, for convenience, we are using the same notation for the form a^* and the cohomology class it represents. Since

$$\mathcal{M}_k \subset \mathcal{BG} \tag{12.49}$$

by an appropriate restriction we get an element of $H^6(M \times \mathcal{M}_k)$ and finally use integration to divide out by a^*, thereby obtaining

$$\int_M c_2(E_\mathcal{G}) \wedge a^* \in H^2(\mathcal{M}_k) \tag{12.50}$$

The map m can now be defined by

$$\begin{aligned} m : H_2(M) &\longrightarrow H^2(\mathcal{M}_k) \\ a &\longmapsto \int_M c_2(E_\mathcal{G}) \wedge a^* \end{aligned} \tag{12.51}$$

We use m to define $q_d(M)$ by setting $d = \dim \mathcal{M}_k/2$ and writing

$$\begin{aligned} q_d(M) : H_2(M) \times \cdots \times H_2(M) &\longrightarrow \mathbf{Z} \\ a_1 \times \cdots \times a_d &\longmapsto \int_{\overline{\mathcal{M}}_k} m(a_1) \wedge \cdots \wedge m(a_d) \end{aligned} \tag{12.52}$$

where $\overline{\mathcal{M}}_k$ denotes a compactification of the moduli space. We see that the $q_d(M)$ are symmetric integer valued polynomials of degree d in $H^2(M)$, i.e. $q_d(M) \in Sym^d(H_2(M))$; also, since $d = \dim \mathcal{M}_k/2$, we now understand why $\mathcal{M}_k$ must be even dimensional.

The Donaldson invariants are, *a priori*, not very easy to calculate since they require detailed knowledge of the instanton moduli space. However, if M is a complex algebraic surface, a positivity argument shows that

$$q_d(M) \neq 0, \qquad \text{for } d \geq d_0 \tag{12.53}$$

---

[5] If M has dimension n and $i : X \to M$ represents the inclusion of the oriented r-dimensional subset X in M then the Poincaré dual of X is the element $[\omega]$ of $H^{n-r}(M)$ defined by $\int_X i^*\alpha = \int_M \alpha \wedge \omega$ for all $[\alpha] \in H^r(M)$.

with d_0 some integer—in other words the $q_d(M)$ are all non-zero when d is large enough. Conversely, if M can be written as the connected sum [6]

$$M = M_1 \# M_2 \qquad (12.54)$$

where M_1 and M_2 both have $b_2^+ > 0$ then

$$q_d(M) = 0, \qquad \text{for all } d \qquad (12.55)$$

Witten has shown how to obtain the $q_d(M)$ as correlation functions in a BRST-supersymmetric topological field theory. This theory is the relativistic generalisation of that described at the end of § 2. We shall only describe some of its principal features; for a full account, cf. Witten [5]. The action S for the theory is given by

$$S = \int_M d^4x \sqrt{g}\, tr\, \{\ \frac{1}{4}F_{\mu\nu}F^{\mu\nu} + \frac{1}{4}F_{\mu\nu}^* F^{\mu\nu} + \frac{1}{2}\phi D_\mu D^\mu \lambda + iD_\mu \psi_\nu \chi^{\mu\nu}$$
$$- i\eta D_\mu \psi^\mu - \frac{i}{8}\phi[\chi_{\mu\nu},\chi^{\mu\nu}] - \frac{i}{2}\lambda[\psi_\mu,\psi^\mu] - \frac{i}{2}\phi[\eta,\eta] - \frac{1}{8}[\phi,\lambda]^2\ \}$$
$$(12.56)$$

where $F_{\mu\nu}$ is the curvature of a connection A_μ and $(\phi,\lambda,\eta,\psi_\mu,\chi_{\mu\nu})$ are a collection of fields introduced in order to construct the right supersymmetric theory; ϕ and λ are both spinless while the multiplet $(\psi_\mu,\chi_{\mu\nu})$ contains the components of a 0-form, a 1-form and a self-dual 2-form respectively. The significance of this choice of multiplet is that the anti-instanton version of the instanton deformation complex 8.104 contains precisely these fields. A quantum number U is assigned to these fields with values $(2,-2,-1,0,1,-1)$ for the fields $(\phi,\lambda,\eta,A_\mu,\psi_\mu,\chi_{\mu\nu})$ respectively. This quantum number is not conserved, because of the existence of instantons; however, it is conserved $mod\,8$.

Even though S contains a metric its correlation functions are independent of the metric g so that S can still be regarded as a topological field theory. This can be shown to follow from the fact that both S and its associated energy momentum tensor $T \equiv (\delta S/\delta g)$ can be written as BRST commutators $S = \{Q,V\}$, $T = \{Q,V'\}$ for suitable V and V'—cf. Witten [5].

With this theory it is possible to show that the correlation functions are independent of the gauge coupling and hence we can evaluate them in

---

[6] The connected sum $X\#Y$ of two closed n dimensional manifolds X and Y is obtained by removing an n-dimensional disc from each of them and then sewing the two resulting manifolds together along their newly created boundaries.

a small coupling limit. In this limit the functional integrals are dominated by the classical minima of S, which for A_μ are just the instantons

$$F_{\mu\nu} = -F^*_{\mu\nu} \qquad (12.57)$$

We also need ϕ and λ to vanish for irreducible connections. If we expand all the fields around the minima up to quadratic terms and do the resulting Gaussian integrals, the correlation functions may be formally evaluated. Let us consider a correlation function

$$< P >= \int \mathcal{DF} \exp[-S] P(\mathcal{F}) \qquad (12.58)$$

where $\mathcal{F}$ denotes the collection of fields present in S and $P(\mathcal{F})$ is a polynomial in the fields. Now S has been constructed so that the zero modes in the expansion about the minima are the tangents to the moduli space $\mathcal{M}_k$; thus, if the $\mathcal{DF}$ integration is expressed as an integral over modes, all the non-zero modes may be integrated out first leaving a *finite dimensional* integration over $\dim \mathcal{M}_k$. The Gaussian integration over the non-zero modes is a Boson–Fermion ratio of determinants, a ratio which supersymmetry constrains to be ∓ 1 since Bosonic and Fermionic eigenvalues are equal in pairs. This amounts to expressing $< P >$ as

$$< P >= \int_{\mathcal{M}_k} P_n \qquad (12.59)$$

where P_n is an n-form over $\mathcal{M}_k$ and $n = \dim \mathcal{M}_k$. The only non-vanishing correlation functions are those for which $P(\mathcal{F})$ has U quantum number equal to $\dim \mathcal{M}_k$. This is due to the fact that the zero modes in ψ_μ and A_μ are the tangents to the moduli space and therefore impart to its integration measure a U weight $-n$; thus to have a non-zero integral $P(\mathcal{F})$ must have U weight n. If the original polynomial $P(\mathcal{F})$ is chosen in the correct way then calculation of $< P >$ reproduces evaluation of the Donaldson polynomials.

A modification of the connected sum operation can also be used to obtain a relation between the $q_d(M)$ and Floer homology. The idea is to cut M into two non-closed pieces M^+ and M^-, we write

$$M = M^+ \cup_h M^- \qquad (12.60)$$

where the symbol $\cup_h$ means that the boundaries of M^+ and M^- are *homology 3-spheres* rather than being ordinary 3-spheres. Such a decomposition is always possible if the intersection form of M splits according to

$q(M) = q(M^+) \oplus q(M^-)$ and both M^+ and M^- have $b_2^+ > 0$ (Freedman and Taylor [1]).

Let the homology spheres which form the boundaries of M^+ and M^- be N^+ and N^-, then we can consider their Floer homology $HF_*(N^+)$ and $HF_*(N^-)$. Now, for some given topological charge k, we consider instantons on the 4-manifolds M^+ and M^-. Just as was the case in § 2, boundary conditions are required since the $M^{\mp}$ are not closed. The boundary conditions are just the specifying of a connection on the boundaries $N^{\mp}$. The solution set for the two boundary value problems are denoted by $C^{\mp}$; should a solution on M^+ smoothly match up to a solution on M^- then we have a conventional instanton on the closed manifold M. In this way we see that

$$C^+ \cap C^- = \mathcal{M}_k \qquad (12.61)$$

The boundary conditions allow us to construct two Floer homology classes

$$[C^{\mp}] \in HF_*(N^{\mp}) \qquad (12.62)$$

By their construction $N^{\mp}$ inherit opposite orientations from M and so the $HF_*(N^{\mp})$ are dual to one another; Donaldson shows that the pairing of these classes using Poincaré duality gives the invariant $q_d(M)$, that is

$$[C^+] \bullet [C^-] = q_d(M) \qquad (12.63)$$

with $\bullet$ representing the pairing of the homology cycles. The simplest example of this occurs when d takes its lowest value, namely, zero. In this case the Donaldson invariant becomes just an integer and, referring to 12.63, shows that to evaluate $q_d(M)$ we must set $d = \dim \mathcal{M}_k/2$. Thus the topological charge k must satisfy

$$8k - 3(1 + b_2^+) = 0 \qquad (12.64)$$

and the moduli space, being zero dimensional, reduces to a discrete set of points. Delving further into 12.63 shows that $q_0(M)$ is just a signed sum over the elements in this discrete set of instantons, the signs being determined to be those of the intersection pairing of the Floer homology. Recall that in the quantum field theory the calculation of $q_d(M)$ is the calculation of a correlation function $< P >$ of U weight $\dim \mathcal{M}_k$; when $d = \dim \mathcal{M}_k/2 = 0$ this reduces to the evaluation of the partition function Z which is thereby expressed as an algebraic sum over instantons of the form

$$Z = \sum_i (-1)^{n_i} \qquad (12.65)$$

with i labelling the ith instanton and $n_i = 0$ or 1 determining the sign of its contribution to Z. Since the intersection form of an n-dimensional manifold is defined using the homology in the middle dimension $(n/2)$, we can view the Floer groups as behaving like the 'middle dimensional' homology of the space $\mathcal{A}/\mathcal{G}$.

The Floer–Donaldson relationship 12.63 can be read in either direction: one make deductions about Floer homology or Donaldson invariants. For example, if M is a complex algebraic surface we know that $q_d(M) \neq 0$, for $d \geq d_0$; but this implies that the homology spheres $N^\mp$ occurring in the decomposition $M = M^+ \cup_h M^-$ have non-trivial Floer homology groups in $HF_*(N^\mp)$. On the other hand, should the homology 3-spheres in the decomposition coincide with a *genuine* 3-sphere—i.e. $M^+ \cup_h M^- \equiv \hat{M}^+ \# \hat{M}^-$ with $\hat{M}^\mp$ the obvious closed manifolds—then the $HF_*(N^\mp)$ are automatically trivial and the $q_d(M)$ are forced to vanish; of course, this is just what was asserted in 12.55.

A further point about 12.63 is that it can be regarded as suggesting that an extension of the definition of $q_d(M)$ to manifolds with boundary (such as $M^\mp$). If this is done then 12.63 shows that the relative Donaldson invariants for a non-closed manifold are not integer valued, but rather take values in the Floer groups of its boundary. We can get some idea of what these relative polynomials look like by applying the decomposition $M = M^+ \cup_h M^-$ to the product

$$\underbrace{H_2(M) \times \cdots \times H_2(M)}_{d \text{ factors}} \tag{12.66}$$

Using $H_2(M) = H_2(M^+) \oplus H_2(M^-)$ on this product permits us to deduce that the various spaces of symmetric polynomials are related by

$$Sym^d(M) = \sum Sym^i(H_2(M^+)) Sym^{d-i}(H_2(M^-)) \tag{12.67}$$

This suggests that we introduce a set of polynomials $q_d^i(M^+)$ in the variables

$$\underbrace{H_2(M^+) \times \cdots \times H_2(M^+)}_{i \text{ factors}} \tag{12.68}$$

and similarly for M^-. These polynomials are the relative invariants and, by combining 12.63 and 12.67, we see that they must take values in the Floer groups $HF_*(M^\mp)$ and obey an evaluation rule of the form

$$q_d(M) = \sum_{i=0}^{d} q_d^i(M^+) \bullet q_d^{d-i}(M^-) \tag{12.69}$$

The Donaldson invariants are infinite in number although it is presumably possible that there exist relations among them. It is important to realise that the Donaldson invariants are *independent* of the metric required to define the instantons and to construct their moduli space $\mathcal{M}_k$. In addition, the $q_d(M)$ are invariant, up to a sign, under orientation preserving diffeomorphisms of M. When the orientation of M is reversed there is no obvious relationship between the two sets of Donaldson invariants. The $q_d(M)$ are *differential* topological invariants rather than topological invariants; this means that they have the potential to distinguish homeomorphic manifolds which have distinct diffeomorphic structures. An example where the $q_d(M)$ are used to show that two homeomorphic manifolds are not diffeomorphic can be found in Ebeling [1].

§ 4. Knots and knot invariants

This section contains some background information about knots which is needed for our discussion below of the knot theoretic properties of Chern–Simons theory, cf. also Crowell and Fox [1], Rolfsen [1] and Birman [1].

Knots are embeddings of a circle S^1 into a three dimensional space; in this section we shall take this space to be $\mathbf{R}^3$ or its compactification S^3. The relation between the embedding and the 'tangled' nature of the knot should be appreciated: All knots are homeomorphic copies of S^1, it is the embedding in $\mathbf{R}^3$ which tangles them up; indeed all knots may be untied in higher dimensions. More precisely, if we replace $\mathbf{R}^3$ by $\mathbf{R}^4$ then all knots are trivial. More generally we can consider the embedding of S^m in $\mathbf{R}^n$ and, if $n > 3(m+1)/2$, any knotted m-sphere in $\mathbf{R}^n$ can be unknotted; the embedding of other spaces in $\mathbf{R}^3$ can also be investigated (Kyle [1]). Reverting to knots, let us consider several of them at a time. In particular we shall take p disjoint circles and embed them in $\mathbf{R}^3$; some of these knots may be enmeshed together like the links of a chain and such an object is referred to as a *link L*, say, with p components. Hence a knot is a link with only one component.

Knots are classified into *types* in the following way: If K_1 and K_2 are two knots then K_1 is said to be *equivalent* to K_2 if there is a homeomorphism $\alpha : \mathbf{R}^3 \to \mathbf{R}^3$ under which K_1 is mapped into K_2; we write

$$\alpha K_1 = K_2 \tag{12.70}$$

This clearly defines an equivalence relation on the set of all knots and the elements of each equivalence class are called knots of the same type.

A somewhat finer classification of knots is that by *isotopy type:* Let

$$\{\alpha_t\}, \qquad t \in [0, 1] \tag{12.71}$$

be a family of homeomorphisms of $\mathbf{R}^3$ where $\alpha_t(x)$ is continuous in t and x and α_0 is the identity; then, if $\alpha_1 K_1 = K_2$, K_1 and K_2 are said to be of the same isotopy type.

Evidently knots of the same isotopy type are also equivalent; however the converse is false. This is because the continuity in t of $\{\alpha_t\}$ causes all the α_t to have the same orientation properties, namely they all preserve orientation since α_0, being the identity, does so. Hence isotopic[7] knots K_1 and K_2 satisfy $\alpha K_1 = K_2$ where α *preserves orientation*, but, in general, $\alpha K_1 = K_2$ does not imply that α is orientation preserving.

If a knot is *equivalent* to a polygonal knot—i.e. a knot made of straight line segments—it is called *tame*, if not it is called *wild*. We shall only consider smooth knots and these are all tame.

The classification of knots proceeds by constructing invariants of knot type. One of the simplest of these is the knot group. Let K be a knot and let $\mathbf{R}^3 - K$ be its complement in $\mathbf{R}^3$; in general this set has non-trivial topology and its fundamental group is the knot group, i.e. the knot group of K is

$$\pi_1(\mathbf{R}^3 - K) \tag{12.72}$$

Another knot invariant is the Alexander polynomial $\Delta(t)$ (cf. Crowell and Fox [1]). The Alexander polynomial is not strictly speaking a polynomial, it can contain negative as well as positive powers of t; such polynomials are referred to as Laurent polynomials or L-polynomials. Given a knot K its Alexander polynomial $\Delta(t)$ is not unique, rather it is only determined up to a factor $\mp t^n$ for some positive or negative n. However, $\Delta(t)$ has the properties

$$\Delta(1) = \mp 1, \qquad \Delta(1/t) = t^n \Delta(t) \tag{12.73}$$

for some *even* integer n; also the coefficients of $\Delta(t)$ are all integers. This lack of uniqueness permits $\Delta(t)$ to be adjusted so that it has no negative powers and has a positive constant term; this is called the normalised form of $\Delta(t)$ and is unique. When $\Delta(t)$ is written in normalised form the integer n in 12.73 is positive and is called the degree of $\Delta(t)$; the degree is also equal to the difference between the largest positive power and the smallest negative power in any representation of $\Delta(t)$. For example, the familiar clover leaf or trefoil knot has

$$\Delta(t) = t^2 - t + 1 \tag{12.74}$$

---

[7] Note that the definition of isotopy is similar to the more common one of homotopy. The crucial difference is that, for an isotopy, all of the members of the family $\{\alpha_t\}$ are invertible as well as continuous; for a homotopy invertibility is not required.

in normalised form. But we could also write

$$\Delta(t) = \frac{1}{t^2} - \frac{1}{t} + 1 \quad \text{or} \quad \Delta(t) = t - 1 + \frac{1}{t} \qquad (12.75)$$

which coincide with 12.74 on multiplication by t^2 and t respectively. The last form of $\Delta(t)$ has the property that it is symmetric under interchange of t and t^{-1} and satisfies $\Delta(0) = 1$; the Alexander polynomial for any link can be written in this form and when we wish to refer to this particular normalisation we shall write $\Delta^{Sym}(t)$.

The *mirror image* of a knot is obtained by changing its $\mathbf{R}^3$ coordinates from (x, y, z) to $(x, y, -z)$. The clover leaf and its mirror image are equivalent but have differing isotopy types; however, their Alexander polynomials are the same.

A more powerful knot and link invariant is the Jones polynomial. This polynomial is an integer L-polynomial and it originated in the theory of finite dimensional von Neumann algebras (Jones [1,2]). It is also an invariant of isotopy type and is able to distinguish the clover leaf from its mirror image. If we denote the Jones polynomial for a general link L by $V_L(t)$ then, if L is the clover leaf and $\tilde{L}$ its mirror image, we have (Jones [1])

$$V_L(t) = t + t^3 - t^4, \qquad V_{\tilde{L}}(t) = \frac{1}{t} + \frac{1}{t^3} - \frac{1}{t^4} \qquad (12.76)$$

$V_L(t)$ has some interesting properties extending those given in 12.73 for the Alexander polynomial Δ_L. These are the following: if L is a link with p components and mirror image $\tilde{L}$ then

$$V_L(-1) = \Delta_L^{Sym}(-1), \qquad V_L(1) = (-2)^{p-1}, \qquad V_{\tilde{L}} = V_L(1/t) \qquad (12.77)$$

and if L is just a knot K we have

$$V_K(e^{2\pi i/3}) = 1, \qquad \left. \frac{dV_K(t)}{dt} \right|_{t=1} = 0 \qquad (12.78)$$

These properties can be easily checked for the case of the clover leaf above. The Jones polynomial has been generalised to a homogeneous polynomial in three variables by Freyd *et al.* [1]; this polynomial—known as the Homfly polynomial after the initials of its authors—can reduce, in special cases, to the Alexander polynomial and the Jones polynomial. Also, both for the Jones and Homfly polynomials there exist distinct links on which these polynomials have the same values, cf. Birman [2] and Kanenobu [1].

The Alexander polynomial for any link can be computed inductively as we now describe. Suppose a link has n crossings then we can relate its Alexander polynomial to that of another link with only $(n-1)$ crossings. This is done via the skein relation: Let L_+, L_- and L_0 be three links which are identical except for the interior of a small disc where they are as displayed below

$$L_+ \qquad\qquad L_- \qquad\qquad L_0 \tag{12.79}$$

Notice that if the $L_{\mp}$ have n crossings then L_0 has $(n-1)$ crossings. The Alexander polynomials of these links are related by the simple equation

$$\Delta_{L_+}(t) - \Delta_{L_-}(t) + (t^{1/2} - t^{-1/2})\Delta_{L_0}(t) = 0 \tag{12.80}$$

Iteration of this relation expresses any $\Delta_L(t)$ in terms of the Alexander polynomial for a finite number of unlinked, unknotted circles. Thus it is only the Alexander polynomial for these latter that we need to calculate.

In a similar way Jones [1] has shown that $V_L(t)$ satisfies

$$tV_{L_+}(t) - t^{-1}V_{L_-}(t) + (t^{1/2} - t^{-1/2})V_{L_0}(t) = 0 \tag{12.81}$$

sometimes written diagrammatically as

$$t \;\;\diagdown\!\!\!\!\diagup\;\; - \;\; t^{-1} \;\;\diagup\!\!\!\!\diagdown\;\; + \;\; (t^{1/2} - t^{-1/2}) \;\; \Big) \; \Big(\;\; = 0 \tag{12.82}$$

Hence $V_L(t)$ is also determined by its values on a finite number of unlinked, unknotted circles. If we now take this example so that L consists of p unlinked, unknotted circles then $V_L(t)$ is given by

$$V_L(t) = \left\{ -\frac{(t - t^{-1})}{(t^{1/2} - t^{-1/2})} \right\}^{p-1} \tag{12.83}$$

§ 5. Chern–Simons theory and knots

Witten [6,7] has shown how to determine $V_L(t)$ from certain correlation functions of the Chern–Simons topological field theory introduced in 12.37. In this section we discuss Chern–Simons theory and knots; the following section will deal with the calculation of $V_L(t)$ itself. Witten's work has the great benefit of providing an intrinsically three dimensional definition of the Jones polynomial, something which is otherwise lacking. In fact many knot calculations are carried out by various projections from three into two dimensions: this can create a technical problem in that differing knots may have the same projection, and such calculations must be shown to be independent of the particular projection used. An interesting feature of the quantum field theory of the Jones polynomial calculation is that, as well as the topological Chern–Simons theory, the machinery of conformal field theory is an indispensable part of the construction.

From the end of the last section it follows that to verify that a given function is the Jones polynomial we must check that it obeys the skein relation 12.81, and show that it takes the same values as the Jones polynomial on links consisting of unlinked, unknotted, circles.

The correlation functions which determine the Jones polynomial are those consisting of Wilson lines. A Wilson line $W(R, C)$ is a function determined by a closed curve C and a G-connection A where

$$W(R, C) = tr \, P \exp \left[\int_C A \right] \qquad (12.84)$$

In this notation R stands for an irreducible representation of G, tr is the trace in the R representation and P is the familiar path ordering symbol. Clearly C can be a knot and, if we consider p Wilson lines, we have a p-component link L. The normalised correlation function associated with these p Wilson lines is

$$< W(R_1, C_1) \cdots W(R_p, C_p) >= \frac{1}{Z(M)} \int \mathcal{D}\mathcal{A}$$
$$W(R_1, C_1) \cdots W(R_p, C_p) \exp \left[-\frac{ik}{4\pi} \int_M tr(A \wedge dA + \frac{2}{3} A \wedge A \wedge A) \right]$$
$$(12.85)$$

where $Z(M)$ is the partition function. As we have observed before the Chern–Simons action is purely a differential topological quantity; introducing a link L in the guise of Wilson lines does not require any non-topological data. In this way we see that the Wilson correlation functions $< W(R_1, C_1) \cdots W(R_p, C_p) >$ have a chance of being purely topological; of course, even if this is true, we still need them to be both non-zero and non-trivial. This will turn out to be so.

As preliminary evidence for the non-trivial topological nature of the correlation functions $< W(R_1, C_1) \cdots W(R_p, C_p) >$ we follow Witten [6,7] and discuss two special cases: the first is an Abelian example where $G = U(1)$, the second is where G is non-Abelian but the gauge coupling is small.

Suppose, then, that $G = U(1)$. This renders the Chern–Simons action quadratic in A and the correlation functions are therefore

$$< W(R_1, C_1) \cdots W(R_p, C_p) >=$$
$$\frac{1}{Z(M)} \int \mathcal{D}A \, W(R_1, C_1) \cdots W(R_p, C_p) \exp\left[-\frac{ik}{4\pi} \int_M tr(A \wedge dA)\right]$$
(12.86)

Since $G = U(1)$ is Abelian a representation R is just a map of the form $\theta \mapsto \exp[in\theta]$, $n \in \mathbf{Z}$, and the path ordering and the trace may be dispensed with. However, before dispensing with the trace, it is necessary to realise that differing representations can have 'traces' whose normalisations differ but have a rational ratio. Thus a Wilson line is now given by

$$W(n, C) = \exp\left[in \int_C A\right], \quad n \in \mathbf{Z} \qquad (12.87)$$

and the correlation function is

$$< W(n_1, C_1) \cdots W(n_p, C_p) >=$$
$$\frac{1}{Z(M)} \int \mathcal{D}A \, W(n_1, C_1) \cdots W(n_p, C_p) \exp\left[-\frac{ik}{4\pi} \int_M (A \wedge dA)\right]$$
(12.88)

Now we commit ourselves to the choice $M = S^3$ since this will enable us to proceed at once to an explicit calculation; we shall comment on the situation for other M later in the chapter.

The quadratic action, together with the exponential dependence on A of a Wilson line, allows the entire integrand to be written as a Gaussian after completing the square. The calculation of the functional integral rests just on the calculation of a Green's function which, for S^3, is an elementary computation. The result is that

$$< W(n_1, C_1) \cdots W(n_p, C_p) >=$$
$$\exp\left[\frac{i}{4k}\epsilon_{ijk} \sum_{l,m=1}^{p} n_l n_m \int_{C_l} dx^i \int_{C_m} dy^j \frac{(x-y)^k}{|x-y|^3}\right] \qquad (12.89)$$

with x^i and y^j local coordinates on the knots C_l and C_m. The basic integral in 12.89 is the linking number $L(C_l, C_m)$ of Gauss. $L(C_l, C_m)$ is one of the oldest invariants in knot theory and originated, at least in part, in a study of Ampère's law in electromagnetic theory, cf. Gauss [1]. Its definition is

$$L(C_l, C_m) = \frac{\epsilon_{ijk}}{4\pi} \int_{C_l} dx^i \int_{C_m} dy^j \frac{(x-y)^k}{|x-y|^3} \qquad (12.90)$$

and, from Ampère's law, where precisely this integral occurs, it follows that $L(C_l, C_m)$ is an integer.

A difficulty with the linking number arises when $l = m$ because then the *self-linking* integral $L(C_l, C_l)$ has singularities due to points where $x - y = 0$. Witten regularises $L(C_l, C_l)$ by giving the knot a framing: this is the replacement of the curve C_l by a ribbon of positive width; the boundary of this ribbon consists of two disjoint curves, one of which we identify with C_l, the other curve is arbitrary and is denoted by C'_l. Having done this the self-linking number of C_l is defined to be the ordinary linking number of C_l and C'_l, that is we define

$$L(C_l, C_l) = L(C_l, C'_l) \tag{12.91}$$

A little thought shows that $L(C_l, C'_l)$ is not unique. There are many such ribbons we can make with boundaries C_l and C'_l and some of these ribbons will have many more twists than others. This lack of uniqueness in the framing of a knot is a familiar feature of knot theory. It does not cause as much trouble as might be expected provided one always knows how the particular knot data being computed changes if a different framing is specified. For example, in the present case, if we express the correlation function in terms of $L(C_l, C_m)$, we obtain

$$< W(n_1, C_1) \cdots W(n_p, C_p) > = \exp\left[\frac{i\pi}{k} \sum_{l,m} n_l n_m L(C_l, C_m)\right] \tag{12.92}$$

Now if the framing of one of the C_l's is changed this is detected in the RHS of 12.92 by a change in the number of twists in the ribbon. Supposing the number of extra twists to be t, the self-linking $L(C_l, C_l)$ increases by t and the RHS of 12.92 becomes

$$\exp\left[\frac{i\pi}{k} n_l^2 t\right] \exp\left[\frac{i\pi}{k} \sum_{l,m} n_l n_m L(C_l, C_m)\right] \tag{12.93}$$

Thus the rule for the transformation of the correlation function under a change of framing is

$$< W(n_1, C_1) \cdots W(n_p, C_p) > \mapsto \exp\left[\frac{i\pi}{k} n_l^2 t\right] < W(n_1, C_1) \cdots W(n_p, C_p) > \tag{12.94}$$

With this first piece of topological evidence uncovered we move on to examine the small coupling limit.

The gauge coupling g is usually present in the curvature $F(A)$ as a coefficient of the quadratic term; using local coordinates one writes

$$F(A) = \frac{1}{2}F_{\mu\nu}^{a}T^{a}dx^{\mu} \wedge dx^{\nu} = \frac{1}{2}(\partial_{\mu}A_{\nu}^{a} - \partial_{\nu}A_{\mu}^{a} + igf^{abc}T^{c}A_{\mu}^{b}A_{\nu}^{c})dx^{\mu} \wedge dx^{\nu}$$
(12.95)

It is more convenient to rescale A_{μ}^{a} by a factor g, rendering the coupling simply an overall factor multiplying the curvature. The Chern–Simons action is then

$$-\frac{ik}{4\pi g^2} \int_M tr(A \wedge dA + \frac{2}{3}A \wedge A \wedge A)$$
(12.96)

We can now see that the small coupling limit $g \to 0$ is the same as the large k limit $k \to \infty$ and, from now on, we shall treat them as equivalent; also, since small coupling can be achieved by taking k large we shall set $g = 1$ and just vary k.

Take the partition function

$$Z(M) = \int \mathcal{D}A \, \exp\left[-\frac{ik}{4\pi} \int_M tr(A \wedge dA + \frac{2}{3}A \wedge A \wedge A)\right]$$
(12.97)

For k large the growing oscillations in the integrand cause the functional integral to be dominated by the stationary points of the action, which we know are the flat connections

$$F(A) = 0$$
(12.98)

Hence, in the small coupling limit, the principle of stationary phase leads us to expect that $Z(M)$ can be expressed as a sum over flat connections. The standard procedure to evaluate in this limit is to expand the action about its stationary points up to quadratic terms; this converts the functional integral into the exponential of a Gaussian which can be expressed in terms of determinants. Since we are working with a gauge theory we will also have to fix a gauge.

Let $S(A)$ denote the action with the coefficient $-ik/4\pi$ factored out. Then, if A_f is a flat connection, we perturb A according to

$$A = A_f + a$$

$$\Rightarrow S(A) = S(A_f) + 2\int_M tr\left(F(A_f) \wedge a\right) + \int_M tr\left(a \wedge d_{A_f}a\right) + 0(a^3)$$

$$= S(A_f) + \int_M tr\left(a \wedge d_{A_f}a\right) + 0(a^3)$$
(12.99)

since A_f is flat. Next we choose a metric on M so that we can impose the gauge condition

$$d_{A_f}^* a = 0 \qquad (12.100)$$

of 8.82. This condition can be encoded into the functional integral for $Z(M)$ by the addition of a Lagrange multiplier field ϕ; we will also have to include a Faddeev–Popov ghost term which acts as a Jacobian. The result of all this is to modify the expression for the partition function to

$$Z(M) = \sum_{A_f} \exp[-ikS(A_f)/4\pi] \int \mathcal{D}a \mathcal{D}\phi \mathcal{D}c \mathcal{D}\bar{c}$$

$$\exp\left[-ik/4\pi \int_M \{tr\,(a \wedge d_{A_f} a) + 2tr\,(\phi \wedge d_{A_f}^* a) + tr\,(d_{A_f}\bar{c} \wedge d_{A_f} c)\}\right]$$

$$(12.101)$$

where we note that c is a 0-form and ϕ is a 3-form. These forms are, of course, matrix valued and are actually sections of bundles constructed from the flat bundle E, say, on which the connection A_f is defined. The appropriate spaces of sections are $\Omega^i(M, E) = \Gamma(M, E \otimes \wedge^i T^*M)$. This formula for $Z(M)$ can be written more compactly using inner products on these forms. To do this we consider the odd forms by themselves and, on the odd forms, we define the operator D by

$$D : \Omega^1(M, E) \oplus \Omega^3(M, E) \longrightarrow \Omega^1(M, E) \oplus \Omega^3(M, E)$$
$$(a, \phi) \longmapsto (*d_{A_f} + d_{A_f}*)(a, \phi) \qquad (12.102)$$

With respect to the vector space decomposition $\Omega^1(M, E) \oplus \Omega^3(M, E)$ we can write D more usefully in the block matrix form

$$D = \begin{pmatrix} *d_{A_f} & d_{A_f}* \\ d_{A_f}* & 0 \end{pmatrix}, \quad \text{where} \quad D(a, \phi) = \begin{pmatrix} *d_{A_f} & d_{A_f}* \\ d_{A_f}* & 0 \end{pmatrix} \begin{pmatrix} a \\ \phi \end{pmatrix} \equiv Dv \qquad (12.103)$$

This linear algebra enables us to write

$$< v, Dv > = -\int_M tr\,(v \wedge *Dv)$$

$$= -\int_M tr\,(a \wedge *^2 d_{A_f} a + a \wedge *d_{A_f} * \phi + \phi \wedge *d_{A_f} * a)$$

$$= -\int_M tr\,(a \wedge d_{A_f} a + 2\phi \wedge d_{A_f}^* a)$$

and $< c, d_{A_f}^* d_{A_f} c > = -\int_M tr\,(d_{A_f}\bar{c} \wedge d_{A_f} c)$

$$(12.104)$$

The partition function is now expressible as

$$Z(M) = \sum_{A_f} \exp[-ikS(A_f)/4\pi]$$

$$\int \mathcal{D}a\mathcal{D}\phi\mathcal{D}c\mathcal{D}\bar{c}\exp\left[\frac{ik}{4\pi}(<v,Dv> + <c,d^*_{A_f}d_{A_f}c>)\right] \qquad (12.105)$$

The functional integral can be written in terms of determinants using zeta functions and we find that

$$\int \mathcal{D}a\mathcal{D}\phi\mathcal{D}c\mathcal{D}\bar{c}\exp\left[\frac{ik}{4\pi}(<v,Dv> + <c,d^*_{A_f}d_{A_f}c>)\right] = C\frac{det\,(d^*_{A_f}d_{A_f})}{\sqrt{det\,D}} \qquad (12.106)$$

where C is a constant to be discussed in a moment and we comment that the ghost determinant has no square root because of the presence of $\bar{c}$ as well as c. Now we come to the constant C which is used to absorb the dependence on the factor $(ik/4\pi)$; let us call this factor μ for short. This is usually a routine matter because, as far as the zeta function is concerned, it amounts to finding the μ-dependence in $det\,(\mu A)$ where A is the elliptic operator under study. It is easy to work out from the definition $det\,A = \exp[-\zeta'_A(0)]$ that $det\,(\mu A) = \mu^{\zeta_A(0)}det\,A$. However, this assumes that μA is a positive operator and, in our case, μ is pure imaginary. Thus we can absorb the dependence on $(k/4\pi)$ into a constant C but the i-dependence requires some more work. It is not difficult to see that the i-dependence contributes a phase factor to C involving the η-function of the first order operator D. The details are of some independent interest and can be found in Witten [7]; we just observe that the phase factor is of the form $\exp[i\pi\eta(0)/2]$ and hereafter we also absorb this phase factor into the constant C.

The partition function $Z(M)$ has now simplified to

$$Z(M) = \sum_{A_f} C\exp[-ikS(A_f)/4\pi]\frac{det\,(d^*_{A_f}d_{A_f})}{\sqrt{det\,D}} \qquad (12.107)$$

Now the same ratio of determinants occuring in 12.107 also occurs in an invariant known as the Ray–Singer analytic torsion (cf. Ray and Singer [1,2]), thus we can regard this formula for $Z(M)$ as further evidence for the deformation invariance of the Chern–Simons theory.

Before leaving this topic we provide a bit more detail on the relation between the analytic torsion and the Chern–Simons theory. In the Abelian case this relation was established by Schwarz [1]. First of all, if we have a bundle with a flat connection such as E, then the analytic torsion is the number $T(M, E)$ where

$$\ln T(M, E) = \sum_0^n (-1)^p p \ln \det \Delta_p^E, \quad n = \dim M \qquad (12.108)$$

The notation Δ_p^E stands for the Laplacian on E-valued p-forms and so Δ_p^E is just the operator $(d_{A_f}^* d_{A_f} + d_{A_f} d_{A_f}^*)$ acting on $\Omega^p(M, E)$. When $n = 3$ we find that

$$T(M, E) = (\det \Delta_1^E)^{-1} (\det \Delta_2^E)^2 (\det \Delta_3^E)^{-3} \qquad (12.109)$$

But, since the $*$ operator commutes with the Laplacians, $\det \Delta_p^E = \det \Delta_{n-p}^E$ and so we have

$$T(M, E) = (\det \Delta_0^E)^{-3} (\det \Delta_1^E) \qquad (12.110)$$

Now if we square our operator D above, a short calculation shows that we get

$$D^2 = \begin{pmatrix} *d_{A_f} & d_{A_f}* \\ d_{A_f}* & 0 \end{pmatrix} \begin{pmatrix} *d_{A_f} & d_{A_f}* \\ d_{A_f}* & 0 \end{pmatrix} = \begin{pmatrix} \Delta_1^E & 0 \\ 0 & \Delta_3^E \end{pmatrix} \qquad (12.111)$$

$$\Rightarrow \det D^2 = (\det \Delta_1^E)(\det \Delta_3^E)$$

Hence, for the ratio of determinants appearing in the partition function, we have

$$\begin{aligned} \frac{\det (d_{A_f}^* d_{A_f})}{\sqrt{\det D}} &= \frac{(\det \Delta_0^E)}{\{(\det \Delta_1^E)(\det \Delta_3^E)\}^{1/4}} \\ &= \frac{(\det \Delta_0^E)^{3/4}}{(\det \Delta_1^E)^{1/4}} \qquad (12.112) \\ &= T(M, E)^{-1/4} \end{aligned}$$

and so

$$Z(M) = \sum_E C \exp[-ikS(E)/4\pi] T(M, E)^{-1/4} \qquad (12.113)$$

where we have replaced the symbol A_f by its associated flat bundle E. It is worth noting that the final expression for the partition function is independent of the metric g introduced to carry out the gauge fixing; this is explicitly checked in Schwarz [1]. A technical difficulty also arises in defining the $det \, \Delta_p^E$ if any of these Laplacians have zero eigenvalues: by Hodge theory this means that some of the cohomology groups $H^p(M; E)$ are non-trivial. We shall just restrict ourselves to situations where these cohomology groups vanish.

§ 6. Chern–Simons theory and the Jones polynomial

We now wish to describe the Chern–Simons method for calculating the Jones polynomial $V_L(t)$. To accomplish this task the attractive topological properties of the Chern–Simons action discussed in the previous section need to be supplemented by something further. The main ingredient of this supplement is the machinery of two dimensional conformal field theory. Some insight into how the conformal field theory enters can be obtained as follows. Suppose M is a 3-manifold M inside which there are various knots. If we consider some two dimensional (Riemann) surface Σ inside M then, near Σ, M may look like $\Sigma \times \mathbf{R}$ and this surface will, in general, be punctured by some of the knots. Thus the presence of the knots suggests that we consider, not just Σ, but Σ with a collection of punctures. But Riemann surfaces with punctures constitute the primary data for a conformal field theory; so we are led to a quest for a relation between the conformal field theory on Σ and knots inside M.

The quantisation of the Chern–Simons action suggested by this picture can now be pursued. For example, if we choose the $A_0 = 0$ gauge on $\Sigma \times \mathbf{R}$ then it becomes natural to regard the phase space as the moduli space $\mathcal{M}_F$ of equivalence classes of flat connections on Σ. This space has a natural symplectic structure and is a compact Kähler manifold; thus quantising on this phase space should give finite dimensional quantum Hilbert spaces H_Σ or $H_{\Sigma_{p,n}}$. These spaces are the sections of the kth power of the determinant line bundle over $\mathcal{M}_F$. The conformal blocks described in chapter 11 are also spaces of sections and, in the axiomatic approach to conformal theories, vector bundles on the moduli space $\mathcal{M}_F$ also occur. The spaces H_Σ can now be identified with the $V(\Sigma)$ of the conformal theory on Σ. A fuller discussion can be found in Witten [6], cf. also Hitchin [1].

To begin our calculations we need a slightly more compact notation for our Wilson line correlation functions. We introduce the quantity $\widetilde{W}(M, \{R\})$ and write

$$< W(R_1, C_1) \cdots W(R_p, C_p) > = \frac{\widetilde{W}(M, \{R\})}{Z(M)}, \quad \text{where} \quad \{R\} = R_1, \ldots, R_p$$

$$(12.114)$$

The symbol $\{R\}$ serves to record the representation content of the Wilson lines and from now on the group G is taken to be $SU(2)$. If we wish to be more explicit we may also write out $\{R\}$ in full; for example, if $p = 2$, we could write

$$\widetilde{W}(M, R_1, R_2) \qquad (12.115)$$

When $p = 0$ we shall write simply

$$\widetilde{W}(M) \qquad (12.116)$$

which we observe is equal to the partition function $Z(M)$. Finally the normalised correlation function is written as $W(M, \{R\})$ where

$$W(M, \{R\}) = \frac{\widetilde{W}(M, \{R\})}{Z(M)} \qquad (12.117)$$

Our first goal will be to derive the skein relation

$$tV_{L_+}(t) - t^{-1}V_{L_-}(t) + (t^{1/2} - t^{-1/2})V_{L_0}(t) = 0 \qquad (12.118)$$

which, in the quantum field theory approach, amounts to the statement that any three vectors in a two dimensional vector space are linearly dependent. To start this derivation we take our manifold M with knots inside and decompose it as a connected sum

$$M = M_1 \# M_2 \qquad (12.119)$$

where it is required that none of the knots in M pass through the two sphere joining M_1 and M_2. This requirement means that we have two sets of knots one in M_1 and the other in M_2; we denote these by $\{R^{(1)}\}$ and $\{R^{(2)}\}$ respectively. Next we wish to show that

$$Z(S^3)\widetilde{W}(M, \{R\}) = \widetilde{W}(M_1, \{R^{(1)}\})\widetilde{W}(M_2, \{R^{(2)}\}) \qquad (12.120)$$

Now we consider the state of affairs just before the connected sum operation joins M_1 to M_2. M_1 has a disc excised from it giving a manifold $\check{M}_1$ with $\partial \check{M}_1 = S^2$; when a manifold has a boundary an associated functional integral is heuristically a wave function on this boundary or, more precisely, an element of a Hilbert space H_{S^2} defined on this boundary. This space H_{S^2} is that for the conformal field theory on S^2. A similar remark applies to $\check{M}_2$ except that the orientation of its boundary is opposite to that of $\check{M}_1$, which has the consequence that the two Hilbert spaces are dual to one another. If we denote the two vectors in the Hilbert spaces by V_1 and V_2 then we have

$$\widetilde{W}(M, \{R\}) = < V_1, V_2 > \qquad (12.121)$$

where the inner product is induced by the dual pairing of the Hilbert spaces. We can think of this equation as expressing a factorisation of probability amplitudes similar to that encountered in string theories. This connected sum process can be applied to S^3 itself in the case where S^3 has *no knots* inside. This gives the equation

$$Z(S^3) = <U_1, U_2> \tag{12.122}$$

where U_1 and U_2 are two more vectors in H_{S^2}. But S^2 has no marked points and so corresponds to the trivial representation of G. Thus we know from 11.75 that H_{S^2} is one dimensional, so that the inner product is like ordinary multiplication, and this allows to make the elementary observation that

$$<U_1, U_2><V_1, V_2> = <V_1, U_1><U_2, V_2> \tag{12.123}$$

Since $<V_1, U_1>$ corresponds to $\check{M}_1$ 'evolving' into M_1, and the same applies to M_2, the RHS is easily seen to be

$$\widetilde{W}(M_1, \{R^{(1)}\})\widetilde{W}(M_2, \{R^{(2)}\}) \tag{12.124}$$

and so we have accomplished our task and shown that

$$Z(S^3)\widetilde{W}(M, \{R\}) = \widetilde{W}(M_1, \{R^{(1)}\})\widetilde{W}(M_2, \{R^{(2)}\}) \tag{12.125}$$

which we rewrite in the form

$$\frac{\widetilde{W}(M, \{R\})}{Z(S^3)} = \frac{\widetilde{W}(M_1, \{R^{(1)}\})}{Z(S^3)} \frac{\widetilde{W}(M, \{R^{(2)}\})}{Z(S^3)} \tag{12.126}$$

We can now contemplate decomposing M many times, yielding

$$M = M_1 \# \cdots \# M_p \tag{12.127}$$

where we still insist that no knots pass through any of the joins. Induction then shows that

$$\frac{\widetilde{W}(M, \{R\})}{Z(S^3)} = \frac{\widetilde{W}(M_1, \{R^{(1)}\})}{Z(S^3)} \cdots \frac{\widetilde{W}(M, \{R^{(p)}\})}{Z(S^3)} \tag{12.128}$$

This result can immediately be applied to the case where $M = S^3$ and $\{R\}$ consists of p unlinked, unknotted circles; in that case it will certainly be possible to decompose S^3 into exactly p components each of which is both a copy of S^3 and contains only one circle. This reduces the calculation for p unlinked, unknotted circles to that for a single unknotted circle; we have

$$\frac{\widetilde{W}(S^3, \{R\})}{Z(S^3)} = \frac{\widetilde{W}(S^3, R_1)}{Z(S^3)} \cdots \frac{\widetilde{W}(S^3, R_p)}{Z(S^3)} \tag{12.129}$$

or $\quad W(S^3, \{R\}) = W(S^3, R_1) \cdots W(S^3, R_p)$

Now we must get to the skein relation itself; this is done by abandoning our earlier restriction that no knots may pass through the join in the connected sum $M_1 \# M_2$. To this end let us decompose $M = M_1 \# M_2$ as before but this time we suppose that the knots have produced 4 punctures in the S^2 joining $\breve{M}_1$ to $\breve{M}_2$; we further require all 4 representations at these punctures to be the 2-dimensional defining representation of $SU(2)$. This means that the Hilbert space $H_{S^2, R, R, R, R} \equiv H_{S^2, 4}$ of the 4-fold punctured S^2 has dimension 2—this can be seen by noting that the physical Hilbert space must be $SU(2)$-invariant and the trivial representation occurs precisely twice in the tensor product $R \otimes R \otimes R \otimes R$. Now suppose that we carry out such a decomposition three times and ordain things so that the only difference between the three decompositions is that, when we look into M_2, 2 of the 4 strands are arranged[8] according to the three skein configurations

$$\begin{array}{ccc} L_+ & L_- & L_0 \end{array} \tag{12.130}$$

This gives us three equations like 12.121; using $+, -$ and 0 for over, under and zero crossing we have

$$M = M_1 \# M_2^+, \quad M = M_1 \# M_2^-, \quad M = M_1 \# M_2^0$$
$$\widetilde{W}(M, \{R^+\}) = <V_1, V_2^+>, \quad \widetilde{W}(M, \{R^-\}) = <V_1, V_2^->, \tag{12.131}$$
$$\widetilde{W}(M, \{R^0\}) = <V_1, V_2^0>$$

But, because $H_{S^2, 4}$ is only two dimensional, the three vectors V_2^+, V_2^- and V_2^0 are linearly dependent, that is

$$\alpha^+ V_2^+ + \alpha^- V_2^- + \alpha^0 V_2^0 = 0$$
$$\Rightarrow \alpha^+ \widetilde{W}(M, \{R^+\}) + \alpha^- \widetilde{W}(M, \{R^-\}) + \alpha^0 \widetilde{W}(M, \{R^0\}) = 0 \tag{12.132}$$

---

[8] We choose 4 strands rather than 2 because, to be completely general, we allow each of the 2 strands displayed in the skein relation to belong to a separate Wilson loop; so these 2 loops must possess both an entry and exit point on S^2, thus we have 4 strands and a 4-fold punctured S^2.

and this is the skein relation.

The values of the constants $\alpha^{\mp}$ and α^0 can be found by reference to conformal field theory. The point is that judiciously chosen diffeomorphisms of the punctured S^2 can change the position of the punctures so that we can pass between the links R^+, R^- and R^0. Then, as far as the Hilbert space is concerned, such diffeomorphisms are matrices acting on $H_{S^2,4}$. In fact Witten shows that if we introduce the 2×2-matrix B such that

$$V_2^0 = BV_2^+ \tag{12.133}$$

then

$$V_2^- = BV_2^0 = B^2 V_2^+ \tag{12.134}$$

This means that, up to an inessential overall constant, $\alpha^{\mp}$ and α^0 are determined by the matrix B. In the context of conformal field theories the matrix B was extensively studied by Moore and Seiberg [1] and using this work we find that [9]

$$
\begin{aligned}
\alpha^+ &= -\exp\left[\frac{2\pi i}{k+2}\right], \quad \alpha^- = \exp\left[-\frac{2\pi i}{k+2}\right] \\
\alpha^0 &= \exp\left[\frac{\pi i}{k+2}\right] - \exp\left[-\frac{\pi i}{k+2}\right]
\end{aligned}
\tag{12.135}
$$

If we introduce the variable

$$t = \exp\left[-\frac{2\pi i}{k+2}\right] \tag{12.136}$$

then we have

$$t\widetilde{W}(M,\{R^+\}) - t^{-1}\widetilde{W}(M,\{R^-\}) + (t^{1/2} - t^{-1/2})\widetilde{W}(M,\{R^0\}) = 0 \tag{12.137}$$

Finally we are now in a position to calculate $\widetilde{W}(M,\{R\})$ explicitly for the case when $M = S^3$. By 12.129 we only need to calculate $\widetilde{W}(S^3,\{R\})$ for a single unknotted circle. To do this we use the skein relation 12.137 with $\{R^+\}$ representing the single unknotted circle. It is elementary to verify that this choice has the consequence that $\{R^+\} = \{R^-\}$ while $\{R^0\}$ consists

---

[9] This calculation also involves changing the framing of some of the links and we recommend reference to Witten [6] for the details.

of two unlinked, unknotted circles; hence, using the normalised correlation functions, we have

$$tW(S^3, \{R^+\}) - t^{-1}W(S^3, \{R^-\}) + (t^{1/2} - t^{-1/2})W(S^3, \{R^0\}) = 0$$
$$\Rightarrow (t - t^{-1})W(S^3, \{R^+\}) + (t^{1/2} - t^{-1/2})\{W(S^3, \{R^+\})\}^2 = 0$$
$$\Rightarrow W(S^3, \{R^+\}) = -\frac{(t - t^{-1})}{(t^{1/2} - t^{-1/2})}$$

$$(12.138)$$

It follows from 12.129 that for a link $L \equiv \{R\}$ consisting of p unlinked, unknotted circles we have

$$W(S^3, \{R\}) = \left\{ -\frac{(t - t^{-1})}{(t^{1/2} - t^{-1/2})} \right\}^p \qquad (12.139)$$

Thus if we analytically continue t to real values and write

$$V_L(t) = -\frac{(t^{1/2} - t^{-1/2})}{(t - t^{-1})} W(S^3, \{R\}) \qquad (12.140)$$

then $V_L(t)$ is the Jones polynomial since it satisfies the skein relation 12.118 and agrees with 12.83 when L consists of p unlinked, unknotted circles.

Finally we must describe the transformation in $\widetilde{W}(M, \{R\})$ under a change in framing. Let $M = M_1 \cup_\Sigma M_2$ represent a decomposition of M obtained by cutting into two so that the boundaries of M_1 and M_2 are the Riemann surface Σ. Suppose, then, that M has been cut into these two pieces and a framed Wilson line passes through Σ at a point P. If we apply a $2\pi t$ Dehn twist around P then the framing of C will change by an amount t. Hence the procedure for changing the framing of C is to cut M as instructed and then to apply the Dehn twist to $P \in \partial M_1$ and then rejoin M_1 and M_2. The effect of the Dehn twist is detected as a linear transformation on the space H_Σ and, from conformal field theory, we can calculate that this transformation is just scalar multiplication by the factor $\exp[2\pi i t h_R]$, where h_R is the conformal weight of the primary field in the R representation. Summarising, the transformation law of $\widetilde{W}(M, \{R\})$ is $\widetilde{W}(M, \{R\}) \mapsto \exp[2\pi i t h_R]\widetilde{W}(M, \{R\})$.

§ 7. Surgery and the Jones polynomial

The Jones polynomial was originally defined for links in $\mathbf{R}^3$ or S^3 but, when quantum field theory is used there is no need to restrict the manifold containing the links to be S^3. All we need to do is to construct the Chern–Simons theory on the three manifold M of interest. There is a well developed surgery theory for three manifolds and, in particular, it is known that after enough surgery any manifold M can be transformed into S^3. Witten used this idea both to generalise $V_L(t)$ to an arbitrary manifold and also to give an illuminating recomputation of the Jones polynomial for S^3. As well as considering a general manifold M we shall consider a general G; thus G is no longer restricted to be $SU(2)$.

The basic procedure is to combine surgery with the conformal block pairing of the previous section. The rather special surgery employed is the following: Let C be a curve inside M which we thicken into a tubular neighbourhood, thereby obtaining a solid torus M_2, say; next we cut out M_2 leaving a piece M_1 behind, then, before gluing the pieces back together, we act with a diffeomorphism K on the boundary of M_2. Note that $\partial M_2 \simeq T^2$. When we glue things back together the result is the manifold M^K and we denote the surgery by writing

$$M^K = M_1 \cup_{T^2} M_2^K \tag{12.141}$$

The action of the diffeomorphism K on the Hilbert space H_{T^2} associated with $\partial \check{M}_2$ is that of a linear transformation. Thus the vector V_2 of 12.121 becomes KV_2 and so the formula for the correlation functions of links inside M^K is

$$\widetilde{W}(M^K, \{R\}) = < V_1, KV_2 > \tag{12.142}$$

Thus if we can compute with the surgery matrix K we can do calculations on links in M^K.

To facilitate computations with K we introduce a basis $\{V_2^0, \ldots, V_2^n\}$ for H_{T^2} so that $\dim H_{T^2} = n + 1$. Since H_{T^2} is the Hilbert space for a conformal block coming from $\overline{LG}$, the $(n+1)$ primary fields each correspond to a representation of G; we choose the basis $\{V_2^0, \ldots, V_2^n\}$ so that each element corresponds to a Wilson line. Note that the trivial representation is always one of the representations occurring above and we adjust our notation so that V_2^0 gives the trivial representation. We recall that, on selecting the trivial representation, the Wilson line $tr\, P \exp[\int_C A]$ collapses to the identity, i.e. the knot associated to the Wilson line disappears.

To work with K we must know its action of the basis of H_{T^2}; in this connection it is useful to define K_i^j by

$$KV_2^i = K_j^i V_2^j \tag{12.143}$$

Now let us consider some examples. Let M be a manifold containing no Wilson lines and with correlation function given by

$$\widetilde{W}(M, \{R\}) = < V_1, V_2 > \tag{12.144}$$

But, if there are no knots inside M, then $\widetilde{W}(M, \{R\})$ is just the partition function $Z(M)$ and the vector V_2 can be taken to be the 'identity' basis element V_2^0. In this way we can write

$$Z(M) =< V_1, V_2^0 > \qquad (12.145)$$

Now if we subject M to a surgery using K we obtain

$$\begin{aligned} Z(M^K) &=< V_1, KV_2^0 > \\ &=< V_1, K_j^0 V_2^j > \\ &= K_j^0 < V_1, V_2^j > \end{aligned} \qquad (12.146)$$

However, recalling that V_2^j indicates the presence inside M_2 of a Wilson line in the R_j representation, we realise that

$$< V_1, V_2^j >= \widetilde{W}(M, R_j) \qquad (12.147)$$

where $\widetilde{W}(M, R_j)$ means the correlation function for an M containing a single Wilson line in the representation R_j. Thus we have established that

$$Z(M^K) = K_j^0 \widetilde{W}(M, R_j) \qquad (12.148)$$

and we note that the LHS is evaluated on M^K while the RHS is evaluated on M; in addition, the LHS contains no Wilson line but the RHS does. We shall see below that this formula 12.148 can be used to recompute the Jones polynomial for S^3.

Another example may be obtained by starting with a Wilson line already inside M_2 and then doing the surgery. Let there be an unknotted Wilson line inside the torus M_2 and parallel to its boundary. This means that $V_2 = V_2^i$ for some i; thus the new surgery formula is

$$\begin{aligned} \widetilde{W}(M^K, R_i) &=< V_1, KV_2^i > \\ &= K_j^i < V_1, V_2^j > \\ &= K_j^i \widetilde{W}(M, R_j) \end{aligned} \qquad (12.149)$$

To carry out a complete surgery calculation we must commit ourselves to some specific manifolds M and M^K; we shall do this shortly but first we want to describe a useful piece of Hamiltonian formalism that will be required during the calculations.

Let Σ be a Riemann surface propagating statically through time according to the Hamiltonian $\mathcal{H}$ of the Chern–Simons theory. If the time is periodic and represented by an S^1 then the Chern–Simons partition function of the 3-manifold $\Sigma \times S^1$ is simply

$$Z(\Sigma \times S^1) = tr\,(\exp[i\mathcal{H}t]) \qquad (12.150)$$

But the Chern–Simons theory, being topological, has $\mathcal{H} = 0$ so that the partition function reduces to the statement

$$Z(\Sigma \times S^1) = tr\,(I) = \dim H_\Sigma \qquad (12.151)$$

If Σ has marked points with associated representations $\{R\}$ then, as Σ evolves round S^1, these points trace out unlinked, unknotted circles and the correlation function changes to

$$\widetilde{W}(\Sigma \times S^1, \{R\}) = \dim H_{\Sigma, \{R\}} \qquad (12.152)$$

More generally, our three dimensional space-time need not be globally a product of space and time, it may be just locally so. One can achieve this by letting Σ evolve into diffeomorphic copies of itself as t progresses: then after a time interval $[0, 1]$ a point $(x, 0)$ in space-time has become $(Kx, 1)$ where K is a diffeomorphism of Σ. Finally, to make time periodic we glue the points $(x, 0)$ and $(Kx, 1)$ together giving a 3-manifold $\Sigma \times_K S^1$ which is a bundle over S^1. The partition function is still a trace but we now have

$$\begin{aligned} Z(\Sigma \times_K S^1) &= tr\,(K \exp[i\mathcal{H}t]) \\ &= tr\,(K) \end{aligned} \qquad (12.153)$$

We now begin our surgery calculation by choosing $M = S^2 \times S^1$ and $M^K = S^3$. Hence we must explain how to cut a solid torus out of $M = S^2 \times S^1$ and glue it back to get an S^3. As explained by Witten, one starts with a solid torus T, say, embedded in $\mathbf{R}^3$. Now we subject $\mathbf{R}^3$ to that inversion which interchanges the interior of T with its exterior; this will leave the boundary $\partial T \simeq T^2$ fixed and it also shows that $\mathbf{R}^3 \cup \{\infty\}$ can be constructed by gluing together two solid tori. It may be of some help to think, loosely, of $\mathbf{R}^3 \cup \{\infty\}$ as being made up of two solid tori linked together, as if belonging to a chain. In this way we can see that, if the radii of one of the tori are represented by the pair (a, b), those of the other are represented by the pair (b, a). This further means that the two solid tori are joined by gluing them together along their boundaries; both boundaries are ordinary tori T^2 and, before gluing them together, one boundary is acted on by a diffeomorphism K which interchanges its radii. Now, referring to chapter 5, this interchanging property identifies K as the modular transformation $S \in SL(2, \mathbf{Z})/\{\mp I\}$. But $\mathbf{R}^3 \cup \{\infty\} \simeq S^3$ and so S^3 is the join of two solid tori.

Next view the solid torus T as $D \times S^1$ where D is a two dimensional disc; if we take two solid tori $D \times S^1$ and $D' \times S^1$ then we can join them by gluing together the two discs D and D' along their boundaries, but, since gluing together D and D' gives an S^2, the resulting 3-manifold is $S^2 \times S^1$.

Hence we have described two ways of gluing solid tori together with the result being either S^3 or $S^2 \times S^1$.

The surgery that we require is got by cutting $S^2 \times S^1$ into two solid tori and then acting with the inversion on one of them before gluing the pieces back together. The result will then be an S^3 and this is the surgery we desire. We have also identified the diffeomorphism K which acts on $\partial T \simeq T^2$: it is the modular transformation S.

Thus, with $M = S^2 \times S^1$, $K = S$ and $M^K = S^3$, 12.148 and 12.149 give us

$$Z(S^3) = S^0_j \widetilde{W}(S^2 \times S^1, R_j)$$

$$\text{and} \quad \widetilde{W}(S^3, R_i) = S^0_j \widetilde{W}(S^2 \times S^1, R_i, R_j)$$

$$\Rightarrow W(S^3, R_i) = \frac{\widetilde{W}(S^3, R_i)}{Z(S^3)} = \frac{S^0_j \dim H_{S^2, R_i, R_j}}{S^0_j \dim H_{S^2, R_j}} \qquad (12.154)$$

where in the last line we used 12.151 and 12.152. However, from 11.75 we know that

$$\dim H_{S^2, R_i} = \delta_{j0}, \qquad \dim H_{S^2, R_i, R_j} = g_{ij} \qquad (12.155)$$

thus

$$W(S^3, R_i) = \frac{S^0_j g_{ij}}{S^0_j \delta_{j0}} = \frac{S_{0i}}{S_{00}} \qquad (12.156)$$

But the matrix elements S_{ij} of S were studied in Gepner and Witten [1] and, for level k representations of $\overline{LG}$ with $G = SU(2)$, which we now assume, they showed that

$$S_{ij} = \sqrt{\frac{2}{k+2}} \sin\left(\frac{(i+1)(j+1)\pi}{k+2}\right) \qquad (12.157)$$

Now for our Jones polynomial calculations the representation R_i was the 2-dimensional defining representation of $SU(2)$; this corresponds to $i = 1$ in the above notation, hence we have found that, for a single unknotted circle,

$$
\begin{aligned}
W(S^3, R_i) &= \frac{\sin(2\pi/(k+2))}{\sin(\pi/(k+2))} \\
&= -\frac{(t - t^{-1})}{(t^{1/2} - t^{-1/2})}
\end{aligned}
\tag{12.158}
$$

in agreement with 12.138.

When $G \neq SU(2)$ we obtain a generalisation of the original Jones polynomial. If $G = SU(N)$ and N is allowed to vary we obtain a two variable polynomial whose variables are determined by N and k. On adjusting the representations of the Wilson lines to be the defining representation of $SU(N)$ we can recover the Homfly polynomial of Freyd et al. [1].

We have by no means exhausted all the possible topological field theories. We shall just mention two more. In two dimensions there is a topological sigma model (cf. Witten [8], Floer [1] and Gromov [1]) which, when quantised, exhibits quantum ground states that are like Floer groups. In $2 + 1$ dimensions a Chern–Simons topological theory of gravity can be written down (Witten [9]) where the group $G = ISO(2,1)$ with $ISO(2,1)$ denoting the inhomogeneous $SO(2,1)$, i.e. $ISO(2,1)$ is $SO(2,1)$ plus the translations. Since $ISO(2,1)$ is non-compact, the moduli space of flat $ISO(2,1)$-connections is non-compact and the Hilbert spaces are no longer finite dimensional. This is an example of a gravity theory which is generally covariant but possesses no metric; nevertheless, it has non-trivial amplitudes which change the topology of space (Witten [10]). More generally, an axiomatic approach, somewhat analogous to that introduced for conformal field theories, can be given, cf. Atiyah [6].

References

Abraham R. and Marsden J.

1. *Hamiltonian mechanics on Lie groups and Hydrodynamics*, Global analysis: Proc. Symp. Pure Math. 16, edited by: Chern S. S. and Smale S., Amer. Math. Soc., (1970).

Adams R. A.

1. *Sobolev Spaces*, Academic Press, (1975).

Adler S. and Dashen R.

1. *Current algebras and applications to particle physics*, Benjamin, (1968).

Adler S.

1. Axial-vector vertex in spinor electrodynamics, *Phys. Rev.*, **177**, 2426–2438, (1969).

2. *Lectures on Elementary particle physics and Quantum field theory*, Brandeis , edited by: Deser S., Grisaru M. and Pendleton H., MIT press, (1970).

Alvarez O., Singer I. M. and Zumino B.

1. Gravitational Anomalies and the families index theorem, *Commun. Math. Phys.*, **96**, 409–417, (1984).

Alvarez O.

1. Conformal anomalies and the index theorem, *Nucl. Phys.*, **286B**, 175–188, (1987).

Alvarez-Gaumé L. and Ginsparg P.

1. The Structure of Gauge and Gravitational Anomalies, *Ann. Phys.*, **161**, 423–490, (1985).

Alvarez-Gaumé L. and Witten E.

1. Gravitational anomalies, *Nucl. Phys.*, **234B**, 269–330, (1983).

Alvarez-Gaumé L., Gomez C. and Vafa C.

1. Strings in the operator formalism, *Nucl. Phys.*, **B303**, 455–521, (1988).

Atiyah M. F. and Bott R.
1. A Lefschetz fixed point formula for elliptic complexes I, *Ann. Math.*, **86**, 374–407, (1967).
2. A Lefschetz fixed point formula for elliptic complexes II, *Ann. Math.*, **88**, 451–491, (1968).
3. *The index theorem for manifolds with boundary*, Differential Analysis (Bombay Colloquium), Oxford University Press, (1964).
4. The Yang–Mills equations over Riemann surfaces, *Phil. Trans. Roy. Soc. Lond. A*, **308**, 523–615, (1982).

Atiyah M. F., Bott R. and Patodi V. K.
1. On the heat equation and the index theorem, *Invent. Math.*, **19**, 279–330; (errata **28**, 227–280, (1975).), (1973).

Atiyah M. F. and Hitchin N. J.
1. Low energy scattering of Non-Abelian monopoles, *Phys. Lett.*, **107A**, 21–25, (1985).
2. *The Geometry and Dynamics of Magnetic Monopoles*, Princeton University Press, (1988).

Atiyah M. F., Hitchin N. J. and Singer I. M.
1. Self-duality in four dimensional Riemannian geometry, *Proc. Roy. Soc. Lond. A.*, **362**, 425–461, (1978).

Atiyah M. F. and Jones J. D. S.
1. Topological Aspects of Yang–Mills Theory, *Commun. Math. Phys.*, **61**, 97–118, (1978).

Atiyah M. F., Patodi V. K. and Singer I. M.
1. Spectral asymmetry and Riemannian geometry, *Bull. Lond. Math. Soc.*, **5**, 229–234, (1973).
2. Spectral asymmetry and Riemannian geometry I, *Math. Proc. Camb. Philos. Soc.*, **77**, 43–69, (1975).
3. Spectral asymmetry and Riemannian geometry II, *Math. Proc. Camb. Philos. Soc.*, **78**, 405–432, (1975).
4. Spectral asymmetry and Riemannian geometry III , *Math. Proc. Camb. Phil. Soc.*, **79**, 71–99, (1976).

Atiyah M. F. and Segal G.
1. The Index of Elliptic Operators II, *Ann. Math.*, **87**, 531–545, (1968).

Atiyah M. F. and Singer I. M.
1. The index of elliptic operators on compact manifolds, *Bull. Amer. Math. Soc.*, **69**, 422–433, (1963).
2. The Index of Elliptic Operators I, *Ann. Math.*, **87**, 485–530, (1968).

3. The Index of Elliptic Operators III, *Ann. Math.*, **87**, 546–604, (1968).

4. The Index of Elliptic Operators IV, *Ann. Math.*, **93**, 119–138, (1971).

5. The Index of Elliptic Operators V, *Ann. Math.*, **93**, 139–149, (1971).

6. Index theory for skew-adjoint Fredholm operators, *Inst. Hautes Études Sci. Publ. Math.*, **37**, 305–326, (1969).

7. Dirac operators coupled to vector potentials, *Proc. Nat. Acad. Sci. U. S. A.*, **81**, 2597–2600, (1984).

Atiyah M. F.

1. *K-Theory*, Benjamin, (1967).

2. K-theory and Reality, *Quart. Jour. Math. (Oxford)*, **7**, 367–386, (1966).

3. Instantons in Two and Four Dimensions, *Commun. Math. Phys.*, **93**, 437–451, (1984).

4. *Magnetic Monopoles in Hyperbolic Space*, Vector Bundles on Algebraic Varieties, edited by: , Oxford University Press, (1987).

5. *New invariants of 3 and 4 dimensional manifolds*, Symposium on the mathematical heritage of Hermann Weyl, edited by: Wells R. O., Amer. Math. Soc., (1988).

6. Topological quantum field theories, *Inst. Hautes Études Sci. Publ. Math.* , **68**, 175–186, (1989).

Bardeen W. A.

1. Anomalous Ward identities in spinor field theories, *Phys. Rev.*, **184**, 1848–1859, (1969).

Belavin A. A. and Knizhnik V. G.

1. Algebraic geometry and the geometry of quantum strings, *Phys. Lett.*, **168B**, 201–206, (1986).

Belavin A. A., Polyakov A. M. and Zamolodchikov A. B.

1. Infinite conformal symmetry in two-dimensional quantum field theory, *Nucl. Phys.* , **B241**, 333–380, (1984).

Bell J. and Jackiw R.

1. A PCAC puzzle: $\pi^0 \rightarrow 2\gamma$ in the σ-model, *Nuovo Cimento*, **60A**, 47–61, (1969).

Bernard C. W., Christ N. H., Guth A. H. and Weinberg E. J.

1. Pseudoparticle parameters for arbitrary gauge groups, *Phys. Rev.*, **D16**, 2967–2977, (1977).

Birman J.

1. *Braids, links and mapping class groups*, Princeton University Press, (1974).

2. On the Jones polynomial of closed 3-braids, *Invent. Math.*, **81**, 287–294, (1985).

Bismut J.-M. and Freed D. S.

1. The analysis of elliptic families: I metrics and connections on determinant bundles., *Commun. Math. Phys.*, **106**, 159–176, (1986).

2. The analysis of elliptic families: II Dirac operators, eta invariants, and the holonomy theorem, *Commun. Math. Phys.*, **107**, 103–163, (1986).

Bogomolny E. B.

1. Stability of Classical solutions, *Sov. J. Nucl. Phys.*, **24**, 861–870, (1976).

Bonora L. and Cotta-Ramusino P.

1. Some Remarks on BRS transformations, anomalies and the cohomology of the Lie algebra of the group of gauge transformations, *Comm. Math. Phys.*, **87**, 589–603, (1983).

Borel A. and Weil A.

1. Représentations linéaires et espaces homogènes Kählerians des groupes de Lie compacts, *Séminaire Bourbaki (exposé par J.-P. Serre)*, **100**, 1–8, (1954).

Bott R.

1. The stable homotopy of the classical groups, *Ann. Math.*, **70**, 313–337, (1959).

2. Homogeneous vector bundles, *Ann. Math.*, **66**, 203–248, (1957).

3. An application of Morse theory to the topology of Lie groups, *Bull. Soc. Math. France*, **84**, 251–281, (1956).

4. Lectures on Morse Theory, Old and New, *Bull. Amer. Math. Soc.*, **7**, 331–358, (1982).

Bott R. and Tu L. W.

1. *Differential Forms in Algebraic Topology*, Springer-Verlag, New York, (1982).

Cappelli A., Itzykson C. and Zuber J-B.

1. Modular invariant partition functions in two dimensions, *Nucl. Phys.*, **B280**, 445–465, (1987).

2. The A-D-E classification of minimal and $A^{(1)}$ conformal invariant theories, *Commun. Math. Phys.*, **113**, 1–26, (1987).

Cardy J. L.

1. Operator content of two-dimensional conformally invariant theories, *Nucl. Phys.*, **B270**, 186–204, (1986).

Cerf J.

1. La stratification naturelle des espaces de fonctions differentiables réelles et le théorème de la pseudo-isotopie, *Inst. Hautes Études Sci. Publ. Math.*, **39**, 5–173, (1970).

Chakrabarti A.

1. Instanton chains with multimonopole limits: Lax pairs for non-axially symmetric cases, *Phys. Rev.*, **D28**, 989–1000, (1983).
2. Construction of hyperbolic monopoles, *Jour. Math. Phys.*, **27**, 340–348, (1986).

Chern S.

1. *Complex Manifolds without Potential Theory, ed. 2*, Springer-Verlag, (1979).

Connes A.

1. Non-Commutative Differential Geometry, *Inst. Hautes Études Sci. Publ. Math.*, **62**, 257–360, (1986).
2. The Action Functional in Non-Commutative Geometry, *Commun. Math. Phys.*, **117**, 673–684, (1988).

Connes A. and Rieffel M.

1. Yang–Mills for non-commutative two tori, *Contemp. Math.*, **62**, 237–266, (1987).

Conway J. H. and Norton S. P.

1. Monstrous Moonshine, *Bull. Lond. Math. Soc.*, **11**, 308–339, (1979).

Crowell R. H. and Fox R. H.

1. *An Introduction to Knot theory*, Springer-Verlag, (1977).

D'Hoker E. and Phong D. H.

1. The geometry of string perturbation theory, *Rev. Mod. Phys.*, **60**, 917–1066, (1988).

Deser S., Jackiw R. and Templeton S.

1. Three dimensional massive gauge theories, *Phys. Rev. Lett.*, **48**, 975–, (1982).

Donaldson S. K.

1. An Application of Gauge Theory to Four Dimensional Topology , *J. Diff. Geom.*, **18**, 279–315, (1983).
2. Nahm's Equations and the classification of Monopoles, *Commun. Math. Phys.*, **96**, 387–407, (1984).
3. *The Geometry of 4-Manifolds*, Proc. of the International Congress of Mathematicians, Berkeley 1986, edited by: Gleason A. M., Amer. Math. Soc., (1987).

4. Polynomial invariants for smooth four manifolds, *Topology*, **29**, 257–315, (1990).

Ebeling W.
1. An example of two homeomorphic, nondiffeomorphic complete intersection surfaces, *Invent. Math.*, **99**, 651–654, (1990).

Eguchi T., Gilkey P. B. and Hanson A. J.
1. Gravitation, Gauge Theories and Differential Geometry, *Phy. Rep.*, **66**, 213–393, (1980).

Eilenberg S. and Steenrod N.
1. *Foundations of algebraic topology*, Princeton University Press, (1952).

Faddeev L. D.
1. Operator anomaly for the Gauss law, *Phys. Lett.*, **145B**, 81–84, (1984).

Floer A.
1. Morse theory for fixed points of symplectic diffeomorphisms, *Bull. A. M. S.*, **16**, 279–281, (1987).
2. An Instanton Invariant for 3-Manifolds, *Commun. Math. Phys.*, **118**, 215–240, (1988).

Forgács P., Horváth Z. and Palla L.
1. Exact fractionally charged self-dual solution, *Phys. Rev. Lett.*, **46**, 392–394, (1981).

Freed D. S.
1. Determinants, Torsion and Strings, *Commun. Math. Phys.*, **107**, 483–513, (1986).

Freed D. S. and Uhlenbeck K. K.
1. *Instantons and Four-Manifolds*, Springer-Verlag, (1984).

Freedman M. H.
1. The topology of 4-dimensional manifolds, *Jour. Diff. Geom.*, **17**, 357–453, (1982).

Freedman M. and Taylor L.
1. Λ-splitting 4-manifolds, *Topology*, **16**, 181–184, (1977).

Freyd P., Yetter D.; Hoste J.; Lickorish W. B. R., Millet K.; and Ocneanu A.
1. A new polynomial invariant of knots and links, *Bull. A. M. S.*, **12**, 239–246, (1985).

Friedan D. and Shenker S.
1. The analytic geometry of two-dimensional conformal field theory, *Nucl. Phys.*, **B281**, 509–545, (1987).

2. The integrable analytic geometry of quantum string, *Phys. Lett.*, **175B**, 287–296, (1986).

Friedan D., Martinec E. and Shenker S.
1. Conformal invariance, supersymmetry and string theory, *Nucl. Phys.*, **B271**, 93–165, (1986).

Friedan D., Qiu Z. and Shenker S. H.
1. *Conformal Invariance, Unitarity and two dimensional critical exponents*, Vertex operators in mathematics and physics, edited by: Lepowsky J., Mandelstam S. and Singer I. M., Springer-Verlag, (1985).
2. Conformal Invariance, Unitarity and Critical exponents in two dimensions, *Phys. Rev. Lett.*, **52**, 1575–1578, (1984).

Fujikawa K.
1. Path integral for gauge theories with Fermions, *Phys. Rev.*, **D21**, 2848–2858, (1980).

Gauss K. F.
1. Zur mathematischen theorie der electrodynamischen Wirkungen (1833), *Werke. Königlichen Gesellschaft der Wissenschaften zu Göttingen*, **5**, 605, (1877).

Gel'fand I. M. and Fuks D. B.
1. Cohomologies of the Lie algebra of the vector fields on the circle, *Funct. Anal. Appl.*, **2**, 92–93, (1968).

Gel'fand I. M. and Shilov G. E.
1. *Generalised Functions, vol. 1*, Academic Press, (1964).

Gell-Mann M. and Ne'eman Y.
1. *The eightfold way*, Benjamin, (1964).

Gepner D. and Witten E.
1. String theory on group manifolds, *Nucl. Phys.*, **B278**, 493–549, (1986).

Goddard P., Kent A. and Olive D.
1. Virasoro algebras and coset space models, *Phys. Lett.*, **152B**, 88–92, (1985).
2. Unitary representations of the Virasoro and super Virasoro algebras, *Commun. Math. Phys.*, **103**, 105–119, (1986).

Goddard P. and Olive D.
1. Kac-Moody and Virasoro algebras in relation to Quantum Physics, *Int. Jour. Mod. Phys.*, **1A**, 303–414, (1986).

Gompf R.
1. Three exotic $\mathbf{R}^4$'s and other anomalies, *J. Diff. Geom.*, **18**, 317–328, (1983).

  2. An infinite set of exotic $\mathbf{R}^4$'s, *J. Diff. Geom.*, **21**, 283–300, (1985).

Green M. B. and Schwarz J. H.
  1. Anomaly cancellations in supersymmetric $D = 10$ gauge theory and superstring theory, *Phys. Lett.*, **149B**, 117–122, (1984).

Green M. B., Schwarz J. H. and Witten E.
  1. *Superstring Theory vol. 1*, Cambridge University Press, (1987).
  2. *Superstring Theory vol. 2*, Cambridge University Press, (1987).

Gribov V. N.
  1. Quantisation of non-Abelian gauge theories, *Nucl. Phys.*, **B139**, 1–19, (1978).

Griffiths P. and Harris J.
  1. *Principles of Algebraic Geometry*, John Wiley, (1978).

Gromov M.
  1. Pseudo-holomorphic curves in symplectic manifolds, *Invent. Math.*, **82**, 307–347, (1985).

Hamilton R.
  1. The inverse function theorem of Nash and Moser, *Bull. Amer. Math. Soc.*, **7**, 65–222, (1982).

Hirzebruch F.
  1. *Topological Methods in Algebraic Geometry*, Springer-Verlag, (1965, (ed. 3) 1966).

Hitchin N. J.
  1. Flat connections and geometric quantisation, *Commun. Math. Phys.*, **131**, 347–380, (1990).

Husemoller D.
  1. *Fibre Bundles*, Springer-Verlag, (1975).

Jaffe A. and Taubes C. H.
  1. *Vortices and Monopoles*, Birkhäuser, Boston, (1980).

Jones V. F. R.
  1. A Polynomial invariant for knots via Von Neumann algebras, *Bull. Amer. Math. Soc.*, **12**, 103–111, (1985).
  2. Hecke algebra representation of braid groups and link polynomials, *Ann. Math.*, **126**, 335–388, (1987).

Kac V. G.
  1. *Infinite dimensional Lie algebras*, Cambridge University Press, (1985).

Kanenobu T.

1. Examples on polynomial invariants of knots and links, *Math. Ann.*, **275**, 555–572, (1986).

Kervaire M. and Milnor J.

1. Groups of Homotopy spheres: I, *Ann. Math.*, **77**, 504–537, (1963).

Kirby R. C.

1. *The Topology of 4-Manifolds*, Springer-Verlag SLNM 1374, (1989).

Kirby R. C. and Siebenmann L. C.

1. On the triangulation of manifolds and the Hauptvermutung, *Bull. Amer. Math. Soc.*, **75**, 742–749, (1969).

2. *Foundational essays on topological manifolds, smoothings, and triangulations*, Ann. Math. Studies, Princeton University Press, (1977).

Klingenberg W.

1. *Lectures on closed geodesics*, Springer-Verlag, (1978).

Kuiper N. H.

1. *A short history of triangulation and related matters*, Proc. of bicentennial congress Wiskundig Genootschap, part I, edited by: Baayen P. C., van Dulst D. and Oosterholf J., Mathematisch Centrum, Amsterdam, (1979).

Kyle R. H.

1. Embeddings of Möbius bands in 3-dimensional space, *Proc. Roy. Irish Acad.*, **57A**, 131–136, (1955).

2. Branched covering spaces and the quadratic forms of a link, *Ann. Math.*, **59**, 539–548, (1954).

3. Branched covering spaces and the quadratic forms of links II, *Ann. Math.*, **69**, 686–699, (1959).

Lepowsky J., Mandelstam S. and Singer I. M. (eds.)

1. *Vertex operators in mathematics and physics*, Springer-Verlag, (1985).

Manin Y. I.

1. The partition function of the Polyakov string can be expressed in terms of theta functions, *Phys. Lett.*, **172B**, 184–185, (1986).

Manton N. S.

1. A remark on the scattering of BPS monopoles, *Phys. Lett.*, **110B**, 54–56, (1982).

Mason G.

1. Finite groups and Modular functions, *Proc. of Symposia in Math.*, **47**, 181–210, (1987).

Mickelsson J.
1. *Current algebras and groups*, Plenum, (1989).
2. On a relation between massive Yang–Mills theories and dual string models, *Lett. Math. Phys.*, **7**, 45–50, (1983).

Milnor J.
1. On manifolds homeomorphic to the 7-sphere, *Ann. Math.*, **64**, 399–405, (1956).
2. *Remarks on infinite dimensional Lie Groups*, Relativity, Groups and Topology II, edited by: De Witt B. S. and Stora R., North Holland, Amsterdam, (1984).
3. *Morse Theory*, Princeton University Press, (1963).

Milnor J. W. and Stasheff J. D.
1. *Characteristic Classes*, Princeton University Press, (1974).

Mitter P. K. and Viallet C. M.
1. On the bundle of connections and the gauge orbit manifold in Yang–Mills theory, *Commun. Math. Phys.*, **79**, 455–472, (1981).

Moise E.
1. Affine structures on 3-manifolds, *Ann. Math.* , **56**, 96–114, (1952).

Moore G. and Seiberg N.
1. Polynomial equations for rational conformal field theories, *Phys. Lett.*, **212B**, 451–460, (1988).

Morse M.
1. *Calculus of variations in the large*, Amer. Math. Soc. Colloq. Publ., (1934).

Nash C. and Sen S.
1. *Topology and Geometry for Physicists*, Academic Press, (1983).

Nash C.
1. Geometry of Hyperbolic Monopoles, *Jour. Math. Phys.*, **27**, 2160–2164, (1986).
2. *Relativistic Quantum Fields*, Academic Press, (1978).
3. A Complex Anomaly, *Phy. Lett.*, **B184**, 239–241, (1987).
4. Sheaf Cohomology and functional integration, *Int. Jour. Mod. Phys. A*, **4**, 4919–4928, (1989).
5. Singularly perturbed Chern–Simons theory, *J. Math. Phys.*, **31**, 2258–2262, (1990).

Palais R. S.
1. Morse Theory on Hilbert Manifolds, *Topology*, **2**, 299–349, (1963).
2. Lusternik-Schnirelman theory on Banach Manifolds, *Topology*, **5**, 115–132, (1966).

3. *Critical point theory and the mini-max principle. Proc. Symp. Pure Math.* **15** , Amer. Math. Soc., (1970).

Palais R. et al.

1. *Seminar on the Atiyah–Singer index theorem*, Ann. Math. Stud. Princeton Univ. Press, (1965).

Peetre J.

1. Une caractérisation abstraite des opérateurs différentiels, *Math. Scand.*, **7**, 211–218, (1959).
2. Rectification à l'article Une caractérisation abstraite des operateurs differentiels, *Math. Scand.*, **8**, 116–120, (1960).

Peter F. and Weyl H.

1. Die Vollständigkeit der primitiven Darstellungen einer geschlossenen kontinuierlichen Gruppe, *Math. Ann.*, **97**, 737–755, (1927).

Polyakov A. M.

1. Quantum geometry of Bosonic strings, *Phys. Lett.*, **103B**, 207–210, (1981).

Pressley A. and Segal G.

1. *Loop Groups*, Oxford University Press, (1986).

Quillen D.

1. Determinants of Cauchy–Riemann operators over Riemann surfaces, *Funct. Anal. and Appl.*, **19**, 31–33, (1985).
2. Superconnections and the Chern character, *Topology*, **24**, 89–95, (1985).

Quinn F.

1. Ends III, *Jour. Diff. Geom.*, **17**, 503–521, (1982).

Rado T.

1. Über den Begriff der Riemannanschen Fläche , *Acta Litt. Scient. Univ. Szegd.*, **2**, 101–121, (1925).

Ray D. B. and Singer I. M.

1. R-torsion and the Laplacian on Riemannian manifolds, *Adv. in Math.*, **7**, 145–201, (1971).
2. Analytic Torsion for complex manifolds, *Ann. Math.*, **98**, 154–177, (1973).

Rohlin V. A.

1. New results in the theory of 4 dimensional manifolds, *Dok. Akad. Nauk. U. S. S. R.*, **84**, 221–224, (1952).

Rolfsen D.

1. *Knots and Links*, Publish or Perish, (1976).

Schonfeld J.
1. A mass term for three dimensional gauge fields, *Nucl. Phys.*, **B185**, 157–171, (1981).

Schwarz A. S.
1. The partition function of degenerate quadratic functional and Ray–Singer invariants, *Lett. Math. Phys.*, **2**, 247–252, (1978).

Seeley R. T.
1. Complex Powers of an Elliptic Operator, *Proc. Symp. in Pure Math., Amer. Math. Soc.*, **10**, 288–307, (1967).

Segal G.
1. Equivariant K-theory, *Inst. Hautes Études Sci. Publ. Math.*, **34**, 129–151, (1968).
2. Unitary Representations of some infinite dimensional groups, *Commun. Math. Phys.*, **80**, 301–42, (1981).
3. Faddeev's anomaly in Gauss's law, *Oxford preprint*, (1985).
4. *Two dimensional conformal field theories and modular functors*, I. A. M. P. Congress, Swansea, 1988, edited by: Davies I., Simon B. and Truman A., Institute of Physics, (1989).

Segal G. and Wilson G.
1. Loop Groups and equations of K dV type, *Inst. Hautes Études Sci. Publ. Math.*, **61**, 5–65, (1985).

Singer I. M.
1. Some Remarks on the Gribov ambiguity, *Commun. Math. Phys.*, **60**, 7–12, (1978).

Smale S.
1. Generalised Poincaré's conjecture in dimensions greater than four, *Ann. Math.*, **74**, 391–406, (1961).

Spanier E. H.
1. *Algebraic Topology*, McGraw-Hill, (1966).

Spivak M.
1. *A Comprehensive Introduction to Differential Geometry, vol. 1*, Publish or Perish, (1979).

Steenrod N.
1. *The Topology of Fibre Bundles*, Princeton University Press, (1970).

Sugawara H.
1. A field theory of currents, *Phys. Rev.*, **170**, 1659–1662, (1968).

Taubes C. H.
1. Self-dual Yang–Mills connections on non-self-dual 4-manifolds, *Jour. Diff. Geom.*, **17**, 139–170, (1982).

2. Stability in Yang–Mills Theories, *Commun. Math. Phys.*, **91**, 235–263, (1983).

3. A framework for Morse theory for the Yang–Mills functional, *Invent. Math.*, **94**, 327–402, (1988).

4. Min-Max theory for the Yang–Mills–Higgs Equations, *Commun. Math. Phys.*, **97**, 473–540, (1985).

5. Casson's invariant and gauge theory, *Jour. Diff. Geom.*, **31**, 547–599, (1990).

Treves F.

1. *Basic Linear Partial Differential Equations*, Academic Press, New York, (1975).

Tsuchiya A. and Kanie Y.

1. Vertex operators in the conformal field theory on CP^1 and monodromy representations of the braid group, *Lett. Math. Phys.*, **13**, 303–312, (1987).

Vafa C.

1. Operator formulation on Riemann Surfaces, *Phy. Lett.*, **B190**, 47–54, (1987).

Verlinde E.

1. Fusion Rules and Modular transformations in 2-d conformal field theory, *Nucl. Phys.*, **B300**, 360–376, (1988).

Virasoro M. A.

1. Subsidiary conditions and ghosts in dual-resonance models, *Phys. Rev.*, **D1**, 2933–2936, (1970).

Weinberg E.

1. Parameter counting for multi-monopole solutions, *Phys. Rev.*, **D20**, 936–944, (1979).

Wells R. O.

1. *Differential Analysis on Complex Manifolds*, Springer-Verlag, New York, (1980).

Whitehead G. W.

1. *Elements of Homotopy Theory*, Springer-Verlag, (1978).

Whittaker E. T. and Watson G. N.

1. *A Course of Modern Analysis*, Cambridge University Press, (1920).

Wilson K. G.

1. Non-Lagrangian models of current algebra, *Phys. Rev.*, **179**, 1499–1512, (1969).

Witten E.

   1. Non-Abelian Bosonisation in two dimensions, *Commun. Math. Phys.*, **92**, 455–72, (1984).

   2. Supersymmetry and Morse theory, *J. Diff. Geom.*, **17**, 661–692, (1982).

   3. An $SU(2)$ anomaly, *Phys. Lett.*, **117B**, 324–328, (1982).

   4. Global gravitational anomalies, *Commun. Math. Phys.*, **100**, 197–229, (1985).

   5. Topological quantum field theory, *Commun. Math. Phys.*, **117**, 353–386, (1988).

   6. *Quantum field theory and the Jones polynomial*, I. A. M. P. Congress, Swansea, 1988, edited by: Davies I., Simon B. and Truman A., Institute of Physics, (1989).

   7. Quantum field theory and the Jones polynomial, *Commun. Math. Phys.*, **121**, 351–400, (1989).

   8. Topological sigma models, *Commun. Math. Phys.*, **118**, 411–466, (1988).

   9. $2+1$ dimensional gravity as an exactly soluble system, *Nucl. Phys.*, **B311**, 46–78, (1988/89).

   10. Topology changing amplitudes in $2+1$ dimensional gravity, *Nucl. Phys.*, **B323**, 113–140, (1990).

Zamolodchikov A. B.

   1. Exact solutions of conformal field theory in two dimensions and critical phenomena, *Rev. in Math. Phys.*, **1**, 197–234, (1990).

   2. Irreversibility of the flux of the renormalisation group in a 2d field theory, *JETP Lett.*, **43**, 730–732, (1986).

Index

Printed and bound by CPI Group (UK) Ltd, Croydon, CR0 4YY

03/10/2024

01040426-0003